13 listed on abebooks
New up to $290 -
Used $125 - 225

Transient Electronics

Transient Electronics

Pulsed Circuit Technology

Paul W. Smith
Fellow of Pembroke College, Oxford, UK

JOHN WILEY & SONS, LTD

Telephone (+44) 1243 779777

Email (for orders and customer service enquiries): cs-books@wiley.co.uk
Visit our Home Page on www.wileyeurope.com or www.wiley.com

Other Wiley Editorial Offices

John Wiley & Sons Inc., 111 River Street, Hoboken, NJ 07030, USA

Jossey-Bass, 989 Market Street, San Francisco, CA 94103-1741, USA

Wiley-VCH Verlag GmbH, Boschstr. 12, D-69469 Weinheim, Germany

John Wiley & Sons Australia Ltd, 33 Park Road, Milton, Queensland 4064, Australia

John Wiley & Sons (Asia) Pte Ltd, 2 Clementi Loop #02-01, Jin Xing Distripark, Singapore 129809

John Wiley & Sons Canada Ltd, 22 Worcester Road, Etobicoke, Ontario, Canada M9W 1L1

British Library Cataloguing in Publication Data

A catalogue record for this book is available from the British Library

ISBN 0-471-97773-X

Typeset in 10/12 Times by Thomson Press (India) Ltd, New Delhi
Printed and bound in Great Britain by Biddles Ltd, Guildford and King's Lynn
This book is printed on acid-free paper responsibly manufactured from sustainable forestry in which at least two trees are planted for each one used for paper production.

Contents

Preface

The analysis of the transient response of electrical and electronic circuits to any transient input signal is a rather more difficult subject than the analysis of the AC response of such circuits when excited by sinusoidal signal sources. Fortunately the use of the Laplace transform method, developed separately by Oliver Heaviside in his operational calculus, has proved to be a very powerful tool, more so than many people realise, for carrying out this type of analysis. The method effectively transforms a difficult problem, based on the classical solution of linear differential equations, into a much simpler one that even first year undergraduates can tackle. However, in a few cases, the classical method can prove to be more efficient particularly in the transient analysis of circuits containing electrical components whose values change with time.

The book therefore starts with a detailed chapter on the Laplace transform method together with a introductory section on the use of the classical method. The chapter gives an insight into the origins of the Laplace transform method so that interested readers can get some sort of understanding of the way in which differential equations, that may be difficult to solve directly, can be converted into simple algebraic equations that are much easier to solve. The use of the method is illustrated by many worked examples and it is recommended that those new to the subject work through the examples to achieve competence in its application. The Laplace transform method is used heavily throughout the later stages of the book as the standard method for analysing the transient response of the many components and circuits described. This chapter is then followed up by a second chapter on transmission lines which is written in such a way as to explain how the Laplace transform method may also be applied to the transient analysis of the response of transmission lines to transient signals. These two chapters then provide the foundation for the rest of the book which is devoted to specific electrical components, circuits and circuit techniques which are used to generate and transform short electrical pulses with pulse duration's down to a few hundreds of picoseconds.

Chapters 3 and 4 are then devoted to the subject of pulse forming using transmission lines (chapter 3) and line simulating *LC* ladder networks (chapter 4). The transient response of conventional wound transformers is described in chapter 5 which is then followed by a chapter on the more recently developed family of transmission line transformers. Chapter 7 deals with the design of pulse generators which are based on the discharge of energy either stored in capacitors or inductors and includes a detailed description of the Marx generator, perhaps the most important generator used in pulsed power systems. Finally chapter 8 introduces the exciting new field of nonlinear pulse generators. The use of nonlinear components has led to the development of a whole new family of pulse generating circuits whose performance, particularly in terms of speed, can far out exceed that of circuits which are restricted to the use of linear components.

It is hoped that this book will provide readers with a comprehensive guide to the most important pulse generating circuits and components that have been reported so far. It can be quite difficult to find information on many pulsed circuit techniques as much of the work has been published in the form of internal research reports, often at defence research establishments, or in books and papers that can only be discovered in the world's largest engineering and physics libraries. For this reason there are extensive lists of references at the end of each chapter so that more detailed information on particular circuits and components can be found relatively easily.

There are very few books that have been written specifically on transient electronics and pulsed electrical circuits. Most notable are the books by Glasoe and Lebacqz, *Pulse Generators* and Lewis and Wells, *Millimicrosecond Pulse Techniques*. Both books are now very old having been written around 50 years ago. A later book by Zepler and Nichols, *Transients in Electronic Engineering* is also worth noting as it also deals specifically with the transient analysis of electrical circuits although it does not include chapters on pulse generating circuits and components that are to be found in this book.

This book, therefore, is designed to give an up-to-date approach to the subjects of transient electronics and pulse generator circuits and is, in part, based on the vast amount of research work carried out over the last 50 years or so in the field of pulsed power technology. This work has been primarily directed towards the development of pulsed electrical circuits capable of generating short electrical pulses at very high power levels for applications mostly in experimental physics and defence. Much of the source material comes from the Proceedings of the International IEEE Pulsed Power Conferences (started in 1977), the Proceedings of the IEEE Power Modulator Symposia (started in 1950) and the IEE Pulsed Power Colloquia (started by the author in 1991).

The book is written so that it should be of use both to undergraduates in electrical and electronic engineering (chapters 1 and 2) and, in particular, to all researchers in pulsed power technology. It should also be of value to engineers who need to know about transient analysis and pulse generation, such as aerospace engineers (lightning and EMP protection), the defence community (electric guns, flash X-rays, etc.), radar engineers (pulsed and impulse radars), and computer engineers (computer protection from transient signals etc.).

In writing this preface, I find it rather amusing to think that I am again writing about the characteristics of a variety of electrical circuits when the first piece of work I wrote on the subject was a project on electrical circuits written at the age of 11 at Whitehorse Road Primary School, Croydon. Some 40 years on I am still devoted to the subject and the book is written as a result of over 30 years research activity in pulsed power. For an experimentalist in physics or electrical engineering, pulsed power technology is arguably the most exciting (quite literally!) field to work in. The world's biggest lasers, plasma experiments, electric guns, particle accelerators are critically dependent on pulsed power technology and simply would not exist without the pioneering research carried out in the field.

Over the years of my research career I have met and worked with many fine physicists and engineers world-wide whose friendship and generosity have made this book possible. However, before thanking those who have helped me to produce this book, I would like to start by expressing my deep gratitude to Jim Holbrook whose lecture course on Network Analysis and Synthesis, that I undertook at Southampton University in 1971, is probably the most valuable course an electrical or electronic engineer could take. His book entitled *Laplace Transforms for Electronic Engineers* clearly illustrates his profound understanding of the Laplace transform method and his ability to communicate this understanding in the most digestible way. A "must buy" for all electrical engineers!

I should also like to thank my colleagues and friends at the former EEV Co. Ltd. (now Marconi Applied Technologies) and, in particular, Peter Maggs, Chris Neale, Colin Pirrie and the late Hugh Menown, for their support and sponsorship over many years. Without their generosity and patronage it would have been impossible to carry out much of the work described in this book. I was also fortunate enough to spend time in the late 70s with the late Charlie Martin and his group at AWE, Aldermaston. Charlie, regarded by many as the Father of pulsed power technology, generously devoted much of his time to the training of new workers in pulsed power technology. His highly individual and unconventional approach to research was very stimulating and a very interesting and amusing account of his career is to be found in the book edited by Martin, Guenther and Kristiansen entitled *J. C. Martin on Pulsed Power.* I should also like to acknowledge the contribution made to my own research by the many postgraduate students and post doctoral research workers who have been part of my research group over the years. Much of this research appears in this book and I should like to thank, in particular, Colin Wilson, Miles Turner, Andy Erickson, Greg Branch, Martin Brown and Osvaldo Rossi for their contribution to the work that is written up in this book. Finally I should like to thank Joanna Ashbourn, John Allen, Nigel Seddon and Peter Choi for agreeing to proof read this book and for their valuable comments.

Paul W. Smith

ACKNOWLEDGEMENTS

The author would like to thank the IEEE for granting permission to use figures from the following papers:

Figures 8.14 and 8.15 are reproduced with permission from Brown, H. P. and Smith P. W. "High Power, Pulsed Soliton Generation at Radio and Microwave Frequencies" Proceedings of the 11th IEEE Pulsed Power Conference, Baltimore (1977) 346–354, © 1977 IEEE.

Figures 8.10 and 8.11 are reproduced with permission from Wilson C. R., Turner M. M. and Smith P. W. "Pulse Sharpening in a Uniform LC Ladder Network Containing Nonlinear Ferroelectric Capacitors" *IEEE Trans. on Electron Devices,* **38** (1991) 767–771, © 1991 IEEE.

The author would also like to thank the American Institute of Physics for permission to reproduce figures from the following papers:

Figures 6.21–6.24 are reproduced with permission from Graneau P. N., Rossi J. O., Brown M. P. and Smith P. W. "A High-voltage Transmission-line Pusle Transformer with Very Low Droop". *Rev. Sci. Instrum.* **67**(7) (1996) 2630–2635, © 1996 American Institute of Physics.

Figures 6.13–6.20 are reproduced with permission from Graneau P. N., Rossi J. O., and Smith P. W. "The Operation and Modelling of Transmission Line Transformers using a Referral Method". *Rev. Sci. Instrum.* **70** (1999) 3180–3185, © 1999 American Institute of Physics.

The author would further like to thank the UK Institute of Physics for permission to reproduce figures from the following paper:

Figures 8.17–8.19 and Figure 8.21 reproduced with permission from Branch G. and Smith P. W. "Fast-rise-time Electromagntic Shock Waves in Nonlinear, Ceramic Dielectrics". *J. Phys. D: Phys.* **29** (1996) 2170–2178, © 1996 Institute of Physics.

1

Mathematical Techniques for Pulse and Transient Circuit Analysis

1.1 INTRODUCTION

The analysis of the transient response of pulsed circuits requires a comprehensive knowledge and understanding of the mathematical methods that can be used. In this chapter the most important mathematical tools are explained. It is not the purpose of this chapter to give a complete description of the techniques of electrical circuit analysis, as there are plenty of good texts available on this topic [1, 2, 3]. It will be assumed, however, that the reader is competent in the basic techniques of network analysis, i.e. the application of Kirchhoff's laws to circuits, the laws of Thévenin and Norton and the principle of superposition. Also desirable is a working knowledge of the technique of signal flow graphs and the associated use of Mason's reduction formula [4] as this can often reduce the labour involved in analysing the transient behaviour of multi-component circuits. It will also be assumed that the reader has an adequate background in mathematical techniques, and in particular is familiar with complex number theory, the solution of integro-differential equations, Fourier analysis and series, and basic matrix and determinental methods. Again there are many texts which cover such topics, but the books by Stephenson [5], Wylie and Barrett [15] and Jeffrey [14] may prove to be among the most useful.

This chapter is largely concerned with the Laplace transform method as developed by Oliver Heaviside. The technique is of prime importance to any electrical engineer concerned with the transient behaviour of electrical circuits [6]. The method has an elegant simplicity for this type of analysis and a very wide range of application. Its application to the transient analysis of circuits involving transmission lines is of particular importance and will be dealt with in the next chapter. Although the growing use of circuit analysis programmes such as PSPICE [7] and MICROCAP [8] can provide a convenient and useful way of analysing the transient behaviour of pulsed circuits, a solid grounding in the application of the Laplace technique to such circuits must be regarded as essential.

1.2 THE CLASSICAL METHOD

Before introducing the Laplace transform method, it is instructive to analyse the transient behaviour of a relatively simple circuit using an integro-differential equation set up using

Kirchhoff's laws. This so called Classical Method [9] can easily become very cumbersome, and in some cases gives equations that are impossible to solve especially if the analysis results in two or more simultaneous integro-differential equations. If a homogeneous solution can be found, arbitrary constants must at some stage be introduced, whose values can only be deduced at the end of the analysis by reference to the initial circuit conditions.

The basis of the method is to use the current voltage relationships for resistors, capacitors and inductors in either algebraic, differential or integral form. These are respectively

$$i(t) = \frac{v(t)}{R} \qquad \text{or} \qquad v(t) = Ri(t) \tag{1.1}$$

$$i(t) = C\frac{dv(t)}{dt} \qquad \text{or} \qquad v(t) = \frac{1}{C}\int_0^t i(\tau)\,d\tau + V(0) \tag{1.2}$$

$$i(t) = \frac{1}{L}\int_0^t v(\tau)\,d\tau + I(0) \qquad \text{or} \qquad v(t) = L\frac{di(t)}{dt} \tag{1.3}$$

where the constants of integration $V(0)$ and $I(0)$ represent the voltage to which the capacitor is charged and the current flowing in the inductor, respectively, at time $t=0$.

Example 1.1

An arbitrary voltage source is switched on to the series combination of a resistor, an inductor and a capacitor as shown in Figure 1.1. Derive expressions for the voltage on the capacitor and the current flowing in the circuit as functions of time if the arbitrary source is chosen to be a DC source of potential V volts. How would the analysis proceed if the source was other than DC?

Applying Kirchhoff's voltage law and using Equations (1.1)–(1.3)

$$v(t) = L\frac{di(t)}{dt} + i(t)R + \frac{1}{C}\int_0^t i(\tau)\,d\tau + V(0) \tag{1.4}$$

Since the capacitor is initially uncharged and the voltage source is DC, this simplifies to

$$V = L\frac{di(t)}{dt} + i(t)R + \frac{1}{C}\int_0^t i(\tau)\,d\tau \tag{1.5}$$

Differentiating with respect to time gives

$$0 = L\frac{d^2i(t)}{dt^2} + \frac{di(t)}{dt}R + \frac{i(t)}{C} \tag{1.6}$$

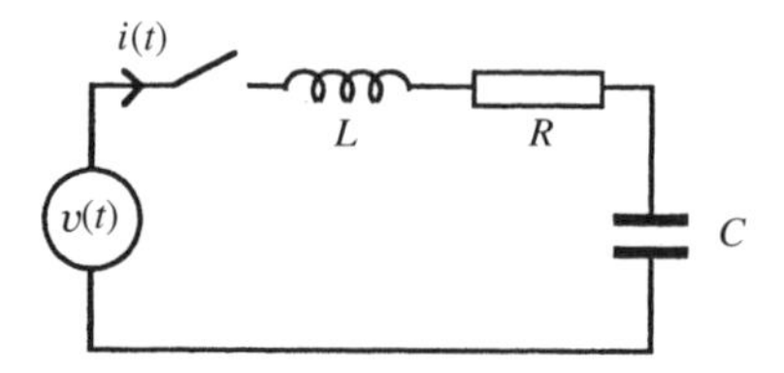

Figure 1.1 Circuit for Example 1.1

Since this is a linear second-order homogeneous differential equation with constant coefficients, a solution of the form

$$i(t) = e^{mt} \tag{1.7}$$

is attempted and gives, on substitution,

$$\left(Lm^2 + Rm + \frac{1}{C}\right)e^{mt} = 0 \tag{1.8}$$

Equation (1.7) must be a solution to Equation (1.6) when m is a root of

$$Lm^2 + Rm + \frac{1}{C} = 0 \tag{1.9}$$

The roots are given by

$$m_1, m_2 = \frac{-R \pm \sqrt{R^2 - 4\dfrac{L}{C}}}{2L} = -\alpha \pm \sqrt{\alpha^2 - \omega_0^2} = -\alpha \pm j\omega_d \tag{1.10}$$

where

$$\alpha = \frac{R}{2L}, \quad \omega_0 = \sqrt{\frac{1}{LC}} \quad \text{and} \quad \omega_d^2 = \omega_0^2 - \alpha^2 \tag{1.11}$$

ω_0 and ω_d are the natural and damped frequencies of the circuit and it is assumed that $\omega_0^2 > \alpha^2$. Thus the general solution to Equation (1.6) is

$$i(t) = A_1 e^{m_1 t} + A_2 e^{m_2 t} \tag{1.12}$$

To evaluate the constants A_1 and A_2 it is noted that just before closure of the switch in the circuit there was no current flowing, therefore at $t = 0_-$, $i(t) = 0$ and $A_1 = -A_2$. Also, at a time just after the switch has been closed, the impedance of the inductor will be very large compared to that of the resistor and capacitor. Therefore

$$V = L\left(\frac{di(t)}{dt}\right)_{t=0_+} \tag{1.13}$$

where the notations 0_- and 0_+ refer to times that occur at an infinitesimally short time before and after the switch is closed at $t=0$, respectively.

Applying this relationship to Equation (1.12), after differentiation and substitution for m_1 and m_2 from Equation (1.10), gives

$$\frac{V}{L} = A_1(-\alpha + j\omega_d) + A_2(-\alpha - j\omega_d) = 2j\omega_d A_1 \tag{1.14}$$

Thus Equation (1.12) becomes

$$i(t) = \frac{V}{2j\omega_d L}\left(e^{-(\alpha - j\omega_d)t} - e^{-(\alpha + j\omega_d)t}\right) = \frac{V}{\omega_d L} e^{-\alpha t} \sin \omega_d t \tag{1.15}$$

This solution corresponds to the case where the circuit is said to be underdamped and the solution is partly oscillatory due to the sine term. In the case where $\omega_0^2 < \alpha^2$ the roots of Equation (1.9) are both real and may be written in the form

$$m_1, m_2 = -\alpha \pm \beta \qquad \text{where} \quad \beta^2 = \alpha^2 - \omega_0^2 \tag{1.16}$$

In this case the solution to Equation (1.6) can easily be found to be

$$i(t) = \frac{V}{\beta L} e^{-\alpha t} \sinh \beta t \tag{1.17}$$

This represents the overdamped case and the solution for $i(t)$ is now no longer oscillatory. Finally, in the case where $\omega_0^2 = \alpha^2$, the roots of Equation (1.9) are both equal to $-\alpha$. Since the two solutions are no longer independent a different general solution must be used, which takes the form

$$i(t) = (A_1 + A_2 t)e^{-\alpha t} \tag{1.18}$$

Applying the initial conditions, as above, gives values for the constants A_1 and A_2 given by

$$A_1 = 0 \quad \text{and} \quad A_2 = \frac{V}{L} \tag{1.19}$$

The solution to Equation (1.6) is now

$$i(t) = \frac{V}{L} t e^{-\alpha t} \tag{1.20}$$

This is the critically damped solution. Typical plots for $i(t)$ for the three damping conditions are shown in Figure 1.2 shown below.

The voltage on the capacitor $v_c(t)$ can be found by using Equation (1.2) since expressions for the current in the circuit have been derived. For example in the case of an underdamped circuit

$$v_c(t) = \frac{1}{C} \int_0^t \left(\frac{V}{\omega_d L} e^{-\alpha \tau} \sin \omega_d \tau \right) d\tau \tag{1.21}$$

assuming the capacitor was initially uncharged and $v_c(0) = 0$. Carrying out the integration gives

$$v_c(t) = V \left[1 - e^{-\alpha t} \left(\cos \omega_d t + \frac{\alpha}{\omega_d} \sin \omega_d t \right) \right] \tag{1.22}$$

In the situation where the voltage source, which is switched on to the circuit, is other than a DC source the problem then becomes one of finding the solution to Equation (1.5) where

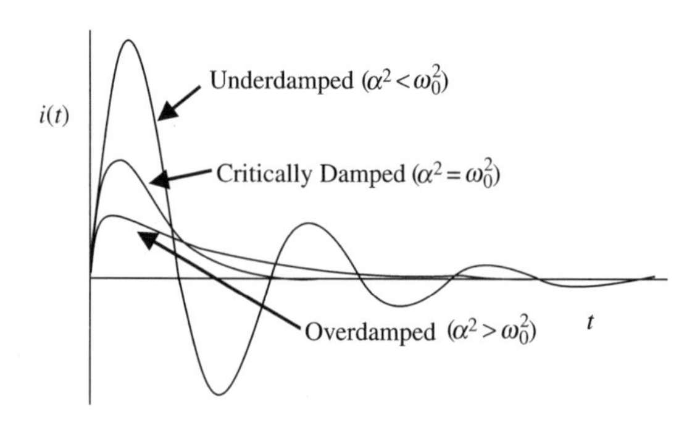

Figure 1.2 Typical current waveforms for an underdamped, critically damped and overdamped series LCR circuit

the source voltage V is replaced by an expression for $v(t)$. After differentiation of the equation, a second-order equation results which is not equal to zero but the derivative of $v(t)$ with respect to time. The general solution of this equation is then the sum of the general solution of the reduced equation (which is the same as Equation (1.6)) i.e. the complementary function and the particular integral of the complete equation. For example if the applied voltage $v(t)$ took the form of a ramp given by

$$v(t) = kt \tag{1.23}$$

then the new form of Equation (1.5) would be

$$v(t) = L\frac{di(t)}{dt} + i(t)R + \frac{1}{C}\int_0^t i(\tau)\,d\tau = kt \tag{1.24}$$

After differentiation this becomes

$$L\frac{d^2i(t)}{dt^2} + \frac{di(t)}{dt}R + \frac{i(t)}{C} = k \tag{1.25}$$

The trial form of the particular integral takes the simple form

$$i(t)_{PI} = A \tag{1.26}$$

Substituting Equation (1.26) into Equation (1.25) gives a value for the constant A of

$$A = kC \tag{1.27}$$

Thus the general solution to Equation (1.25) is given by

$$i(t) = kC + A_3e^{m_1t} + A_4e^{m_2t} \tag{1.28}$$

where the new constants A_3 and A_4 must be evaluated from the initial circuit conditions as before. Since the current in the circuit is initially zero before the switch is closed, then using Equation (1.28) gives

$$kC + A_3 + A_4 = 0 \tag{1.29}$$

As before, by considering the relative sizes of the impedances in the circuit just after the switch has been closed, this results in an equation similar to Equation (1.13)

$$(kt)_{t=0_+} = L\left(\frac{di(t)}{dt}\right)_{t=0_+} \cong 0 \tag{1.30}$$

Applying this relationship to Equation (1.28) gives

$$A_3(-\alpha + j\omega) + A_4(-\alpha - j\omega) = 0 \tag{1.31}$$

From Equations (1.29) and (1.31) the constants A_3 and A_4 are found to be

$$\begin{aligned} A_3 &= \frac{-kC(\alpha + j\omega)}{2j\omega} \\ A_4 &= \frac{-kC(-\alpha + j\omega)}{2j\omega} \end{aligned} \tag{1.32}$$

Thus by substituting these relationships into the general solution Equation (1.28) and rearranging, an expression for the underdamped current $i(t)$ in the circuit can be derived as

$$i(t) = kC\left[1 - e^{-\alpha t}\left(\frac{\alpha}{\omega}\sin\omega t + \cos\omega t\right)\right] \tag{1.33}$$

Clearly, if a less simple function than a ramp for the applied voltage source $v(t)$ had been chosen, the derivation of an expression for the current in the circuit would have been much more laborious, as would have been the case if the initial conditions in the circuit were not zero, i.e. there was an initial current flowing and the capacitor was partially charged prior to switch closure.

Example 1.2

A capacitor charged to a potential V is discharged into a resistor $R(t)$ whose value changes with time. Determine an expression for the current that flows in the circuit $i(t)$ when the capacitor is discharged.

Applying Kirchhoff's voltage law to this simple circuit and using Equations (1.1) and (1.2) gives

$$i(t)R(t) + \frac{1}{C}\int_0^t i(\tau)\,d\tau + V = 0 \tag{1.34}$$

Differentiating this expression with respect to time gives

$$\frac{di(t)}{dt}R(t) + \frac{dR(t)}{dt}i(t) + \frac{i(t)}{C} = 0$$

$$\text{or} \qquad \frac{di(t)}{dt} + \frac{\dfrac{dR(t)}{dt}i(t) + \dfrac{i(t)}{C}}{R(t)} = 0 \tag{1.35}$$

This equation has an integrating factor $\phi(t)$ [5] which is given by

$$\phi(t) = \exp\left(\int \frac{\dfrac{dR(t)}{dt}}{R(t)}\,dt + \int \frac{1}{CR(t)}\,dt\right)$$

$$= R(t)\exp(A)\exp\left(\int \frac{1}{CR(t)}\,dt\right), \quad A \text{ constant} \tag{1.36}$$

Hence the current flowing in the circuit $i(t)$ is given by

$$i(t) = \frac{1}{R(t)}B\exp\left(-\int \frac{1}{CR(t)}\,dt\right) \tag{1.37}$$

Since the current initially flowing in the circuit $i(0)$ depends on the initial value of $R(t)$, i.e. $R(0)$, then the combined constants of integration B can be found to be

$$B = i(0)R(0) = V \tag{1.38}$$

Hence $i(t)$ is given by

$$i(t) = \frac{V}{R(t)} \exp\left(-\int \frac{1}{CR(t)}\, dt\right) \tag{1.39}$$

Clearly the full solution to this type of problem will require $R(t)$ to be defined. It is worth noting that although the Laplace transform method to be described later in this chapter is generally a far easier method for solving time-dependent circuit analysis problems, in this type of problem the integrating factor method turns out to give the easiest solution.

1.3 THE COMPLEX FREQUENCY METHOD

From the above section it is clear that the use of differential equations to find both the natural response of a network and its response to a forcing function (the forced response) can be quite lengthy and labour-intensive. Fortunately there are other much easier methods by which such responses can be derived. The most powerful of these is the use of Laplace transforms, but before introducing the method it is instructive to examine the complex frequency method which helps to explain the operation of the Laplace method, and also introduces the very important concept of complex frequency.

Referring back to Example 1.1, it was seen that the underdamped response of a series LCR circuit to DC voltage source took the form

$$i(t) = \frac{V}{\omega_d L} e^{-\alpha t} \sin \omega_d t \tag{1.40}$$

i.e. an exponentially decaying sinusoid which is a signal type that is very commonly encountered in the solution of a variety of engineering problems. A voltage signal of this type can also be written in a generalised form as

$$v(t) = V_p e^{\sigma t} \cos(\omega t + \theta) \tag{1.41}$$

where V_p is the peak amplitude of the signal and θ is a phase constant. In Figure 1.3 sketches of the signal are shown for the cases in which ω and/or σ are zero and the polarity of σ is changed. From the figure it can be seen that the representation of a generalised voltage signal in the form of Equation (1.41) is both useful and versatile. By the use of Euler's relationship Equation (1.41) can be written in the form

$$v(t) = \mathcal{R}e\left(V_p e^{\sigma t} e^{j(\omega t + \theta)}\right) \tag{1.42}$$

or its complex conjugate. Putting

$$\mathbf{s} = \sigma + j\omega \tag{1.43}$$

Equation (1.42) can be written as

$$v(t) = \mathcal{R}e\left(V_p e^{j\theta} e^{\mathbf{s}t}\right) = \mathcal{R}e\left(\mathbf{V} e^{\mathbf{s}t}\right) \tag{1.44}$$

where

$$\mathbf{V} = V_p e^{j\theta} \tag{1.45}$$

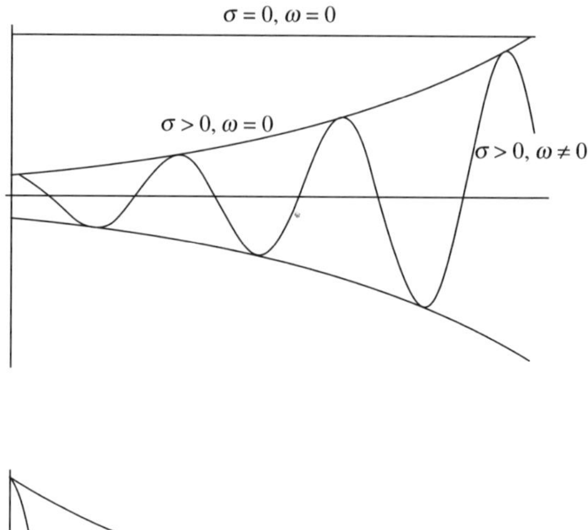

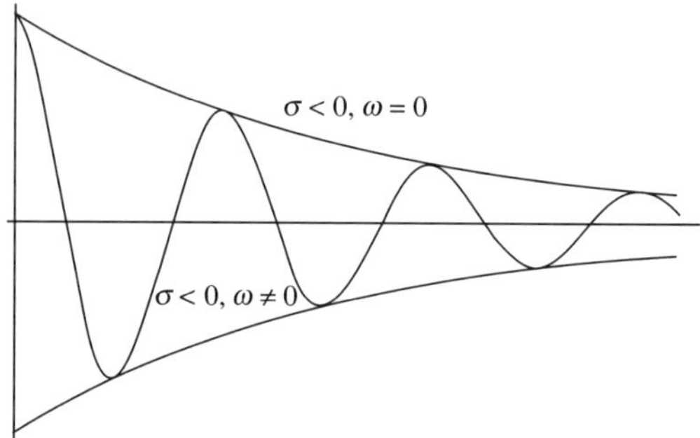

Figure 1.3 Sketches of the generalised voltage signal given by Equation (1.41) with $\theta = 0$

If such a signal is applied as the forcing function to the circuit in Example 1.1, the response of the current in the circuit must take the form

$$i(t) = \mathscr{Re}\left(I_p e^{j\phi} e^{st}\right) = \mathscr{Re}(\mathbf{I}e^{st}), \qquad \text{with} \quad \mathbf{I} = I_p e^{j\phi} \tag{1.46}$$

since there are no non-linear elements in the circuit and the natural frequency and decay rate in the circuit remain the same. The angle $(\phi - \theta)$ represents the difference in phase angle between the applied voltage and the current flowing in the circuit. If the $\mathscr{Re}$ is suppressed, remembering that it must be put back to find the circuit response in the time domain, and the expressions for voltage and current in the circuit (Equations (1.44) and (1.46)) are substituted into Equation (1.4) with $V(0) = 0$, the relationship between current and voltage becomes

$$\mathbf{V}e^{st} = \mathbf{s}L\,\mathbf{I}\,e^{st} + R\mathbf{I}e^{st} + \frac{1}{\mathbf{s}C}\mathbf{I}e^{st} \tag{1.47}$$

or

$$\mathbf{V} = \mathbf{s}L\mathbf{I} + R\mathbf{I} + \frac{1}{\mathbf{s}C}\mathbf{I} \tag{1.48}$$

or

$$\frac{\mathbf{V}}{\mathbf{I}} = \mathbf{Z}(\mathbf{s}) = \mathbf{s}L + R + \frac{1}{\mathbf{s}C} \tag{1.49}$$

This equation does not contain the time variable t but only the complex frequency parameter **s** and takes a much simpler form than the integro-differential Equation (1.4) from which it was derived. Finding the current in the circuit as a function of time is relatively simple if **V** is replaced using Equation (1.44), **s** is defined, and it is remembered that the real part of **I** is required to give the solution, i.e.

$$i(t) = \mathcal{R}e\left\{\frac{\mathbf{V}}{\mathbf{s}L + R + \dfrac{1}{\mathbf{s}C}} e^{\mathbf{s}t}\right\} \tag{1.50}$$

It should be also be remembered, however, that this method of solving integro-differential equations of similar form to Equation (1.4) will only work if the forcing function or stimulus applied to a circuit can be put in complex exponential form.

Example 1.3

A voltage signal $v(t) = 10e^{-5t}\cos(10t + 45°)V$ is applied to the series combination of a $10\,\Omega$ resistor and a 5 H inductor. Derive an expression for the current flowing in the circuit as a function of time.

Expressing the applied voltage as the real part of a complex exponential function gives

$$v(t) = \mathcal{R}e(\mathbf{V}e^{\mathbf{s}t})$$
$$\text{where} \quad \mathbf{V} = 10\angle 45° \quad \text{and} \quad \mathbf{s} = -5 + j10 \tag{1.51}$$

An expression for the voltage and current in the circuit as a differential equation can be derived, as outlined earlier, using the relations given by Equations (1.1) and (1.3), which is

$$v(t) = 10i(t) + 5\frac{di(t)}{dt} \tag{1.52}$$

Applying the complex frequency method to this equation gives

$$10\angle 45° e^{\mathbf{s}t} = 10\mathbf{I}e^{\mathbf{s}t} + 5\mathbf{sI}e^{\mathbf{s}t} \tag{1.53}$$

Hence

$$\mathbf{I} = \frac{10\angle 45°}{10 + 5\mathbf{s}} = \frac{10\angle 45°}{-15 + j50} = 0.19\angle -28.3° A \tag{1.54}$$

Thus the required expression for the current in the circuit is given by

$$i(t) = 0.19e^{-5t}\cos(10t - 28.3°)\text{A} \tag{1.55}$$

1.4 THE LAPLACE TRANSFORM METHOD

It became clear, in the last section, that the solution of the integro-differential equations, which result from the analysis of the response of a simple circuit or network to a stimulus or forcing function, could be considerably simplified if the forcing function could be expressed in a complex exponential form such as that given by Equation (1.44). Using the method, the integro-differential equations were transformed into simple algebraic equations involving the complex frequency variable **s**. However the method would appear to be confined to forcing functions which can be expressed in complex exponential form which is rather restrictive.

To get over this problem it is possible to express any periodic forcing function, that is likely to be applied to a network, in the form of a Fourier series of sine and cosine terms, or a Fourier series of complex exponential terms using the techniques of Fourier Analysis [6]. Since sine and cosine terms can be expressed as either the imaginary or real part of a complex exponential, it becomes clear that the analysis of the behaviour of a given circuit to any forcing function can be achieved by transforming the function into a Fourier series of either type and then analysing the response of the circuit to each term in the series. Provided the circuit is linear, one can apply the principle of superposition and sum the responses to each of the individual terms to get the complete circuit response. This method will of course be very laborious and involve many separate analyses to get an accurate result. In the case of a non-periodic forcing function the Fourier integral

$$F(\omega) = \int_{-\infty}^{\infty} f(t)e^{-j\omega t}\,dt \tag{1.56}$$

can be used to transform the forcing function $f(t)$ into its frequency spectrum $F(\omega)$ and this can then be used to determine a given response. Unfortunately applying the Fourier integral to a number of forcing functions that are commonly in use in circuit analysis fails, because for a function to be Fourier transformable requires

$$\int_{-\infty}^{\infty} |f(t)|\,dt \leq \infty \tag{1.57}$$

which then excludes many functions such as periodic functions or functions involving steps. This problem can be resolved by multiplying such non-convergent functions by a convergence factor CF of the form

$$\mathrm{CF} = e^{-ct} \tag{1.58}$$

where the value of c is chosen to make the function convergent [6]. Thus the Fourier integral in Equation (1.56) becomes

$$F(c, \omega) = \int_{0}^{\infty} f(t)e^{-ct}e^{-j\omega t}\,dt \tag{1.59}$$

where the lower limit of the integral has been changed to reflect the fact that in most, if not all, cases the response of a circuit to a forcing function applied at time $t = 0$ is required. Strictly, as has been mentioned before, the lower limit should be written as 0_- to take into account that $f(t)$ may be an impulse or a step function. Clearly c is not a constant and will vary according to the function that is required to become convergent. Thus if we let $c = \sigma$ and the exponents in Equation (1.59) are combined, the equation can be written as

$$F(\sigma + j\omega) = \mathbf{F}(\mathbf{s}) = \int_{0_-}^{\infty} f(t)e^{-\mathbf{s}t}\,dt \tag{1.60}$$

which is the equation for the direct Laplace transform. As will be seen later, this integral will enable virtually all of the circuit stimuli or forcing functions in the transient analysis of electrical circuits to be dealt with. The inverse transform which allows circuit responses in the complex frequency domain to be converted back to the time domain is given by

$$f(t) = \frac{1}{2\pi j} \int_{\sigma_1 - j\infty}^{\sigma_1 + j\infty} \mathbf{F}(\mathbf{s}) e^{\mathbf{s}t} \, ds \tag{1.61}$$

where σ_1 is a real positive quantity which is greater than the convergence variable σ. This integral is often not easy to perform as it involves contour integration in the complex plane, but as will be explained later much simpler methods exist to find the inverse Laplace transform of a given $\mathbf{F}(\mathbf{s})$.

1.4.1 Application of the Laplace Transform Method

Although the way in which the Laplace transform was introduced may appear a little complicated, in practice it greatly simplifies the problem of analysing the transient response of a circuit or network to the sudden application of an electrical signal (forcing function). Indeed it is arguably the most important technique that an Electrical Engineer must master to be able to cope with such analyses. However its range of application is much broader than at first might be thought, in that the reduction of a given circuit response in the complex frequency domain, by setting either $\sigma = 0$ and $j\omega = 0$, or $\sigma = 0$ automatically gives the DC or continuous AC response of that circuit, respectively. Furthermore most circuit elements or components can in some way be represented by a circuit model involving passive components and current or voltage sources. Thus the behaviour of such elements or components can be analysed when subject to any type of circuit stimulus or signal. In this way the behaviour of semiconductor circuits, electrical machines, transformers, gas discharge devices, etc., can be analysed using relatively straightforward algebraic techniques.

The basic idea of the method is to convert both the electrical circuit or network and the circuit stimulus into their complex frequency equivalent forms. From this the algebraic relationships for voltage or current are derived in terms of $\mathbf{s}$ using standard network techniques and theorems. The resulting expressions are then simplified into a form where the inverse transform can be found, usually with the help of a set of Laplace transform tables. The method has often been compared to the use of logarithms for the multiplication or division of two large numbers. A difficult calculation is essentially made easier by converting the two numbers into logarithms, performing the easier operations of addition or subtraction and then taking the antilogarithm of the result to get the answer to the problem. This then gives the desired response of the circuit in the time domain. Figure 1.4 is a diagram which describes the basis of the method.

The conversion of resistors, capacitors and inductors into their complex frequency impedances can be achieved using the integro-differential relationships for these components as given in Equations (1.1), (1.2) and (1.3), and by the application of a voltage signal of the form given by Equation (1.44). For example, in the case of a capacitor

$$i(t) = C\frac{dv(t)}{dt} \tag{1.62}$$

which becomes

$$\mathscr{Re}(V(\mathbf{s})e^{\mathbf{s}t}) = \frac{1}{\mathbf{s}C}\mathscr{Re}(I(\mathbf{s})e^{\mathbf{s}t}) \tag{1.63}$$

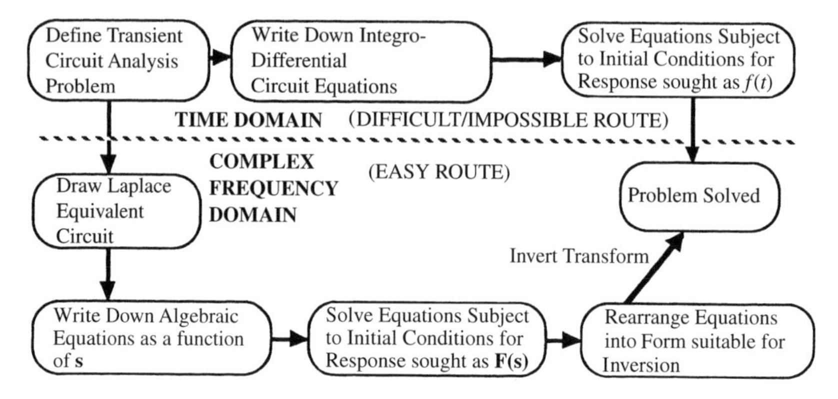

Figure 1.4 The Laplace transform concept

assuming that the capacitor is initially uncharged. If the factor e^{st} is suppressed and the $\mathscr{R}e$ is ignored, the complex frequency impedance of a capacitor becomes

$$\mathbf{Z}(\mathbf{s}) = \frac{1}{\mathbf{s}C} \tag{1.64}$$

By reference to Equations (1.1) and (1.3) it is easily seen that the complex frequency impedances of an inductor and capacitor are $\mathbf{s}L$ and R, respectively.

In order to take the Laplace method further it is necessary to examine the transformation, into the complex frequency domain, of a number of signals (forcing functions) that are likely to be applied to an electrical circuit. These signals are sketched in Figure 1.5.

1.4.2 Laplace Transforms of Some Basic Signals

The unit step

The unit step function is defined as

$$\begin{aligned} u(t) &= 0 \qquad t < 0 \\ &= 1 \qquad t \geq 0 \end{aligned} \tag{1.65}$$

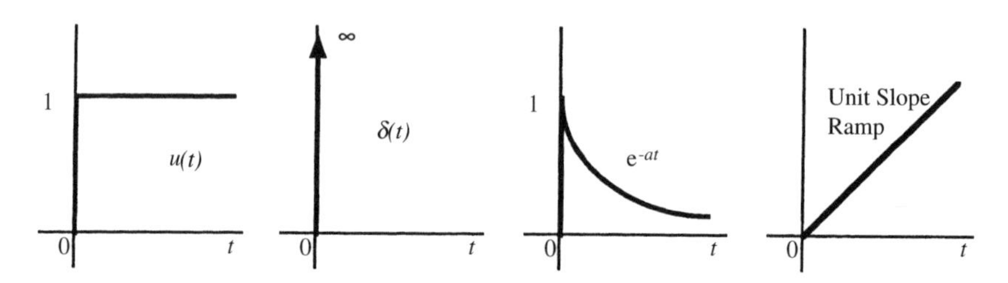

Figure 1.5 Some basic signals

Thus from Equation (1.60) which defines the Laplace transform of a function of time $f(t)$, the transform of the unit step is found to be

$$\begin{aligned}\mathscr{L}\{u(t)\} = \mathbf{F}(\mathbf{s}) &= \int_{0_-}^{\infty} u(t)e^{-\mathbf{s}t}\,dt \\ &= \left[-\frac{1}{\mathbf{s}}e^{-\mathbf{s}t}\right]_{0_-}^{\infty} \\ &= \frac{1}{\mathbf{s}}\end{aligned} \tag{1.66}$$

The delta function

The delta function or unit impulse function may be defined as

$$\left.\begin{aligned}\delta(t) &= 0 \qquad t \neq 0 \\ \int_{-\infty}^{\infty} \delta(t)\,dt &= 1 \\ \text{or,} \quad \int_{0_-}^{0_+} \delta(t)\,dt &= 1\end{aligned}\right\} \tag{1.67}$$

Thus using the definition for the Laplace transformation gives

$$\mathscr{L}\{\delta(t)\} = \mathbf{F}(\mathbf{s}) = \int_{0_-}^{\infty} \delta(t)e^{-\mathbf{s}t}\,dt = 1 \tag{1.68}$$

The exponential function

An exponential function that is suddenly applied a time $t = 0$ is written as $e^{-at}u(t)$. Thus

$$\left.\begin{aligned}\mathscr{L}\{e^{-at}u(t)\} &= \int_{0_-}^{\infty} e^{-(\mathbf{s}+a)t}\,dt \\ &= \left[-\frac{1}{\mathbf{s}+a}e^{-(\mathbf{s}+a)t}\right]_{0_-}^{\infty} \\ \mathbf{F}(\mathbf{s}) &= \frac{1}{\mathbf{s}+a}\end{aligned}\right\} \tag{1.69}$$

The ramp function

As with the sudden application of the exponential function, a ramp function applied at a time $t = 0$ is written as $tu(t)$. Thus

$$\mathscr{L}\{tu(t)\} = \mathbf{F}(\mathbf{s}) = \int_{0_-}^{\infty} te^{-\mathbf{s}t}\,dt = \frac{1}{\mathbf{s}^2} \tag{1.70}$$

1.4.3 Some Properties of the Laplace Transformation

The development of the Laplace transforms of some more difficult signals can be made easier if some of the properties of the Laplace transform are examined. From this it is then possible to construct a table of the transforms of the signals which are most commonly encountered in electrical circuits.

Linearity

The symbol $\mathscr{L}$ which represents the Laplace transform operation is a linear operator, and consequently

$$\mathscr{L}\left\{\sum_i f_i(t)\right\} = \sum_i \mathscr{L}\{f_i(t)\} = \sum_i \mathbf{F}_i(\mathbf{s}) \tag{1.71}$$

Constant product or scaling

This is simply written as

$$\mathscr{L}\{Kf(t)\} = K\mathscr{L}\{f(t)\} = K\mathbf{F}(\mathbf{s}) \tag{1.72}$$

where K is an arbitrary constant.

Differentiation

Given that the Laplace transform of a function of time $f(t)$ exists, what is the Laplace transform of its derivative $f'(t)$? Again, from the definition of the Laplace transformation and using integration by parts

$$\begin{aligned}\mathscr{L}\{f'(t)\} &= \int_{0_-}^{\infty} f'(t)e^{-\mathbf{s}t}\,dt \\ &= [f(t)e^{-\mathbf{s}t}]_{0_-}^{\infty} + \mathbf{s}\int_{0_-}^{\infty} f(t)e^{-\mathbf{s}t}\,dt\end{aligned} \tag{1.73}$$

It is easily seen that the term on the left is just $f(0_-)$ and the term on the right is $\mathbf{s}$ times the Laplace transform of $f(t)$, i.e.

$$\mathscr{L}\{f'(t)\} = \mathbf{sF}(\mathbf{s}) - f(0_-) \tag{1.74}$$

It can also be shown in the case of higher derivatives that

$$\mathscr{L}\{f^n(t)\} = \mathbf{F}^n(\mathbf{s}) = \mathbf{s}^n\mathbf{F}(\mathbf{s}) - \mathbf{s}^{n-1}f(0_-) - \mathbf{s}^{n-2}f'(0_-) - \cdots - f^{n-1}(0_-) \tag{1.75}$$

Differentiation with respect to t in the time domain corresponds to multiplication by $\mathbf{s}$ in the complex frequency domain with initial conditions being taken into account by the terms of the type $f^n(0_-)$. This, in part, helps to explain why integro-differential equations in the time domain, associated with a particular circuit analysis problem, become much simpler to handle in the complex frequency domain.

Integration

In this case it is necessary to find the effect, in the complex frequency domain, of integrating a given time-dependent function. Once again, using integration by parts gives

$$\begin{aligned}\mathscr{L}\left\{\int_{0_-}^{t} f(\tau)\,d\tau\right\} &= \int_{0_-}^{\infty}\left[\int_{0_-}^{t} f(\tau)\,d\tau\right]e^{-\mathbf{s}t}\,dt \\ &= \left[-\frac{e^{-\mathbf{s}t}}{\mathbf{s}}\int_{0_-}^{t} f(\tau)\,d\tau\right]_{0_-}^{\infty} + \frac{1}{\mathbf{s}}\int_{0_-}^{\infty} f(t)e^{-\mathbf{s}t}\,dt \\ &= \frac{\mathbf{F}(\mathbf{s})}{\mathbf{s}} \end{aligned} \tag{1.76}$$

Integration in the time domain thus corresponds to division by **s** in the complex frequency domain.

Change of scale

This is a useful property which can be used to extend the standard table of Laplace transforms

$$\left.\begin{aligned}\mathscr{L}\{f(at)\} &= \int_{0_-}^{\infty} f(at)e^{-\mathbf{s}t}\,dt \\ &= \int_{0_-}^{\infty} f(u)e^{-\mathbf{s}\left(\frac{u}{a}\right)}\,d\left(\frac{u}{a}\right) \\ &= \frac{1}{a}\int_{0_-}^{\infty} f(u)e^{-\mathbf{s}\left(\frac{u}{a}\right)}\,du \\ \mathbf{F}(\mathbf{s}) &= \frac{1}{a}f\left(\frac{\mathbf{s}}{a}\right)\end{aligned}\right\} \tag{1.77}$$

Translation in time

Often it is desirable to apply a second signal to a circuit at some time delay after the application of an initial signal. This can be achieved by translation of the second signal in time such that the function is shifted to a new position on the time axis without any distortion or change in the signal under translation. To explain how this may be done, consider the shifting of a parabola described by $f(t) = t^2$ as shown in Figure 1.6(a) to a new starting point at a time $t = t_1$ as shown in Figure 1.6(c). Writing a new function $f(t) = (t - t_1)^2$ gives the curve shown in Figure 1.6(b) which clearly does not have the correct form. However if $f(t)$ is multiplied by a unit step function which is also shifted to a new starting point at the same time $t = t_1$, the correct definition of the defined function results. To shift the unit step, the definition of the unit step function given by Equation (1.65) is modified to become

$$\begin{aligned} u(t - t_1) &= 0 \quad t < t_{1_-} \\ &= 1 \quad t \geq t_1 \end{aligned} \tag{1.78}$$

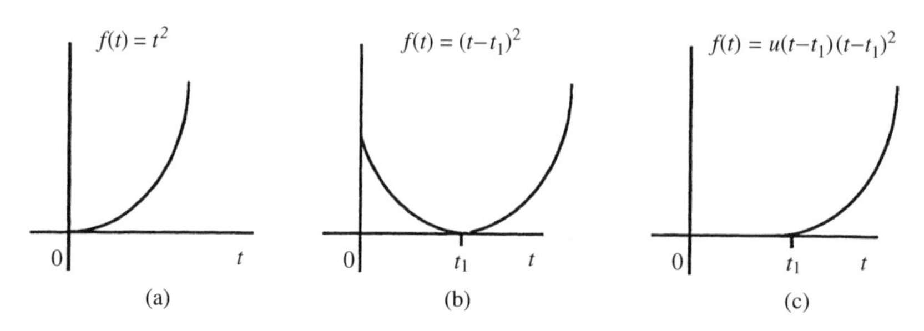

Figure 1.6 Plots of the parabola and its shifted replicas

The effect of this on the Laplace transform of the shifted function can be derived as follows

$$\left.\begin{aligned}\mathscr{L}\{f(t-t_1)u(t-t_1)\} &= \int_{0_-}^{\infty} f(t-t_1)u(t-t_1)e^{-\mathbf{s}t}\,dt \\ &= \int_{t_{1-}}^{\infty} f(t-t_1)e^{-\mathbf{s}t}\,dt \\ &= \int_{0_-}^{\infty} f(T)e^{-\mathbf{s}(T+t_1)}dT, \qquad \text{with} \quad (t-t_1)=T \\ &= e^{-\mathbf{s}t_1}\int_{0_-}^{\infty} f(T)e^{-\mathbf{s}T}\,dT \\ &= e^{-\mathbf{s}t_1}\mathscr{L}\{f(t)\} = e^{-\mathbf{s}t_1}\mathbf{F}(\mathbf{s})\end{aligned}\right\} \tag{1.79}$$

This can also be stated in a slightly altered form as

$$\mathscr{L}\{f(t)u(t-t_1)\} = e^{-\mathbf{s}t_1}\mathscr{L}\{f(t+t_1)\} \tag{1.80}$$

Note that this property is sometimes referred to as the shifting property, which should not be confused with the shifting theorem described in the next section.

Shifting theorem

The shifting theorem is another useful property of Laplace transforms which can help in finding the inverse of Laplace transforms and also the transforms of functions multiplied by exponential terms.

$$\left.\begin{aligned}\text{Let} \quad \mathscr{L}\{f(t)\} &= \mathbf{F}(\mathbf{s}) \\ \text{Then} \quad \mathscr{L}\{e^{-at}f(t)\} &= \int_{0_-}^{\infty} e^{-at}f(t)e^{-\mathbf{s}t}\,dt \\ &= \int_{0_-}^{\infty} f(t)e^{-(\mathbf{s}+a)t}\,dt \\ &= \mathbf{F}(\mathbf{s}+a)\end{aligned}\right\} \tag{1.81}$$

Thus multiplying a function of time $f(t)$ by an exponential term of the form e^{-at} has the effect of adding the variable a to $\mathbf{s}$ in the Laplace transform of $f(t)$.

Using the above properties and theorems together with the Laplace transform integral, a table of transforms can be constructed, which is added as an appendix at the rear of this book. Although the list is far from comprehensive, it does list the transforms which are most commonly encountered in electrical engineering. Note that for functions such as e^{-at}, strictly the function should be multiplied by the unit step function $u(t)$ to show that the function is starting at a time $t = 0$, but for clarity this factor has been left out of the list of transforms. For a more comprehensive list of transforms the reader is referred to the books by Spiegel [10] and Doetsch [11]. In order to illustrate the use of the above concepts and properties two examples now follow.

Example 1.5

Using the table of Laplace transforms and the list of transform properties above find the transforms of: (i) $\sin 2t$, (ii) $\cos 2t$, (iii) e^{-at} sin 2t, (iv) $t^n f(t)$, (v) $g(t) = (t-1)^3$, $t > 1$.

(i) From the transform tables

$$\mathscr{L}\{\sin t\} = \frac{1}{\mathbf{s}^2 + 1}$$

Hence using the change of scale property

$$\mathscr{L}\{\sin 2t\} = \frac{1}{2}\frac{1}{\left(\frac{\mathbf{s}}{2}\right)^2 + 1} = \frac{2}{\mathbf{s}^2 + 4} \tag{1.82}$$

or more simply by putting $\omega = 2$ into the entry for $\sin \omega t$.

(ii) Using the differentiation property and the result from (i)

$$\left.\begin{aligned} \frac{d(\sin 2t)}{dt} &= 2\cos 2t \\ \therefore \quad \mathscr{L}\{\cos 2t\} &= \frac{\mathbf{s}}{2}\frac{2}{\mathbf{s}^2 + 4} = \frac{\mathbf{s}}{\mathbf{s}^2 + 4} \end{aligned}\right\} \tag{1.83}$$

or more simply by putting $\omega = 2$ into the entry for $\cos \omega t$.

(iii) Using the shifting theorem and the entry for $\sin \omega t$

$$\mathscr{L}\{e^{-at}\sin 2t\} = \frac{2}{(\mathbf{s} + a)^2 + 4} \tag{1.84}$$

(iv) The required Laplace transform can be obtained by considering the first and higher derivatives of $\mathbf{F}(\mathbf{s})$ as follows

$$\left.\begin{aligned} \mathbf{F}'(\mathbf{s}) &= \frac{d}{d\mathbf{s}}\int_{0-}^{\infty} f(t)e^{-\mathbf{s}t}\,dt = \int_{0-}^{\infty} (-t)f(t)e^{-\mathbf{s}t}\,dt \\ \therefore \quad \mathbf{F}^{(n)}(\mathbf{s}) &= (-1)^n \mathscr{L}\{t^n f(t)\} \\ \text{or} \quad \mathscr{L}\{t^n f(t)\} &= (-1)^n \mathbf{F}^{(n)}(\mathbf{s}) \end{aligned}\right\} \tag{1.85}$$

where $\mathbf{F}^n(\mathbf{s})$ is the nth derivative of $\mathbf{F}(\mathbf{s})$.

(v) From the tables

$$\mathscr{L}\{t^3\} = \frac{3!}{\mathbf{s}^4} = \frac{6}{\mathbf{s}^4}$$

Therefore using the translation property the required transform is

$$\mathscr{L}\{g(t)\} = \frac{6e^{-\mathbf{s}}}{\mathbf{s}^4} \tag{1.86}$$

Example 1.6

The Laplace transforms of a number of waveforms or signals, including repetitive ones, can be found by using a superposition of a number of elementary functions as described above. By this means find the Laplace transforms of the functions shown in Figure 1.7.

(i) This waveform can be decomposed into a ramp with unit positive slope which starts at time $t = 0$, followed by a ramp with equal but negative slope starting at a time $t = t_1$. Thus the Laplace transform is given by

$$\mathbf{F}(\mathbf{s}) = \frac{1}{\mathbf{s}^2} + \left(-\frac{e^{-\mathbf{s}t_1}}{\mathbf{s}^2}\right) = \frac{1}{\mathbf{s}^2}(1 - e^{-\mathbf{s}t_1}) \tag{1.87}$$

(ii) In this case the waveform must be split into three parts, a ramp with unit slope which starts at a time $t = t_1$, a second ramp with a slope of -2 starting at a time $t = 2t_1$, and a third ramp with unit positive slope starting at time $t = 3t_1$. The Laplace transform is therefore given by

$$\mathbf{F}(\mathbf{s}) = \frac{e^{-\mathbf{s}t_1}}{\mathbf{s}^2} - \frac{2e^{-\mathbf{s}2t_1}}{\mathbf{s}^2} + \frac{e^{-\mathbf{s}3t_1}}{\mathbf{s}^2} = \frac{1}{\mathbf{s}^2}\left(e^{-\mathbf{s}t_1} - 2e^{-\mathbf{s}2t_1} + e^{-\mathbf{s}3t_1}\right) \tag{1.88}$$

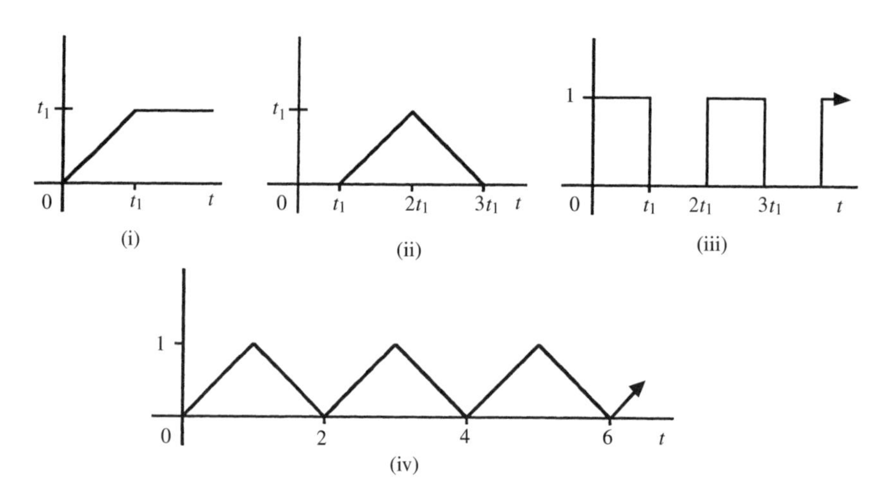

Figure 1.7 Waveforms for Example 1.5

(iii) The repetitive rectangular waveform can be split into a series of alternating positive and negative steps delayed successively by the time t_1. Thus the Laplace transform of the waveform is given by

$$\mathbf{F}(\mathrm{s}) = \frac{1}{\mathrm{s}} - \frac{e^{-\mathrm{s}t_1}}{\mathrm{s}} + \frac{e^{-\mathrm{s}2t_1}}{\mathrm{s}} - \frac{e^{-\mathrm{s}3t_1}}{\mathrm{s}} + \cdots = \frac{1}{\mathrm{s}(1+e^{-\mathrm{s}t_1})} \tag{1.89}$$

Note that series of exponential terms has been simplified into the sum of a geometric progression.

(iv) This repetitive waveform can be split, as with the second example, into a series of ramp functions with positive and negative unity slope. Thus

$$\begin{aligned}
\mathbf{F}(\mathrm{s}) &= \frac{1}{\mathrm{s}^2} - \frac{2e^{-\mathrm{s}}}{\mathrm{s}^2} + \frac{2e^{-2\mathrm{s}}}{\mathrm{s}^2} - \frac{2e^{-3\mathrm{s}}}{\mathrm{s}^2} + \frac{2e^{-4\mathrm{s}}}{\mathrm{s}^2} - \cdots \\
&= \frac{1}{\mathrm{s}^2}(-1 + 2 - 2e^{-\mathrm{s}} + 2e^{-2\mathrm{s}} - 2e^{-3\mathrm{s}} + 2e^{-4\mathrm{s}} \cdots) \\
&= \frac{1}{\mathrm{s}^2}\left(-1 + \frac{2}{1+e^{-\mathrm{s}}}\right) \\
&= \frac{1}{\mathrm{s}^2}\left(\frac{1-e^{-\mathrm{s}}}{1+e^{-\mathrm{s}}}\right) \\
&= \frac{1}{\mathrm{s}^2}\left(\frac{e^{+\mathrm{s}/2} - e^{-\mathrm{s}/2}}{e^{+\mathrm{s}/2} + e^{-\mathrm{s}/2}}\right) \\
&= \frac{1}{\mathrm{s}^2}\tanh\frac{\mathrm{s}}{2}
\end{aligned} \tag{1.90}$$

1.4.4 Finding the Inverse Laplace Transform $\mathscr{L}^{-1}$

Before applying Laplace transforms to circuits it is necessary to look in more detail at the process of finding the inverse of a Laplace transform to return to the time domain. For simple functions in **s**, reference to the set of Laplace transforms, such as the one at the rear of this book, is usually the easiest and most convenient method, having first rearranged the function **F(s)** to be inverted into a form which can be recognised in the tables. For more complicated functions in **s**, which quite often take the form of a rational function of the type **F(s)**=**P(s)**/**Q(s)**, where **P(s)** and **Q(s)** are polynomials in **s** with the degree of **P(s)** usually being less than that of **Q(s)**, the function can be simplified into a series of partial fractions. Alternatively if the **F(s)** to be inverted has a suitable series expansion in **s**, the terms in the expansion can be individually inverted and then summed to find the related $f(t)$. Finally, use can be made of the complex inversion formula given by Equation (1.61). This, in practice, involves carrying out a contour integration in the complex plane. Although this might seem at first sight to be a difficult process, reference to more detailed texts on the use of Laplace transforms, such as that by Holbrook [6], shows that this can, in some cases, be relatively straightforward. However the details of this method lie outside the scope of this book. It is of value to look at the partial fraction method and some examples of the inversion process.

Partial fractions

Often, in the use of the Laplace transform method, the analysis results in a form of **F(s)** of the type

$$\mathbf{F}(\mathbf{s}) = \frac{\mathbf{P}(\mathbf{s})}{\mathbf{Q}(\mathbf{s})} \tag{1.91}$$

where **P(s)** and **Q(s)** are polynomials in **s**, **Q(s)** usually being of higher order than **P(s)**. If, for example, these can be reduced to a series of partial fractions of the form

$$\mathbf{F}(\mathbf{s}) = \frac{\mathbf{P}(\mathbf{s})}{\mathbf{Q}(\mathbf{s})} = \frac{A}{(\mathbf{s}+a_1)} + \frac{B}{(\mathbf{s}+a_2)} + \frac{C}{(\mathbf{s}+a_3)} + \cdots \tag{1.92}$$

Then, by reference to the Laplace transform table this can be simply inverted to

$$f(t) = Ae^{-a_1 t} + Be^{-a_2 t} + Ce^{-a_3 t} + \cdots \tag{1.93}$$

The first stage in the process is to divide both the top and bottom polynomials in the rational function in **s** (**F(s)**) by any coefficient which may be multiplying the highest order term in **s** in **Q(s)**. The roots of **Q(s)** are then found either by inspection or by the use of an appropriate factorisation formula. This may only be possible if numerical values for the coefficients of the various terms in **s** are substituted first.

In the case where the degree of **P(s)** is greater than or equal to that of **Q(s)**, it is necessary to divide the numerator **P(s)** by the denominator **Q(s)** so that the remainder can then be expanded into partial fractions. This process is best illustrated by the following example.

Example 1.7

Find the inverse Laplace transform of

$$\mathbf{F}(\mathbf{s}) = \frac{2\mathbf{s}^2 + 7\mathbf{s} + 16}{\mathbf{s}^2 + 4\mathbf{s} + 8} \tag{1.94}$$

First divide the numerator by the denominator to give

$$\mathbf{F}(\mathbf{s}) = 2 - \frac{\mathbf{s}}{\mathbf{s}^2 + 4\mathbf{s} + 8} \tag{1.95}$$

which can be written as

$$\mathbf{F}(\mathbf{s}) = 2 - \frac{(\mathbf{s}+2) - 2}{(\mathbf{s}+2)^2 + 4} \tag{1.96}$$

The inverse Laplace transform can be found by reference to the tables as

$$f(t) = 2\delta(t) + e^{-2t}(\sin 2t - \cos 2t) \tag{1.97}$$

There are three main methods which can be used to find the partial fraction expansion of rational functions in **s**. These are the method of equating coefficients, the use of the "cover-up" rule, and the use of the Heaviside expansion formula. These methods are now described.

Equating coefficients

This method is really only suitable if the denominator $\mathbf{Q(s)}$ only has a small number of factors and is easily factorised. Again the method is best described by way of an example.

Example 1.8

Find the inverse Laplace transform of

$$\mathbf{F(s)} = \frac{\mathbf{s} + 10}{\mathbf{s}^2 + 5\mathbf{s} + 4} \tag{1.98}$$

The first stage is to factorise the denominator, and split the denominator into partial fractions with unknown coefficients assigned to the numerators of the fractions, i.e.

$$\mathbf{F(s)} = \frac{\mathbf{s} + 10}{(\mathbf{s}+1)(\mathbf{s}+4)} = \frac{A}{(\mathbf{s}+1)} + \frac{B}{(\mathbf{s}+4)} \tag{1.99}$$

The coefficients A and B are known as residues and can be found, in this case, by combining the denominators of the two fractions and equating the resulting numerator to that of the original expression. Thus

$$\mathbf{s} + 10 = A(\mathbf{s}+4) + B(\mathbf{s}+1) \tag{1.100}$$

Equating the coefficients of $\mathbf{s}^0$ and $\mathbf{s}^1$ gives

$$\left.\begin{aligned} A + B = 1 \quad &\text{and} \quad 4A + B = 10 \\ \therefore \quad A = 3 \quad &\text{and} \quad B = -2 \end{aligned}\right\} \tag{1.101}$$

Thus the transform and its inverse become

$$\left.\begin{aligned} F(\mathbf{s}) &= \frac{3}{\mathbf{s}+1} - \frac{2}{\mathbf{s}+4} \\ \text{and} \quad f(t) &= 3e^{-1} - 2e^{-4} \end{aligned}\right\} \tag{1.102}$$

In the case of denominators which have quadratic terms or repeated roots the partial fraction expansions take different forms. For the quadratic case the required expansion should, for example, take the form

$$\begin{aligned} \mathbf{F(s)} &= \frac{P(\mathbf{s})}{(\mathbf{s}+a)[(\mathbf{s}+b)^2 + c^2]} \\ \mathbf{F(s)} &= \frac{A}{(\mathbf{s}+a)} + \frac{B\mathbf{s} + C}{[(\mathbf{s}+b)^2 + c^2]} \end{aligned} \tag{1.103}$$

In the case of a squared quadratic term the expansion should, again for example, take the form

$$\begin{aligned} \mathbf{F(s)} &= \frac{\mathbf{P(s)}}{[(\mathbf{s}+a)^2 + b^2]^2} \\ \mathbf{F(s)} &= \frac{A\mathbf{s} + B}{(\mathbf{s}+a)^2 + b^2} + \frac{C\mathbf{s} + D}{[(\mathbf{s}+a)^2 + b^2]^2} \end{aligned} \tag{1.104}$$

Finally, for the case where the denominator contains repeated terms or roots the expansion should be

$$\begin{aligned}\mathbf{F}(\mathbf{s}) &= \frac{\mathbf{P}(\mathbf{s})}{(\mathbf{s}+a)^n\mathbf{R}(\mathbf{s})} \\ \mathbf{F}(\mathbf{s}) &= \frac{A_1}{(\mathbf{s}+a)^n} + \frac{A_2}{(\mathbf{s}+a)^{n-1}} + \frac{A_3}{(\mathbf{s}+a)^{n-2}} + \cdots + \frac{A_{n-1}}{(\mathbf{s}+a)} + \frac{\mathbf{P}'(\mathbf{s})}{\mathbf{R}(\mathbf{s})}\end{aligned} \tag{1.105}$$

where A_1, A_2, A_3, ... are residues and $\mathbf{P}'(\mathbf{s})/\mathbf{R}(\mathbf{s})$ represents the remaining terms in the expansion.

The "cover-up" method

This is a very useful method of finding the residues associated with a particular partial fraction expansion which involves many factors in the denominator. Consider an expansion of the form

$$\left.\begin{aligned}\mathbf{F}(\mathbf{s}) &= \frac{\mathbf{P}(\mathbf{s})}{\mathbf{Q}(\mathbf{s})} = \frac{A}{(\mathbf{s}+a)} + \frac{B}{(\mathbf{s}+b)} + \frac{C}{(\mathbf{s}+c)} + \frac{D}{(\mathbf{s}+d)} \cdots \\ \therefore \quad \mathbf{P}(\mathbf{s}) &= A(\mathbf{s}+b)(\mathbf{s}+c)(\mathbf{s}+d)\ldots \\ &\quad + B(\mathbf{s}+a)(\mathbf{s}+c)(\mathbf{s}+d)\ldots \\ &\quad + C(\mathbf{s}+a)(\mathbf{s}+b)(\mathbf{s}+d)\ldots \\ &\quad + \cdots\end{aligned}\right\} \tag{1.106}$$

Putting $\mathbf{s} = -a$ into the expression for $\mathbf{P}(\mathbf{s})$ gives

$$\begin{aligned}\mathbf{P}(-a) &= A(-a+b)(-a+c)(-a+d)\ldots \\ \text{or} \quad A &= \frac{\mathbf{P}(-a)}{(-a+b)(-a+c)(-a+d)\ldots} \\ \text{or} \quad A &= (\mathbf{s}+a)\mathbf{F}(\mathbf{s})|_{\mathrm{s}=-\mathrm{a}}\end{aligned} \tag{1.107}$$

To see how this method works in practice and how the term $(\mathbf{s}+a)$, for example, must cancel such that A is not zero, an example follows.

Example 1.9

Find the partial fraction expansion for

$$\mathbf{F}(\mathbf{s}) = \frac{\mathbf{s}^2 + 2\mathbf{s} + 3}{\mathbf{s}^3 + 6\mathbf{s}^2 + 11\mathbf{s} + 6} \tag{1.108}$$

This expression factorises to

$$\mathbf{F}(\mathbf{s}) = \frac{\mathbf{s}^2 + 2\mathbf{s} + 3}{(\mathbf{s}+1)(\mathbf{s}+2)(\mathbf{s}+3)} \tag{1.109}$$

Therefore using the "cover-up" method the residues are found to be

$$\left.\begin{aligned} A &= (\mathbf{s}+1)\mathbf{F}(\mathbf{s})|_{\mathbf{s}=-1} \\ &= (-1+1)\frac{1-2+3}{(-1+1)(-1+2)(-1+3)} \\ &= 1 \\ B &= (\mathbf{s}+2)\mathbf{F}(\mathbf{s})|_{\mathbf{s}=-2} \\ &= -3 \\ C &= (\mathbf{s}+3)\mathbf{F}(\mathbf{s})|_{\mathbf{s}=-3} \\ &= 3 \\ \therefore \quad \mathbf{F}(\mathbf{s}) &= \frac{1}{(\mathbf{s}+1)} - \frac{3}{(\mathbf{s}+2)} + \frac{3}{(\mathbf{s}+3)} \end{aligned}\right\} \tag{1.110}$$

In the case of a denominator with complex roots the "cover-up" method will also work. For example, the denominator of Equation (1.103) will have complex roots if the term c^2 is positive. Consider the function

$$\begin{aligned} \mathbf{F}(\mathbf{s}) &= \frac{\mathbf{P}(\mathbf{s})}{[(\mathbf{s}+b)^2 + c^2)]} \\ &= \frac{\mathbf{P}(\mathbf{s})}{[(\mathbf{s}+b-jc)(\mathbf{s}+b+jc)]} \\ &= \frac{A}{(\mathbf{s}+b-jc)} + \frac{B}{(\mathbf{s}+b+jc)} \end{aligned} \tag{1.111}$$

where it is again assumed that c^2 is positive. Using Equation (1.107) gives

$$\left.\begin{aligned} A &= \frac{\mathbf{P}(-b+jc)}{2jc} \\ B &= \frac{\mathbf{P}(-b-jc)}{-2jc} \end{aligned}\right\} \tag{1.112}$$

from which it can be seen that A and B are complex conjugates and therefore B may be written as $B=A^*$. The inverse Laplace transform of Equation (1.111) is therefore

$$\begin{aligned} f(t) &= \mathscr{L}^{-1}\left[\frac{A}{(\mathbf{s}+b-jc)} + \frac{A^*}{(\mathbf{s}+b+jc)}\right] \\ &= e^{-bt}(Ae^{jct} + A^*e^{-jct}) \\ &= 2e^{-bt}(\mathscr{Re}\{A\}\cos ct - \mathscr{Im}\{A\}\sin ct) \end{aligned} \tag{1.113}$$

The Heaviside expansion formula

This can be regarded as an extension of the above method. Referring back to Equation (1.106), if the reconstituted denominator is differentiated with respect to $\mathbf{s}$,

$\mathbf{Q}'(\mathbf{s})$ is given by

$$\begin{aligned}\mathbf{Q}'(\mathbf{s}) = (\mathbf{s}+b)(\mathbf{s}+c)(\mathbf{s}+d)\ldots\\ +(\mathbf{s}+a)(\mathbf{s}+c)(\mathbf{s}+d)\ldots\\ +(\mathbf{s}+a)(\mathbf{s}+b)(\mathbf{s}+d)\ldots\\ +\cdots\end{aligned} \tag{1.114}$$

If $\mathbf{s}$ is set equal to $-a$ then all but the first term of $\mathbf{Q}'(\mathbf{s})$ vanishes, and therefore

$$A = \frac{\mathbf{P}(-a)}{\mathbf{Q}'(-a)}, \quad B = \frac{\mathbf{P}(-b)}{\mathbf{Q}'(-b)} \quad \text{etc.} \tag{1.115}$$

1.4.5 The Laplace Transform Circuit

Now that the most useful properties and techniques associated with Laplace transforms have been covered, it is possible to show just how effective and simple the application of Laplace transforms to the analysis of the performance of electrical circuits can be. Rather than writing the circuit equations for a particular analysis in the form of integro-differential equations and converting them using the Laplace transforms of differentials and integrals, it is much easier to convert the circuit to be analysed into its "Laplace transform" form. Thus the analysis proceeds automatically in the complex frequency domain rather than in the time domain. To do this it is necessary to re-examine the ways in which electrical components can be represented in the Laplace transform circuit. This is done by reference to the Equations (1.1) to (1.3) which give the voltage/current relationships for resistors, capacitors and inductors. These equations are repeated below for convenience.

$$i(t) = \frac{v(t)}{R} \quad \text{or} \quad v(t) = Ri(t) \tag{1.1}$$

$$i(t) = C\frac{dv(t)}{dt} \quad \text{or} \quad v(t) = \frac{1}{C}\int_0^t i(\tau)\,d\tau + V(0) \tag{1.2}$$

$$i(t) = \frac{1}{L}\int_0^t v(\tau)\,d\tau + I(0) \quad \text{or} \quad v(t) = L\frac{di(t)}{dt} \tag{1.3}$$

The Laplace transforms of these equations can be taken using the expressions derived earlier for the Laplace transforms of the differential and integrals of specific functions of time. Thus these equations become

$$\left.\begin{aligned}\mathbf{I}(\mathbf{s}) = \frac{\mathbf{V}(\mathbf{s})}{R} \quad &\text{or} \quad \mathbf{V}(\mathbf{s}) = \mathbf{I}(\mathbf{s})R\\ \mathbf{I}(\mathbf{s}) = \mathbf{s}C\mathbf{V}(\mathbf{s}) - CV(0) \quad &\text{or} \quad \mathbf{V}(\mathbf{s}) = \frac{\mathbf{I}(\mathbf{s})}{\mathbf{s}C} + \frac{V(0)}{\mathbf{s}}\\ \mathbf{I}(\mathbf{s}) = \frac{\mathbf{V}(\mathbf{s})}{\mathbf{s}L} + \frac{I(0)}{\mathbf{s}} \quad &\text{or} \quad \mathbf{V}(\mathbf{s}) = \mathbf{s}L\mathbf{I}(\mathbf{s}) - LI(0)\end{aligned}\right\} \tag{1.116}$$

From these relationships a number of conclusions can be drawn as to how to represent resistors, capacitors and inductors in the Laplace circuit. In the case of resistors it is clear that they can be represented without change as the current/voltage relationships are unchanged in the complex frequency domain. In the case of a capacitor it can be seen that it can be represented either as an admittance $\mathbf{s}C$ in parallel with a current source $CV(0)$ or as an impedance $1/\mathbf{s}C$ in series with a voltage source $V(0)/\mathbf{s}$. Similarly an inductor can be represented as an admittance $1/\mathbf{s}L$ in parallel with a current source $I(0)/\mathbf{s}$ or as and impedance $\mathbf{s}L$ in series with a voltage source $LI(0)$. Notice how easily the Laplace method deals with initial conditions, at the start of an analysis, in the form of initial voltages or currents with the equations giving the polarities of the sources directly. In most cases it is more convenient to represent a capacitor or an inductor as an impedance with an associated voltage source rather than as an admittance with an associated current source. However this flexibility can be useful in certain circuit analysis problems in an analogous way to the use of Thévenin's and Norton's theorems in mesh and nodal analysis, respectively. In order to give a pictorial explanation of the two possible representations of capacitors and inductors, the appropriate equivalent circuits are drawn in Figure 1.8.

The representation of a transformer can be carried out in a similar way. Figure 1.9 gives both the standard representation of a transformer in the time domain and its equivalent representation in the Laplace circuit or complex frequency domain. With reference to the time domain circuit the two equations linking the currents and voltages in the primary and secondary circuits can be written as

$$\left.\begin{aligned} v_p(t) &= L_p\frac{di_p(t)}{dt} + M\frac{di_s(t)}{dt} \\ v_s(t) &= L_s\frac{di_s(t)}{dt} + M\frac{di_p(t)}{dt} \end{aligned}\right\} \tag{1.117}$$

where the subscripts p and s refer to the primary and secondary circuits.

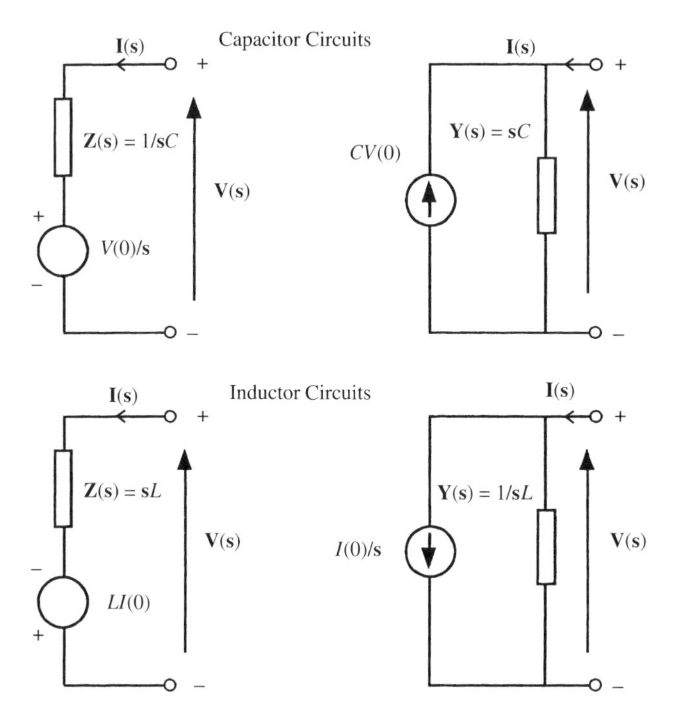

Figure 1.8 Equivalent circuit models for a capacitor and an inductor in the Laplace circuit

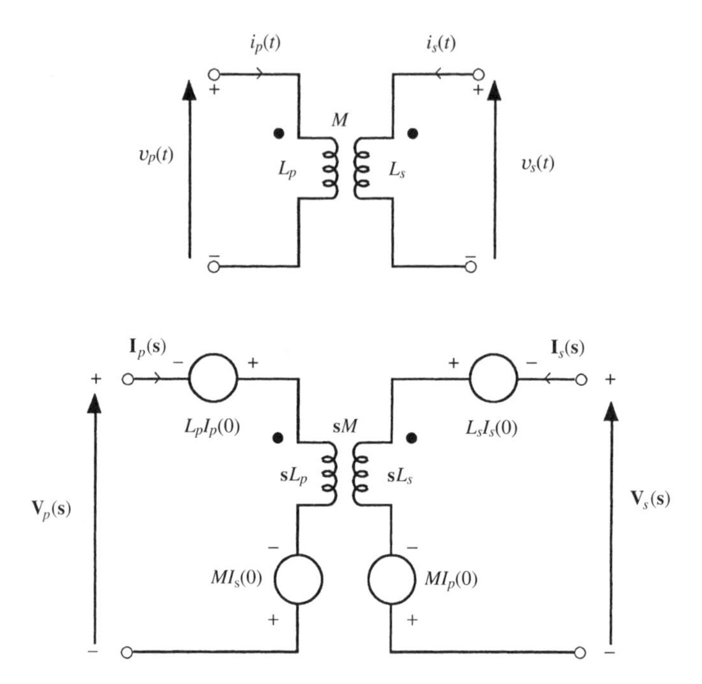

Figure 1.9 Circuit for the transformer in the time domain and its Laplace equivalent

Note that the two dots which are marked on the mutual inductance are there to denote the phase relationships between the primary and secondary circuits. As marked, the windings are in phase, i.e. when the "dotted" end of the primary is positive the "dotted" end of the secondary is also positive and in phase with the primary. The Laplace equivalent circuit can now be derived as before by taking the transform of Equation (1.117). This gives

$$\left.\begin{aligned}\mathbf{V}_p(\mathbf{s}) &= \mathbf{s}L_p\mathbf{I}_p(\mathbf{s}) - L_pI_p(0) + \mathbf{s}M\mathbf{I}_s(\mathbf{s}) - MI_s(0)\\ \mathbf{V}_s(\mathbf{s}) &= \mathbf{s}L_s\mathbf{I}_s(\mathbf{s}) - L_sI_s(0) + \mathbf{s}M\mathbf{I}_p(\mathbf{s}) - MI_p(0)\end{aligned}\right\} \tag{1.118}$$

As with the Laplace transform equations for the capacitor and inductor, the various terms in the equations can be interpreted as impedances and voltage sources whose polarities can be determined directly from the polarities of the related terms in Equation (1.118). Thus the equivalent Laplace circuit form of a transformer can be drawn directly with reference to Equation (1.118). Care though must be taken that, if the directions of current flow or polarity of the applied voltages are changed, the polarity of the voltage sources in the model are properly corrected. Similarly, if the phase relationship of the windings is altered the polarities of the sources must also be changed.

To fully illustrate the use of these equivalent circuits and the Laplace transform method a number of examples now follow.

Example 1.10

A capacitor charged to a positive potential V is discharged into the series combination of an inductor, a resistor, and a capacitor. Determine the current through the circuit as a function of time and the potential on the second capacitor.

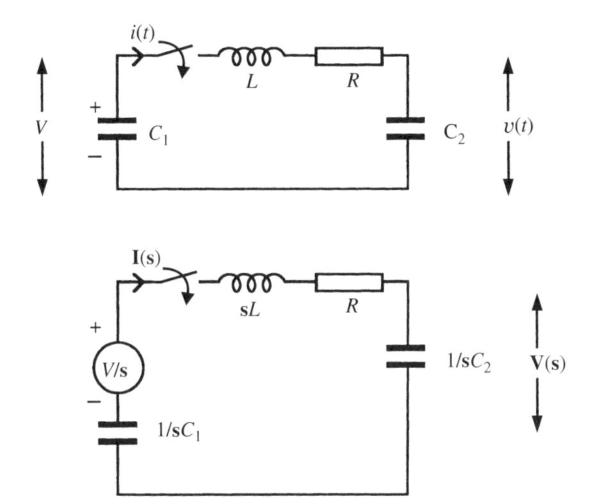

Figure 1.10 The circuit and its Laplace transformed equivalent for Example 1.10

The circuit and its Laplace transform equivalent are shown in Figure 1.10. C_1 is the capacitor to be discharged.

The problem is simplified by combining the two capacitors as they are in series to a new capacitor C given by

$$C = \frac{C_1 C_2}{C_1 + C_2} \tag{1.119}$$

Therefore the problem reduces, in part, to that of Example 1.1 so that a direct comparison between the Laplace method and the classical method can be made. The current in the circuit is easily found to be

$$\begin{aligned} \mathbf{I}(\mathbf{s}) &= \frac{V}{\mathbf{s}} \frac{1}{\mathbf{s}L + R + \dfrac{1}{\mathbf{s}C}} \\ &= \frac{CV}{\mathbf{s}^2 LC + \mathbf{s}CR + 1} \end{aligned} \tag{1.120}$$

In order to put this into a form which is recognisable in the transform tables, the first stage is to divide the numerator and denominator by LC to remove the coefficient of the highest term in $\mathbf{s}$ ($\mathbf{s}^2$).

$$\mathbf{I}(\mathbf{s}) = \frac{V}{L} \frac{1}{\mathbf{s}^2 + \mathbf{s}\dfrac{R}{L} + \dfrac{1}{LC}} \tag{1.121}$$

This expression can then be factorised by putting, as before

$$\alpha = \frac{R}{2L}, \quad \omega_0 = \sqrt{\frac{1}{LC}} \quad \text{and} \quad \omega_d^2 = \omega_0^2 - \alpha^2 \tag{1.122}$$

Therefore Equation (1.121) becomes

$$\begin{aligned}\mathbf{I}(s) &= \frac{V}{L}\frac{1}{(\mathbf{s}+\alpha)^2+\omega_d^2}\\ &= \frac{V}{\omega_d L}\frac{\omega_d}{(\mathbf{s}+\alpha)^2+\omega_d^2}\end{aligned} \tag{1.123}$$

As in Example 1.1 three cases are considered, namely ω_d^2 is positive (underdamping), is zero (α^2 is equal to ω_0^2) (critical damping), and ω_d^2 is negative and can be written as $\omega_d^2 = -\beta^2$ for clarity (overdamping). In these cases Equation (1.123) becomes

$$\left.\begin{aligned}\mathbf{I}(\mathbf{s}) &= \frac{V}{\omega_d L}\frac{\omega_d}{(\mathbf{s}+\alpha)^2+\omega_d^2} & \omega_d^2 > 0\\ \mathbf{I}(\mathbf{s}) &= \frac{V}{L}\frac{1}{(\mathbf{s}+\alpha)^2} & \omega_d^2 = 0\\ \mathbf{I}(\mathbf{s}) &= \frac{V}{\beta L}\frac{\beta}{(\mathbf{s}+\alpha)^2-\beta^2} & \omega_d^2 < 0\end{aligned}\right\} \tag{1.124}$$

By reference to the table of Laplace transforms and with the use of the shifting theorem the inverse transforms of each of these equations are easily found to be

$$\left.\begin{aligned}i(t) &= \frac{V}{\omega_d L}e^{-\alpha t}\sin\omega_d t & \omega_d^2 > 0\\ i(t) &= \frac{Vt}{L}e^{-\alpha t} & \omega_d^2 = 0\\ i(t) &= \frac{V}{\beta L}e^{-\alpha t}\sinh\beta t & \omega_d^2 < 0\end{aligned}\right\} \tag{1.125}$$

To find the potential on C_2 is, again, very straightforward using the integral form for the current through a capacitor (Equation (1.2)) and the integration property of Laplace transforms, i.e.

$$\begin{aligned}v(t) &= \frac{1}{C_2}\int_0^t i(\tau)\,d\tau \qquad (V(0)=0)\\ \therefore\quad \mathbf{V}(\mathbf{s}) &= \frac{\mathbf{I}(\mathbf{s})}{\mathbf{s}C_2}\end{aligned} \tag{1.126}$$

Thus in the case where $\omega_d^2 > 0$ the Laplace transform $\mathbf{V}(\mathbf{s})$ and its inverse become

$$\begin{aligned}\mathbf{V}(\mathbf{s}) &= \frac{V}{\mathbf{s}C_2 L}\frac{1}{(\mathbf{s}+\alpha)^2+\omega_d^2}\\ \mathbf{V}(\mathbf{s}) &= \frac{V}{C_2 L(\alpha^2+\omega_d^2)}\left(\frac{1}{\mathbf{s}}-\frac{(\mathbf{s}+2\alpha)}{(\mathbf{s}+\alpha)^2+\omega_d^2}\right)\\ \mathbf{V}(\mathbf{s}) &= V\frac{C_1}{C_1+C_2}\left(\frac{1}{\mathbf{s}}-\frac{\alpha}{\omega_d}\frac{\omega_d}{(\mathbf{s}+\alpha)^2+\omega_d^2}-\frac{(\mathbf{s}+\alpha)}{(\mathbf{s}+\alpha)^2+\omega_d^2}\right)\\ v(t) &= V\frac{C_1}{C_1+C_2}\left[1-\exp(-\alpha t)\left(\frac{\alpha}{\omega_d}\sin\omega_d t+\cos\omega_d t\right)\right]\end{aligned} \tag{1.127}$$

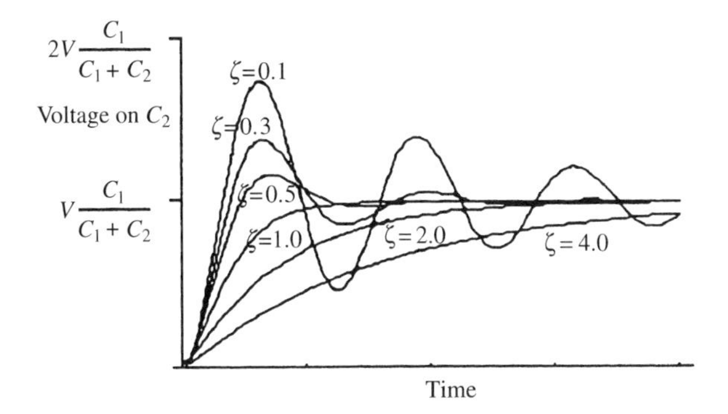

Figure 1.11 Typical voltage waveforms for the potential on C_2 for different damping factors

where the rules for finding partial fractions have been applied and it has been recognised that $\alpha^2 + \omega_d^2 = \omega_0^2$. The voltage on C_2 for the other two cases of interest, namely ω_d^2 is zero (α^2 is equal to ω_0^2) and ω_d^2 is negative, are also straightforward to find.

In some other texts on this topic Equation (1.123) is written in a different form

$$\begin{aligned}\mathbf{I}(\mathbf{s}) &= \frac{V}{L}\frac{1}{\mathbf{s}^2 + 2\mathbf{s}\zeta\omega_0 + \omega_0^2} \\ &= \frac{V}{L}\frac{1}{(\mathbf{s}+\zeta\omega_0)^2 + \omega_0^2(1-\zeta^2)}\end{aligned} \tag{1.128}$$

where ζ is known as the damping factor. Since it is easy to show that $\omega_d^2 = \omega_0^2(1-\zeta^2)$, the three damping conditions correspond to $\zeta < 1$ (underdamping), $\zeta = 1$ (critical damping), and $\zeta > 1$ (overdamping). Figure 1.11 shows typical plots of the potential on C_2 as a function of time for different values of ζ. From this plot it is interesting to note that the fastest rise-time of potential occurs in the underdamped condition where $\zeta < 1$, and that the potential on C_2 can rise to levels above the original potential on C_1. This fact is often put to good use in the circuit technique of resonant charging where the voltage gain can be used to charge a capacitor to a potential which is close to twice that of a charging power supply. If the resistor in the circuit given in Figure 1.10 is assumed to be so small that it can be neglected, then Equation (1.127) with $\alpha = 0$ becomes

$$v(t) = V\frac{C_1}{C_1 + C_2}(1 - \cos\omega_d t) \tag{1.129}$$

Since the maximum value of $(1 - \cos\omega_d t)$ is 2, then if C_1 is very much larger than C_2 the maximum value to which $v(t)$ can rise is almost 2 V. A typical resonant charging circuit is shown in Figure 1.12. Here C_1, which is chosen to be much larger than C_2 and is often

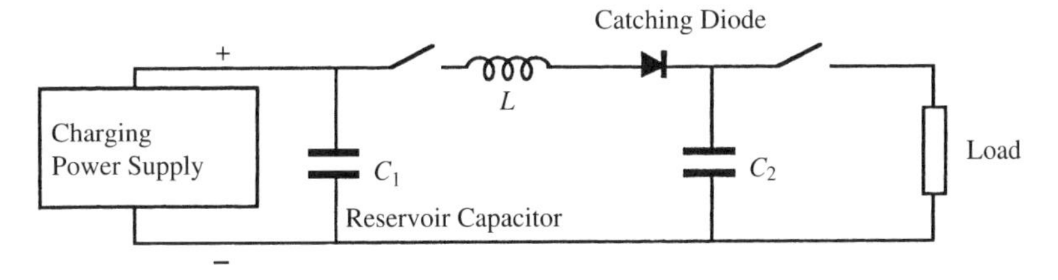

Figure 1.12 Typical resonant charging circuit

referred to as the reservoir capacitor, charges C_2 when the switch on the right is open and the switch on the left is closed. The diode keeps C_2 to the maximum potential and prevents it from falling as $v(t)$ would do through its dependency on the $(1 - \cos \omega_d t)$ term. The diode is often referred to as a catching diode.

Example 1.11

The useable bandwidth of resistive voltage dividers is often restricted by the presence of stray capacitance, which can be represented by a capacitor which lies in parallel with the resistor which forms the top leg of the divider. Show that the bandwidth of the divider can, in theory, be increased to an infinite extent if a second capacitor is placed in parallel with the resistor which forms the bottom leg of the divider, and determine its value.

The Laplace transform circuit for this problem is shown in Figure 1.13. From this circuit the output voltage of the divider in terms of the input can be written as

$$
\begin{aligned}
\mathbf{V}_o(\mathbf{s}) &= \frac{R_2 \| \dfrac{1}{\mathbf{s}C_2}}{R_1 \| \dfrac{1}{\mathbf{s}C_1} + R_2 \| \dfrac{1}{\mathbf{s}C_2}} \mathbf{V}_i(\mathbf{s}) \\
&= \frac{\dfrac{R_2}{\mathbf{s}C_2 R_2 + 1}}{\dfrac{R_1}{\mathbf{s}C_1 R_1 + 1} + \dfrac{R_2}{\mathbf{s}C_2 R_2 + 1}} \mathbf{V}_i(\mathbf{s}) \\
&= \frac{R_2}{R_2 + \dfrac{R_1(\mathbf{s}C_2 R_2 + 1)}{(\mathbf{s}C_1 R_1 + 1)}} \mathbf{V}_i(\mathbf{s})
\end{aligned}
\tag{1.130}
$$

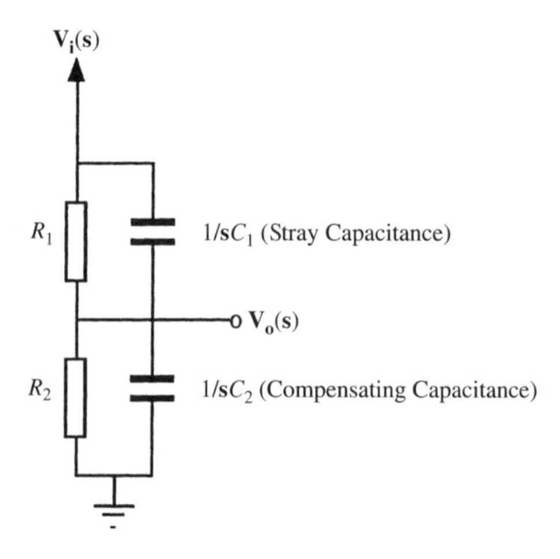

Figure 1.13 Laplace transform circuit for Example 1.11

where the symbol $\|$ represents the parallel combination of each resistor and its parallel capacitor. For the final equation to be independent of frequency would require

$$\begin{aligned} sC_2R_2 + 1 &= sC_1R_1 + 1 \\ \text{or} \qquad C_2 &= \frac{R_1}{R_2}C_1 \end{aligned} \tag{1.131}$$

which gives the required condition and necessary value of C_2 in terms of the other components.

Example 1.12

A 1 μH inductor is to be used as an energy storage element in a pulse generator circuit, as shown in Figure 1.14. The inductor is supplied with current $i(t)$ from a power supply which has an output impedance which is low enough to be neglected. After the current in the inductor has reached 100 kA, switch SW2 closes very rapidly and connects the inductor to a load which comprises a 10 Ω resistor R_2 in parallel with a 10 nF capacitor C_1. At the same time switch SW1 opens, also very rapidly, thereby disconnecting the inductor from the supply. If the inductor has an associated resistance of 0.1 Ω as represented by resistor R_1, derive an expression for the potential on the load as a function of time $v(t)$ and plot the result.

It is often more convenient to assign letters to the component values in this type of problem and then to substitute the numbers back at some later stage. This has been done in the Laplace transform circuit which is also given in Figure 1.14. Since the impedance of the power supply can be neglected, the storage inductor can be directly connected to one side of the load via a voltage source $LI(0)$ due to the initial current flowing in the inductor at the

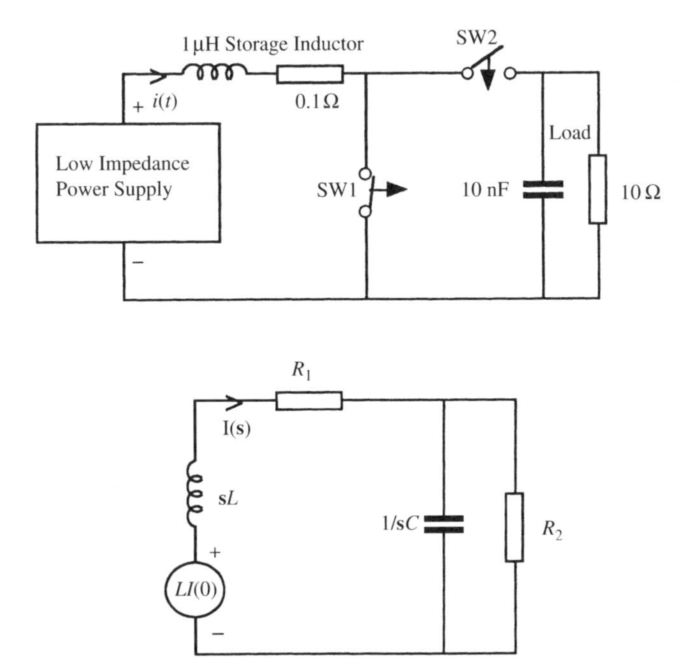

Figure 1.14 Circuit and its Laplace transform equivalent for Example 1.11

start of the analysis. Switches SW1 and SW2 have been replaced by an open circuit and a short circuit since their operation, at the start of the analysis, is assumed to be infinitely fast. From the transformed circuit an expression for the voltage on the load can be derived by considering that the circuit is a potential divider with the load resistor and parallel capacitor acting as the bottom leg of the divider (Impedance Z_2) and the inductor and its equivalent series resistance acting as the top leg (Impedance Z_1). Thus

$$\left.\begin{aligned} Z_1 &= R_1 + \mathbf{s}L \\ Z_2 &= \frac{\dfrac{R_2}{\mathbf{s}C}}{R_2 + \dfrac{1}{\mathbf{s}C}} = \frac{R_2}{\mathbf{s}CR_2 + 1} \\ \mathbf{V}(\mathbf{s}) &= \frac{LI(0)\left(\dfrac{R_2}{\mathbf{s}CR_2 + 1}\right)}{(R_1 + \mathbf{s}L) + \left(\dfrac{R_2}{\mathbf{s}CR_2 + 1}\right)} \\ &= \frac{\dfrac{I(0)}{C}}{\mathbf{s}^2 + \mathbf{s}\left(\dfrac{1}{CR_2} + \dfrac{R_1}{L}\right) + \dfrac{(R_1 + R_2)}{LCR_2}} \end{aligned}\right\} \tag{1.132}$$

This can be factored into the form

$$\left.\begin{aligned} V(\mathbf{s}) &= \frac{I(0)}{C}\frac{\beta}{(\mathbf{s}^2 + \alpha^2) + \beta^2} \\ \text{with} \quad \alpha &= \frac{1}{2CR_2} + \frac{R_1}{2L} \\ \text{and} \quad \beta^2 &= \frac{(R_1 + R_2)}{LCR_2} - \alpha^2 \end{aligned}\right\} \tag{1.133}$$

At this stage it is sensible to insert component values to determine the sign of β^2 so that the correct inverse transform can be taken. From the components given in the example α and β^2 are found to be

$$\left.\begin{aligned} \alpha &= \frac{1}{2 \times 10^{-7}} + \frac{0.1}{2 \times 10^{-6}} \\ &= 5.1 \times 10^6\,\mathrm{s}^{-1} \\ \beta^2 &= \frac{10.1}{10^{-13}} - \alpha^2 \\ &= 7.5 \times 10^{13}\,\mathrm{s}^{-2} \end{aligned}\right\} \tag{1.134}$$

Since β^2 is positive the inverse transform can be found directly by reference to the tables as

$$\left.\begin{aligned} v(t) &= \frac{I(0)}{C\beta}e^{-\alpha t}\sin\beta t \\ \text{with} \quad I(0) &= 100\,\mathrm{kA} \\ \text{and} \quad \beta &= 8.66 \times 10^6\,\mathrm{s}^{-1} \end{aligned}\right\} \tag{1.135}$$

This result is plotted in Figure 1.15.

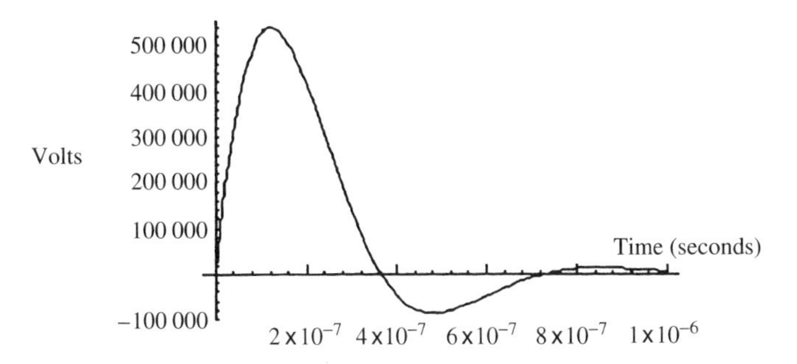

Figure 1.15 Voltage waveform for Example 1.12

Example 1.13

The standard ignition circuit of a car can be represented by the circuit shown in Figure 1.16. The battery of the car has a potential V and an internal resistance of R_1, and is connected to the primary of the ignition coil (a transformer) by switch SW. Once the current from the battery, which flows through the primary winding, has reached its maximum value as determined by its internal resistance, switch SW is opened and the electrical energy stored in the coil is delivered to a spark plug from the secondary winding of the transformer. If the spark plug can be represented by a second resistor R_2, determine the amplitude of the potential applied to the gap as a function of time.

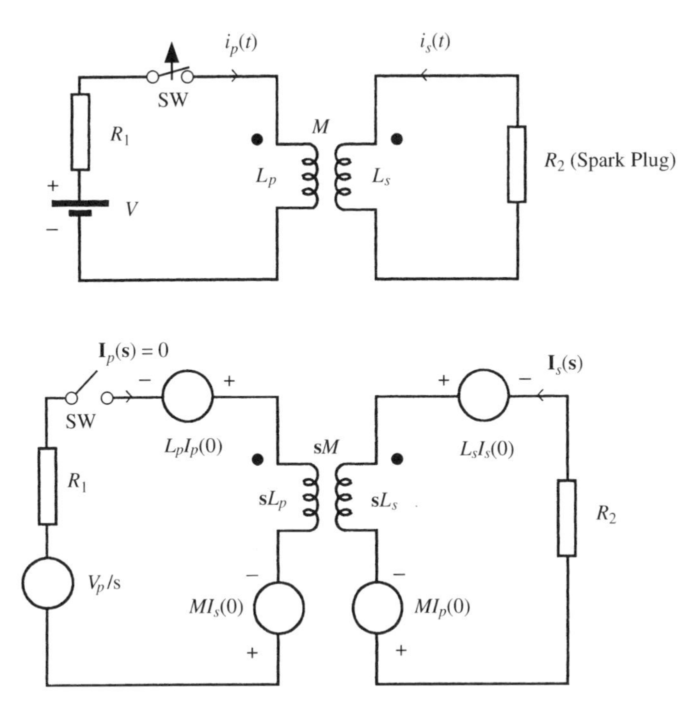

Figure 1.16 Circuits for the ignition circuit problem in Example 1.12

The circuit for this problem and its Laplace transformed equivalent circuit are given in Figure 1.16. At the start of the analysis, the switch is opened. This happens when the current in the primary winding has reached its maximum value of V/R_1. Because the current in the primary circuit has ceased to change there will be no current flowing in the secondary circuit. Hence the initial conditions for the problem are

$$\mathbf{I}_p(0) = \frac{V}{R_1} \quad \text{and} \quad \mathbf{I}_s(0) = 0 \tag{1.136}$$

Analysis of the secondary circuit (by potential divider) gives

$$\begin{aligned} \mathbf{V}_{R_2}(\mathbf{s}) &= -[M\mathbf{I}_p(0) + L_s\mathbf{I}_s(0)]\frac{R_2}{\mathbf{s}L_s + R_2} \\ &= -\frac{MR_2V}{L_sR_1}\frac{1}{\mathbf{s}+\dfrac{R_2}{L_s}} \end{aligned} \tag{1.137}$$

By reference to the table of Laplace transforms the required potential as a function of time is found to be

$$v_{R_2}(t) = -\frac{MR_2V}{L_sR_1}\exp\left(\frac{-R_2t}{L_s}\right) \tag{1.138}$$

1.4.6 System or Transfer Functions

It should be clear from the above section and examples that, in general, the Laplace method can provide an effective means of determining both the transient and continuous AC or DC response of a given circuit subjected to some form of stimulus. In the majority of cases problems are concerned with the response of a circuit in terms of the current or voltage which results in a given part of a circuit when a signal, usually in the form of a voltage signal, is applied to another part of the circuit. The resulting relationship between input signal to response, which is derived, is the system function or transfer function and can be written in the general form

$$\mathbf{R}(\mathbf{s}) = \mathbf{E}(\mathbf{s})\mathbf{H}(\mathbf{s}) \tag{1.139}$$

where $\mathbf{R}(\mathbf{s})$ is the response, $\mathbf{E}(\mathbf{s})$ is the stimulus or excitation and $\mathbf{H}(\mathbf{s})$ is the system function. If $\mathbf{R}(\mathbf{s})$ and $\mathbf{E}(\mathbf{s})$ are either both current signals or both voltage signals then $\mathbf{H}(\mathbf{s})$ is dimensionless and takes the form of a current or voltage transfer function, respectively, and may be written in the form $\mathbf{T}(\mathbf{s})$ where $\mathbf{T}$ stands for transfer. If $\mathbf{R}(\mathbf{s})$ is a current signal and $\mathbf{E}(\mathbf{s})$ is a voltage signal or vice versa then $\mathbf{H}(\mathbf{s})$ takes the form of an admittance function or an impedance function, respectively, and may be written in the form $\mathbf{Y}(\mathbf{s})$ or $\mathbf{Z}(\mathbf{s})$. As will be apparent from the above examples, the general mathematical form of $\mathbf{H}(\mathbf{s})$, independent of its specific function will be the same as given in Equation (1.91) and can be written in the general form

$$\begin{aligned} \mathbf{H}(\mathbf{s}) &= \frac{\mathbf{P}(\mathbf{s})}{\mathbf{Q}(\mathbf{s})} \\ &= \frac{a_n\mathbf{s}^n + a_{n-1}\mathbf{s}^{n-1} + a_{n-2}\mathbf{s}^{n-2} + \cdots + a_0}{b_m\mathbf{s}^m + b_{m-1}\mathbf{s}^{m-1} + b_{m-2}\mathbf{s}^{m-2} + \cdots + b_0} \end{aligned} \tag{1.140}$$

usually with $m > n$. This expression can be further modified, as is standard in the process of finding the inverse transform, by making the coefficient of the highest term in $\mathbf{s}$ equal to unity, and then factorising, i.e.

$$\mathbf{H}(\mathbf{s}) = \frac{K\left(\mathbf{s}^n + \frac{a_{n-1}}{a_n}\mathbf{s}^{n-1} + \frac{a_{n-2}}{a_n}\mathbf{s}^{n-2} + \cdots + \frac{a_0}{a_n}\right)}{\left(\mathbf{s}^m + \frac{b_{m-1}}{b_m}\mathbf{s}^{m-1} + \frac{b_{m-2}}{b_m}\mathbf{s}^{n-2} + \cdots + \frac{b_0}{b_m}\right)}$$

$$= \frac{K(\mathbf{s}+z_n)(\mathbf{s}+z_{n-1})(\mathbf{s}+z_{n-2})\dots(\mathbf{s}+z_0)}{(\mathbf{s}+p_m)(\mathbf{s}+p_{m-1})(\mathbf{s}+p_{m-2})\dots(\mathbf{s}+p_0)} \tag{1.141}$$

Here K is equal to a_n/b_m. From the second expression it is easily seen that $\mathbf{H}(\mathbf{s})$ becomes equal to zero when $\mathbf{s}$ is equal to $-z_n, -z_{n-1}, -z_{n-2}$, etc., and equal to infinity when $\mathbf{s}$ is equal to $-p_m, -p_{m-1}, p_{m-2}$ etc. The p's and z's in the expression are known as the poles and zeros of $\mathbf{H}(\mathbf{s})$ and represent the critical values of the complex frequency $\mathbf{s}$ which determines the transient and steady state behaviour of the circuit or network to which $\mathbf{H}(\mathbf{s})$ relates. Referring back to Equation (1.139), it is clear that the stimulus to a circuit $\mathbf{E}(\mathbf{s})$ can be represented in a similar way. Thus the poles of the system function $\mathbf{H}(\mathbf{s})$ are known as the natural frequencies of the system and the poles of the excitation function or stimulus $\mathbf{E}(\mathbf{s})$ are known as the forced frequencies. Clearly much can be learned about the general form of the response of a system in the time domain to a given stimulus by inspection of both the system and excitation functions. Although a full description of the extent to which this observation can be used to determine a given system response lies outside the scope of this book (interested readers are referred to more specialised texts on this topic [1,12]), a method for determining both the rise-time and delay time of the response of a given network to a given stimulus will be described.

1.4.7 Direct Determination of Rise and Delay Time Response of Networks

Often the rise-time of a pulsed signal on a load is of considerable interest in designing pulse generators as is the time taken for a signal to reach a load after the signal has been injected into a circuit to which the load is attached (delay time). The definition of delay and rise-time is clearly important. Delay time is usually defined as the time taken for the signal at the load to reach half its maximum value, whereas there are a number of ways of defining rise-time. The time taken for the rising edge of a signal to move from 10% to 90% of its final value can be used. Alternatively the slope of the tangent drawn at the half maxiumum value point can be drawn and used to determine rise-time, as shown in Figure 1.17, which shows a typical circuit response to excitation by a unit step. Unfortunately all these definitions are unhelpful in inferring the delay time or rise-time of a given system directly from the system function, and can only be used if the functional variation with time of a signal has been derived. An alternative method for determining such a response directly from the system function of a network has been developed by Elmore [13] which relies on a different definition of rise-time. Referring to Figure 1.17, it can be seen that if the first derivative of the rising edge of the signal response is plotted, an alternative definition of delay and rise-time is possible. In the case of delay time, the time coordinate of the centroid [14] of area of the curve of $v'(t)$

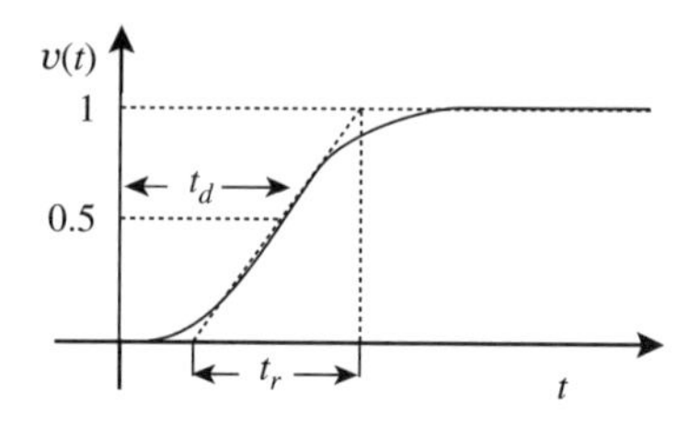

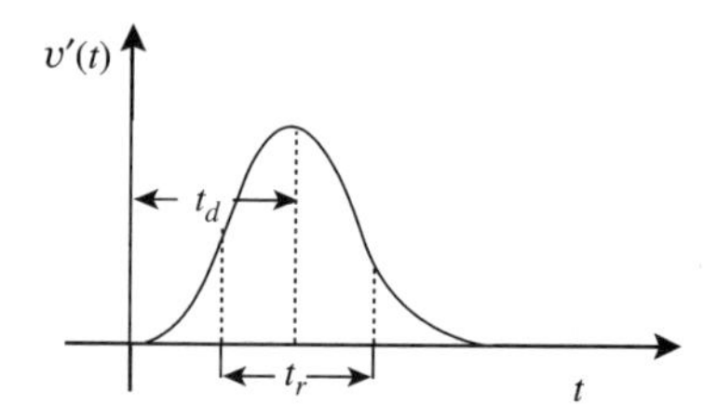

Figure 1.17 Typical response curves illustrating the definitions of rise and delay time

can be used, which is defined as

$$t_d = \frac{\int_0^\infty t v'(t)\,dt}{\int_0^\infty v'(t)\,dt} = \int_0^\infty t v'(t)\,dt \qquad \text{since} \qquad \int_0^\infty v'(t)\,dt = 1 \tag{1.142}$$

The only disadvantage of this definition is that the shape of the response curve must be monotonic, i.e. without oscillation or overshoot, and therefore the method must be restricted to waveforms of this type. A definition of rise-time can be made in a similar way by realising that the radius of gyration k [14] of the area under the curve of $v'(t)$, about a vertical axis drawn at the delay time, must be proportional to the rise-time. Such a radius of gyration is determined by

$$k^2 = \frac{\int_0^\infty (t - t_d)^2 v'(t)\,dt}{\int_0^\infty v'(t)\,dt} = \int_0^\infty (t - t_d)^2 v'(t)\,dt \tag{1.143}$$

Clearly the shape of $v'(t)$ must be known for an accurate estimate of k, and in the original paper by Elmore, which was primarily concerned with rise-time estimation in multi-stage amplifiers, the shape was estimated to be close to that of a Gaussian error curve. Based on this assumption the constant of proportionality would be 2π, and therefore the rise-time in this case can be defined to be

$$t_r = \left\{2\pi\left[\left[\int_0^\infty t^2 v'(t)\,dt - t_d^2\right]\right]\right\}^{1/2} \tag{1.144}$$

To relate these definitions to a given system function, whose rise and delay times are required, it is necessary to first take the general form of the system function **H(s)** as given in

Equation (1.140) and normalise it to $\mathbf{H}'(\mathbf{s})$, as follows

$$\mathbf{H}'(\mathbf{s}) = \frac{\mathbf{H}(\mathbf{s})}{\mathbf{H}(0)}$$

$$\mathbf{H}'(\mathbf{s}) = \frac{a'_n\mathbf{s}^n + a'_{n-1}\mathbf{s}^{n-1} + a'_{n-2}\mathbf{s}^{n-2} + \cdots + 1}{b'_m\mathbf{s}^m + b'_{m-1}\mathbf{s}^{m-1} + b'_{m-2}\mathbf{s}^{m-2} + \cdots + 1} \quad (1.145)$$

where $a'_n = \dfrac{a_n}{a_0}, \quad b'_m = \dfrac{b_m}{b_0}$ etc.

This normalised system function can also be considered as the response $\mathbf{R}(\mathbf{s})$ of the system to a unit impulse or delta function since

$$\mathbf{E}(\mathbf{s}) = \mathscr{L}\{\delta(t)\} = 1 \quad (1.146)$$

This response therefore corresponds to the curve of $v'(t)$ as shown in Figure 1.17, and it can be seen that $\mathbf{H}'(\mathbf{s})$ is simply the Laplace transform of $v'(t)$ given by

$$\mathbf{H}'(\mathbf{s}) = \int_{0_+}^{\infty} v'(t)e^{-\mathbf{s}t}\,dt \quad (1.147)$$

Elmore's method then relies on expanding this integral by expanding the exponential term in a power series in $\mathbf{s}t$, expanding Equation (1.444) for $\mathbf{H}'(\mathbf{s})$ in ascending powers of $\mathbf{s}$ and comparing terms to find that

$$\left.\begin{aligned} t_d &= b'_1 - a'_1 \\ t_r &= \{2\pi[b'^2_1 - a'^2_1 + 2(a'_2 - b'_2)]\}^{1/2} \end{aligned}\right\} \quad (1.148)$$

Although the method can be an extremely useful and simple means for finding both rise and delay times for a given circuit, it must be remembered that the method assumes that the curve of the rising edge of the response must be monotonic and that the differential of this curve is reasonably well approximated by a Gauss error curve.

Example 1.14

Referring to Example 1.10, compare the results of using Elmore's method for rise and delay time to the 10 to 90% rise-time and time to 50% of the maximum potential on capacitor C_2 if $C_1 = C_2 = 2$ μF, $L = 1$ μH, and R is either 5 or 10 Ω.

It is necessary to first check that the component values given lead to an overdamped response. This is done by calculating values for α and ω_0 as determined by Equation (1.122). This gives values of 2.5×10^6 and 5.0×10^6 s^{-1} for α when R is 5 and 10 Ω, respectively, and 10^6 radians s^{-1} for ω_0. Since in both cases α is greater than ω_0, then the circuit is clearly overdamped and the potential on C_2 can easily be shown to be given by

$$\left.\begin{aligned} v(t) &= V\frac{C_1}{C_1+C_2}\left[1 - \exp(-\alpha t)\left(\frac{\alpha}{\beta}\sinh\beta t + \cosh\beta t\right)\right], \\ \text{where} \quad \alpha &= \frac{R}{2L} \quad \text{and} \quad \beta = \sqrt{\alpha^2 - \frac{1}{LC}} \end{aligned}\right\} \quad (1.149)$$

R	10–90% Rise	Time to 50%	t_r	t_d
5 Ω	10.46 μs	3.5 μs	12 μs	5 μs
10 Ω	21.75 μs	7.1 μs	25 μs	10 μs

Figure 1.18 Comparison of rise and delay times for Example 1.10

Since the response is overdamped it must be monotonic, so the criterion for the use of Elmore's method is satisfied. The corresponding system function or impulse response can be derived directly from Equation (1.127) to be

$$\mathbf{V}(\mathbf{s}) = \frac{V}{C_2 L} \frac{1}{(\mathbf{s}+\alpha)^2 - \beta^2} \tag{1.150}$$

This expression can be normalised as follows

$$\begin{aligned} \mathbf{V}'(\mathbf{s}) &= \frac{\mathbf{V}(\mathbf{s})}{\mathbf{V}(0)} = \frac{\alpha^2 - \beta^2}{\mathbf{s}^2 + 2\alpha\mathbf{s} + \alpha^2 - \beta^2} \\ &= \frac{1}{\dfrac{\mathbf{s}^2}{\alpha^2 - \beta^2} + \dfrac{2\alpha\mathbf{s}}{\alpha^2 - \beta^2} + 1} \end{aligned} \tag{1.151}$$

From this expression and by comparison to Equation (1.145) it can be seen that

$$\left.\begin{aligned} a_1' &= 0 \\ a_2' &= 0 \\ b_1' &= \frac{2\alpha}{\alpha^2 - \beta^2} = RC \\ b_2' &= \frac{1}{\alpha^2 - \beta^2} = LC \end{aligned}\right\} \tag{1.152}$$

Therefore the delay and rise-time are given by

$$\begin{aligned} t_d &= RC \\ t_r &= \{2\pi[(RC)^2 - 2LC]\}^{1/2} \end{aligned} \tag{1.153}$$

The values for the 10 to 90% rise-time and the delay time, as determined by the time to reach 50% of the maximum potential on C_2, can calculated by putting appropriate numerical values into Equation (1.148). Similarly, estimates for rise and delay time using Elmore's method can be made using Equation (1.152). The results of these calculations are summarised in Figure 1.18.

Although, in this example, Elmore's method gives a useful indication of rise and delay times, in each case the figures are overestimated.

REFERENCES

[1] Kuo F. F. "Network Analysis and Synthesis". John Wiley & Sons (1966) ISBN 0-471-51116-1

[2] Hayt JR. W. H. and Kemmerly J. E. "Engineering Circuit Analysis". McGraw-Hill (1993) ISBN 0-07-112736-4

[3] Carter G. W. and Richardson A. "Techniques of Circuit Analysis". Cambridge University Press (1972) ISBN 0-521-08435-0
[4] Karni S. "Analysis of Electrical Networks". John Wiley & Sons (1986) ISBN 0-471-84312-1
[5] Stephenson G. "Mathematics Methods for Science Students". Longman Scientific & Technical (1973) ISBN 0-582-44416-0
[6] Holbrook J. G. "Laplace Transforms for Electronic Engineers". Pergamon Press (1966)
[7] "Pspice" is published by Microsim Corporation, Irvine, CA 92718, USA.
[8] "Microcap" is published by Spectrum Software, Sunnyvale, CA 94086
[9] Zepler E. E. and Nichols K. G. "Transients in Electrical Engineering". Chapman and Hall (1971) ISBN 412-10130-0
[10] Spiegel M. R. "Laplace Transforms". Shaum's Outline Series in Mathematics, McGraw-Hill (1965) ISBN 07-060231-x
[11] Doetsch G. "Guide to the Applications of Laplace Transforms". D. Van Nostrand Co. Ltd. (1961)
[12] Maddock R. J. "Poles and Zeros in Electrical and Control Engineering". Holt, Rinehart and Winston (1982) ISBN 0-03-910346-3
[13] Elmore W. C. "The Transient Response of Damped Linear Networks with Particular Regard to Wideband Amplifiers". *J. Appl. Phys.* (1948) **19**, 55–63
[14] Jeffrey A. "Mathematics for Engineers and Scientists". Van Nostrand Reinhold (1982) ISBN 0-442-30728-4
[15] Ray Wylie C. and Barrett L. C. "Advanced Engineering Mathematics" McGraw-Hill (1985) ISBN 0-07-072188-2

2

Transmission Line Theory and Transient Response

2.1 INTRODUCTION

In this chapter the propagation characteristics of transmission lines are described with particular reference to their transient response. Although there are many good books on the subject of transmission line theory (see, for example, the books by Dunlop and Smith [1], Chipman [2], Connor [3] and Matick [4]) few deal specifically with their transient behaviour. Books which to varying degrees deal with this aspect of transmission line behaviour include Zepler and Nichols [5], Rizzi [6], and Lewis and Wells [7].

Transmission lines basically comprise two or more conductors and are usually used to carry electrical signals with frequencies reaching the microwave frequency bands. The important feature of transmission lines is that their cross-sectional configuration and the spacing between the conductors in the line remain constant throughout their length. If an insulator other than air is inserted between the lines, to prevent electrical breakdown between them, then the dielectric constant or permittivity of the material must also remain constant throughout the length of the line. In the cases where the cross-sectional configuration changes with distance along the line or the dielectric constant of the insulator is changed, the amplitude and wavelength of any propagating signal will also change. This effect can be used to build high-frequency transformers and Chapter 6 describes some of the characteristics of this type of transformer. When the line has a uniform physical construction and takes, for example, the form of two straight parallel conductors of fixed length which connect a signal source to a load, it is clear that the instantaneous currents in the two lines, at any point, must be equal in magnitude but flow in opposite directions. Similarly at the intersection of any plane, at a fixed point and transverse to the line, the instantaneous potential on the line takes a unique value which depends on the location of the plane with respect to the signal source or the load.

In principle the propagation characteristics of time-varying electrical signals on a transmission line could be deduced using Maxwell's equations. However the solution of the resulting equations which describe the electrical and magnetic fields between the conductors of the line can be difficult, particularly when the regions containing the fields are bounded by conductors or dielectric insulators which have anything other than very simple shapes. Thus the standard method for analysing the propagation characteristics of transmission lines relies on the use of a lumped element electrical model of the line. The model is constructed by considering the resistance along and between the conductors of the line and the inductance and capacitance along a fixed length of the line. This basic model, for the

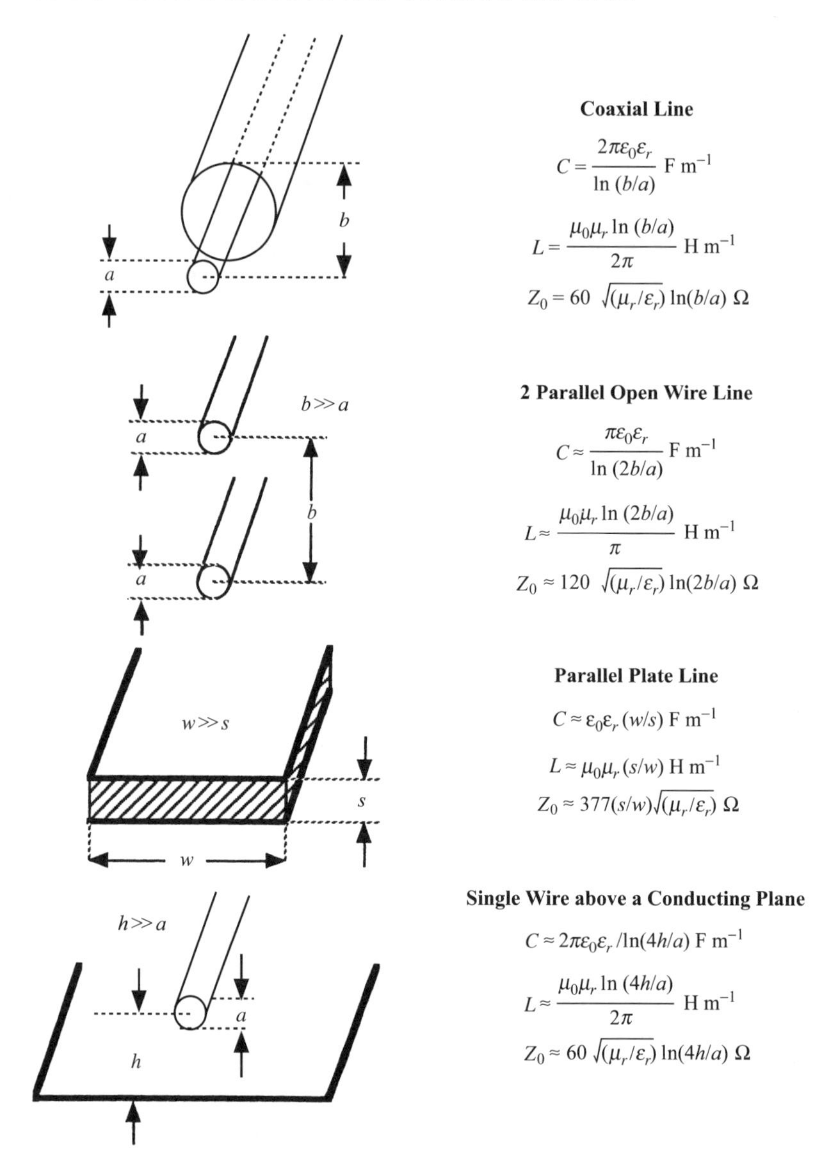

Figure 2.1 Capacitance and inductance per unit length and impedance of 4 commonly encountered transmission line types. Physical dimensions are in metres.

description of the propagation of electrical signals on transmission lines, was developed originally by Oliver Heaviside during the 1880s and the theory presented in the first section of this chapter, as in other books on transmission lines, is based on Heaviside's pioneering work.

Although a wide variety of different types of transmission line can be envisaged, the basic physical structure of four of the most commonly encountered types of transmission line is shown in Figure 2.1. Since the inductance and capacitance of a fixed length of line is of fundamental importance to the model to be described, formulae for these variables for each of the lines are marked in the figure. In the formulae the relative permeability μ_r and relative permittivity ϵ_r of any material which may fill the space between the lines are included. Clearly if the lines are air-insulated then μ_r and $\epsilon_r = 1$. Care should be taken to note that some of the expressions are approximate and the approximations used are noted on the diagram.

2.2 CIRCUIT ANALYSIS OF TRANSMISSION LINES

The purpose of this analysis is to determine the instantaneous values of voltage and current along a uniform length of line when a signal is injected at one end and the line is connected to a terminating load at the other end. To achieve this analysis a lumped element model is used, despite the fact that the electrical parameters of a transmission line are uniformly distributed along its length. For simplicity it is assumed in this analysis that the line consists of just two conductors, although it would not be difficult to adapt the theory to lines which contained more than two conductors. So that a lumped element model can be developed which accurately describes the line it is necessary to consider the behaviour of a fixed length of line whose electrical parameters are restricted as follows.

1. The combined resistance of the two fixed lengths of conductors that form the line R.
2. The total conductance between the two conductors of the line G. Although, with the advent of modern insulating materials, this conductance would be expected to be very small, there are cases where it is important to consider the effect of small leakage currents which may flow between the lines. G also includes the dielectric loss in any dielectric used to insulate the line.
3. Since there is a physical space between the two conductors of the line, current flowing on the lines must set up magnetic flux whose size will depend on the inductance L of the fixed length of line.
4. Similarly, the conductors, together with any insulating dielectric lying between them, will have a capacitance C.

By referring to the data, which is available from manufacturers of standard transmission lines, or by the use of the formulae given in Figure 2.1 and some simple estimates of resistance and conductance, it is possible to arrive at values of these parameters for a standard length of line (usually 1 metre). These values can then be used to determine the values of the electrical parameters R, G, L and C for a very short length of line, considered in the model, by multiplying them by the length of the short section which must, of course, also be in metres. A complete line can then be thought of as a long cascade of these very short lengths of line to which the electrical parameters are assigned as shown in Figure 2.2.

If the length of a very short elemental length of line is given a value Δz then the change in the amplitude of the voltage and current of a signal by passage through this element can be determined. Thus the change in current is given by

$$i(z,t) - i(z+\Delta z,t) = -\Delta i(z,t) = G\Delta z v(z+\Delta z,t) + C\Delta z\frac{\partial v(z+\Delta z,t)}{\partial t} \quad (2.1)$$

and the change in voltage by

$$v(z,t) - v(z+\Delta z,t) = -\Delta v(z,t) = R\Delta z i(z,t) + L\Delta z\frac{\partial i(z,t)}{\partial t} \quad (2.2)$$

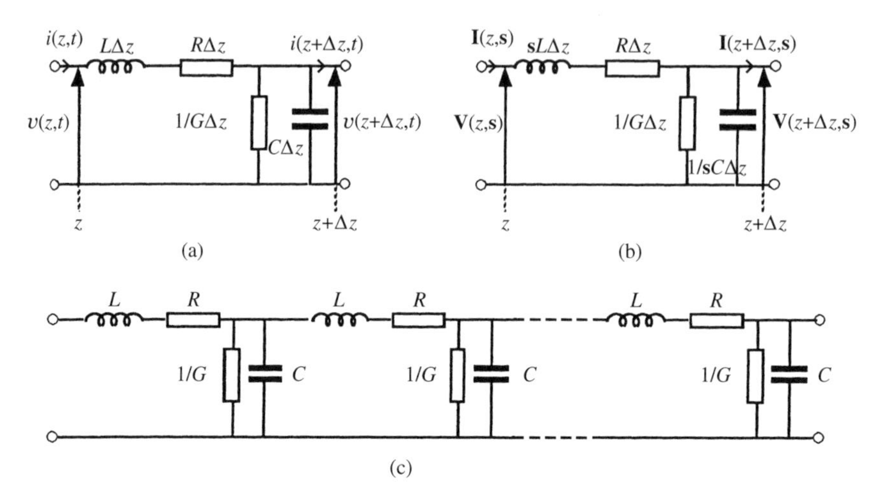

Figure 2.2 Equivalent circuits of a uniform transmission line: (a) an elemental length in the time domain, (b) the Laplace transform circuit representation of an elemental length (in the complex frequency domain), (c) a cascade of elemental lengths to represent a finite length of line

In writing these difference equations it should be remembered that any injected signal will be time-dependent so that, at any point on a line, the current and voltage depend both on line position and time. This will lead to partial differential equations dependent on z and t which are derived from Equations (2.1) and (2.2). For small Δz, use of Taylor's theorem in an expansion about z gives

$$\left.\begin{aligned} i(z+\Delta z,t) &\cong i(z,t)+\frac{\partial i(z,t)}{\partial z}\Delta z \\ v(z+\Delta z,t) &\cong v(z,t)+\frac{\partial v(z,t)}{\partial z}\Delta z \end{aligned}\right\} \tag{2.3}$$

where terms involving $(\Delta z)^2$ and $(\Delta z)^3$, etc., are ignored. Substitution of Equation (2.3) into Equations (2.1) and (2.2) and taking the limit as Δz tends to zero gives

$$\frac{\partial i(z,t)}{\partial z}=-Gv(z,t)-C\frac{\partial v(z,t)}{\partial t} \tag{2.4}$$

$$\frac{\partial v(z,t)}{\partial z}=-Ri(z,t)-L\frac{\partial i(z,t)}{\partial t} \tag{2.5}$$

A solution of these two equations results in expressions for v and i as functions of z and t which depend on the electrical characteristics of the source connected to the line at $z=0$ and the load connected at the end of a fixed length l of the line. To solve these equations the first stage is to use the Laplace transformation (as described in Chapter 1), to get

$$\frac{d\mathbf{I}(z,\mathbf{s})}{dz}=-G\mathbf{V}(z,\mathbf{s})-\mathbf{s}C\mathbf{V}(z,\mathbf{s})+CV(z,0) \tag{2.6}$$

$$\frac{d\mathbf{V}(z,\mathbf{s})}{dz}=-R\mathbf{I}(z,\mathbf{s})-\mathbf{s}L\mathbf{I}(z,\mathbf{s})+LI(z,0) \tag{2.7}$$

These equations could, of course, have been derived directly from the Laplace transform circuit in Figure 2.2(b), without the need to proceed via partial differential equations, with the condition that Δz is allowed to approach zero. If it is assumed that the line is initially uncharged and there is no current flowing in it, then $V(z, 0) = 0$ and $I(z, 0) = 0$ and Equations (2.6) and (2.7) become

$$\frac{d\mathbf{I}(z,\mathbf{s})}{dz} = -(G + \mathbf{s}C)\mathbf{V}(z,\mathbf{s}) \tag{2.8}$$

$$\frac{d\mathbf{V}(z,\mathbf{s})}{dz} = -(R + \mathbf{s}L)\mathbf{I}(z,\mathbf{s}) \tag{2.9}$$

Differentiation of Equation (2.9) with respect to z gives

$$\frac{d^2\mathbf{V}(z,\mathbf{s})}{dz^2} = -(R + \mathbf{s}L)\frac{d\mathbf{I}(z,\mathbf{s})}{dz} \tag{2.10}$$

which results, after substitution of Equation (2.8), in an equation in $\mathbf{V}(z, \mathbf{s})$ only in the form

$$\frac{d^2\mathbf{V}(z,\mathbf{s})}{dz^2} = (R + \mathbf{s}L)(G + \mathbf{s}C)\mathbf{V}(z,\mathbf{s}) \tag{2.11}$$

It is worth noting that this equation would still apply even if the line had been initially charged or there was a current flowing in it, because both $V(z, 0)$ and $I(z, 0)$ would have constant values and their derivatives with respect to z would therefore be zero. Equation (2.11) is usually written in the form

$$\left.\begin{aligned} \frac{d^2\mathbf{V}(z,\mathbf{s})}{dz^2} &= \gamma(\mathbf{s})^2\mathbf{V}(z,\mathbf{s}) \\ \text{where} \quad \gamma(\mathbf{s}) &= \sqrt{(R + \mathbf{s}L)(G + \mathbf{s}C)} \end{aligned}\right\} \tag{2.12}$$

$\gamma(\mathbf{s})$ is known as the propagation constant. An equation for $\mathbf{I}(z, \mathbf{s})$ can also be derived from Equations (2.8) and (2.9) and it takes an identical form to Equation (2.12), i.e.

$$\frac{d^2\mathbf{I}(z,\mathbf{s})}{dz^2} = \gamma(\mathbf{s})^2\mathbf{I}(z,\mathbf{s}) \tag{2.13}$$

The general solutions to Equations (2.12) and (2.13) take the form

$$\left.\begin{aligned} \mathbf{V}(z,\mathbf{s}) &= \mathbf{V}_1(\mathbf{s})\exp(-\gamma(\mathbf{s})z) + \mathbf{V}_2(\mathbf{s})\exp(\gamma(\mathbf{s})z) \\ \mathbf{I}(z,\mathbf{s}) &= \mathbf{I}_1(\mathbf{s})\exp(-\gamma(\mathbf{s})z) + \mathbf{I}_2(\mathbf{s})\exp(\gamma(\mathbf{s})z) \end{aligned}\right\} \tag{2.14}$$

where the parameters $\mathbf{V}_1(\mathbf{s})$, $\mathbf{V}_2(\mathbf{s})$, $\mathbf{I}_1(\mathbf{s})$ and $\mathbf{I}_2(\mathbf{s})$ are arbitrary coefficients whose values are determined by the initial conditions applied to the line. The parameters $\mathbf{I}_1(\mathbf{s})$ and $\mathbf{I}_2(\mathbf{s})$ can be related to $\mathbf{V}_1(\mathbf{s})$ and $\mathbf{V}_2(\mathbf{s})$ by first differentiating the expression for $\mathbf{V}(z, \mathbf{s})$ with respect to z in Equation (2.14) to get

$$\frac{d\mathbf{V}(z,\mathbf{s})}{dz} = \gamma(\mathbf{s})[\mathbf{V}_2(\mathbf{s})\exp(\gamma(\mathbf{s})z) - \mathbf{V}_1(\mathbf{s})\exp(-\gamma(\mathbf{s})z)] \tag{2.15}$$

Substitution from Equations (2.9) and (2.12) results in expression for $\mathbf{I}(z, \mathbf{s})$ of the form

$$\begin{aligned}\mathbf{I}(z, \mathbf{s}) &= \frac{\gamma(\mathbf{s})}{R + \mathbf{s}L}[\mathbf{V}_1(\mathbf{s})\exp(-\gamma(\mathbf{s})z) - \mathbf{V}_2(\mathbf{s})\exp(\gamma(\mathbf{s})z)] \\ &= \sqrt{\frac{G + \mathbf{s}C}{R + \mathbf{s}L}}[\mathbf{V}_1(\mathbf{s})\exp(-\gamma(\mathbf{s})z) - \mathbf{V}_2(\mathbf{s})\exp(\gamma(\mathbf{s})z)] \end{aligned} \tag{2.16}$$

By comparison to Equation (2.14) it is clear that

$$\left.\begin{aligned} &\mathbf{I}_1(\mathbf{s}) = \frac{\mathbf{V}_1(\mathbf{s})}{\mathbf{Z}_0(\mathbf{s})} \quad \text{and} \quad \mathbf{I}_2(\mathbf{s}) = \frac{-\mathbf{V}_2(\mathbf{s})}{\mathbf{Z}_0(\mathbf{s})} \\ \text{where} \quad &\mathbf{Z}_0(\mathbf{s}) = \sqrt{\frac{R + \mathbf{s}L}{G + \mathbf{s}C}} \end{aligned}\right\} \tag{2.17}$$

$\mathbf{Z}_0(\mathbf{s})$ is clearly an impedance and is known as the characteristic impedance of the line.

2.3 CONTINUOUS SINUSOIDAL TRANSMISSION LINE EXCITATION

So far, the analysis of the electrical characteristics of transmission lines has been kept completely general, and expressions for the voltage and current on a line have been given in terms of the complex frequency parameter $\mathbf{s}$ so that the transient response of lines can be determined. However it is helpful, at this stage, to examine the response of transmission lines to continuous sinusoidal excitation before looking at their transient response. This can be done in a straightforward manner by replacing $\mathbf{s}$ by $j\omega$, as explained in Chapter 1, and by assuming that $\mathbf{V}(z, \mathbf{s})$ takes a sinusoidal form given by

$$\mathbf{V}(z, \mathbf{s}) = V(z, j\omega) = \mathcal{Re}\{V(z)e^{j\omega t}\} \tag{2.18}$$

If it is also assumed that the current $\mathbf{I}(z, \mathbf{s})$ takes a similar form, then by substitution Equations (2.8) and (2.9) can be written as

$$\left.\begin{aligned} \frac{dV(z)}{dz} &= -(R + j\omega L)I(z) \\ \frac{dI(z)}{dz} &= -(G + j\omega C)V(z) \end{aligned}\right\} \tag{2.19}$$

Following an identical procedure, as outlined above, to get expressions for $V(z)$ and $I(z)$ under continuous sinusoidal excitation yields the solutions

$$\left.\begin{aligned} &V(z) = V_1\exp(-\gamma z) + V_2\exp(\gamma z) \\ &I(z) = I_1\exp(-\gamma z) + I_2\exp(\gamma z) = \frac{1}{Z_0}[V_1\exp(-\gamma z) - V_2\exp(\gamma z)] \\ \text{with} \quad &\gamma = \sqrt{(R + j\omega L)(G + j\omega C)} \\ \text{and} \quad &Z_0 = \sqrt{\frac{(R + j\omega L)}{(G + j\omega C)}} \end{aligned}\right\} \tag{2.20}$$

Since γ is clearly a complex quantity it is often split into real and imaginary parts and is written as

$$\gamma = \alpha + j\beta \tag{2.21}$$

α is known as the attenuation constant and β as the phase constant. If this form of the propagation constant is substituted into the solution for the voltage on the line $V(z)$ and the time dependency that was assumed in Equation (2.18) is used then a complete expression for the instantaneous voltage on the line at any distance z from the input of the line is given by

$$v(z,t) = \hat{V}_1 \exp(-\alpha z)\mathscr{Re}\left[\exp j(\omega t - \beta z + \phi_1)\right] + \hat{V}_2 \exp(\alpha z)\mathscr{Re}\left[\exp j(\omega t + \beta z + \phi_2)\right] \tag{2.22}$$

In this equation $\hat{V}_1$ and $\hat{V}_2$ are the peak values of the voltage coefficients V_1 and V_2 and ϕ_1 and ϕ_2 are the phase angles of V_1 and V_2 when $t = z = 0$. Inspection of this equation shows that at any point on the line the distance z from the input the voltage is composed of two components. The first term describes a harmonically varying voltage with peak value $\hat{V}_1$ and frequency ω whose amplitude is decreasing exponentially through the term $\exp(-\alpha z)$. Since the phase angle of the voltage signal at the input of the line is unknown the term ϕ_1 must be included, and the way in which the phase angle of the signal changes as it propagates along the line is given by the term $\exp(-\beta z)$. This component is usually referred to as the forward or incident wave. The second component is similar to the first except that it is growing in amplitude with increasing z through the term $\exp(\alpha z)$ or, in other words, it is decreasing in amplitude as it nears the input of the line and it has a different phase angle ϕ_2. This component is referred to as the backward or reflected wave.

The wavelength of the signal on the line can be deduced by examination of either the first or second component. If, for a fixed point in time, the phase angle of the signal is examined at two distinct positions on the line, namely at $z = z_1$ and $z = z_2$, and these positions are separated in distance by a complete wavelength λ, the phase angle of the wave must have changed by 2π radians. Therefore, considering the first component

$$\begin{aligned} [\beta(z_1 + \lambda) + \phi_1] + [\beta z_1 + \phi_1] &= 2\pi \\ \text{or} \qquad \beta &= \frac{2\pi}{\lambda} \end{aligned} \tag{2.23}$$

The propagation velocity of the signal v_p can then be obtained from β and the frequency of the signal f in Hz or ω in radians s^{-1} as follows

$$v_p = f\lambda = \frac{\omega\lambda}{2\pi} = \frac{\omega}{\beta} \tag{2.24}$$

2.3.1 Low Loss and Loss-free Lines

Since the resistive loss both along a transmission line and between its conductors is usually quite small, particularly at low frequencies, the approximations $R \ll \omega L$ and $G \ll \omega C$ can be used to simplify the expressions for both the propagation constant and characteristic impedance of a transmission line. Starting from the expression for the propagation constant γ given in Equation (2.20), rearrangement and expansion by the binomial theorem gives

$$
\begin{aligned}
\gamma &= j\omega\sqrt{LC}\left(1+\frac{R}{j\omega L}\right)^{\frac{1}{2}}\left(1+\frac{G}{j\omega C}\right)^{\frac{1}{2}} \\
&\approx j\omega\sqrt{LC}\left(1+\frac{R}{2j\omega L}\right)\left(1+\frac{G}{2j\omega C}\right) \\
&= j\omega\sqrt{LC}\left(1-\frac{RG}{4\omega^2 LC}-j\frac{R}{2\omega L}-j\frac{G}{2\omega C}\right)
\end{aligned} \tag{2.25}
$$

In the binomial expansion, terms involving R^2 and G^2 and higher orders have been discarded. It is now possible to split the propagation constant γ into its real and imaginary parts to get approximate expressions for the attenuation constant α and the phase constant β, i.e.

$$
\left.
\begin{aligned}
\alpha &\approx \frac{R}{2}\sqrt{\frac{C}{L}}+\frac{G}{2}\sqrt{\frac{L}{C}} \\
\beta &\approx \omega\sqrt{LC}\left(1-\frac{RG}{4\omega^2 LC}\right) \approx \omega\sqrt{LC}
\end{aligned}
\right. \tag{2.26}
$$

In the second approximation for β it has been recognised that the term involving RG is small enough to be neglected, particularly at high frequencies. In the special case where there is no resistive loss in the line, it is said to be loss-free or loss-less. In this case $R = 0$, $G = 0$, $\alpha = 0$ and $\beta = \omega\sqrt{(LC)}$. In both cases β has a linear dependence on frequency; consequently Equation (2.24) shows that sinusoidal signals of different frequencies will have the same phase velocities. If this were not the case, as would be true for lossy lines, phase distortion would occur; as the variation in phase velocity with frequency would cause the shape of signals, which are non-sinusoidal, to change as they propagate on a transmission line.

Using an identical method to obtain a low loss expression for the characteristic impedance gives the approximate relationship

$$
Z_0 = \sqrt{\frac{L}{C}}\left[1-\frac{jR}{2\omega L}-\frac{jG}{2\omega C}\right] \tag{2.27}
$$

Again, at high frequencies the two expressions in brackets will tend to zero, hence

$$
\left.
\begin{aligned}
Z_0 &\approx \sqrt{\frac{L}{C}} \\
\text{and} \qquad \alpha &\approx \frac{R}{2Z_0}+\frac{GZ_0}{2}
\end{aligned}
\right\} \tag{2.28}
$$

Note that, in this particular case, Z_0 has a real value which is independent of frequency.

From the expression for β in Equation (2.26) it is possible to derive an expression for the phase velocity in terms of the inductance and capacitance per unit length of the line L and C, as follows

$$
v_p = \frac{\omega}{\beta} = \frac{1}{\sqrt{LC}} \tag{2.29}
$$

Clearly, if the values of L and C are calculated for a metre length of line, then v_p will be in metres s^{-1}. Similarly an expression for the wavelength of a signal on a line in terms of its free-space wavelength can also be derived. Note, though, that if the line is air insulated, i.e. ϵ_r is unity, then the wavelength on the line will be the same as the free-space wavelength. From Equation (2.24)

$$\left.\begin{aligned} f &= \frac{v_p}{\lambda} = \frac{c}{\lambda_0} \\ \therefore \quad \lambda &= \frac{\lambda_0}{c\sqrt{LC}} \end{aligned}\right\} \tag{2.30}$$

Here c is the velocity of light and λ_0 is the free-space wavelength. If the values for L and C given in Figure 2.1, for any of the different types of line shown, are inserted into Equation (2.29) it is possible to derive expressions for the phase velocity on the line in terms of the relative permeability and permittivity of the medium insulating the line. For example in the case of a coaxial line

$$\left.\begin{aligned} v_p &= \frac{1}{\sqrt{LC}} = \frac{1}{\sqrt{\dfrac{\mu_0\mu_r}{2\pi}\ln\left(\dfrac{b}{a}\right)\dfrac{2\pi\varepsilon_0\varepsilon_r}{\ln\left(\dfrac{b}{a}\right)}}} = \frac{c}{\sqrt{\mu_r\varepsilon_r}} \\ \text{since} \quad c &= \frac{1}{\sqrt{\mu_0\varepsilon_0}} \end{aligned}\right\} \tag{2.31}$$

Similarly the wavelength on the line will be given by

$$\lambda = \frac{\lambda_0}{\sqrt{\mu_r\varepsilon_r}} \tag{2.32}$$

Example 2.1

A standard 50 Ω coaxial transmission line has the following electrical parameters: $R = 20\,\text{m}\Omega\,\text{m}^{-1}$, $G = 5\times10^{-4}$ S m^{-1}, $L = 250$ nH m^{-1}, $C = 100$ pF m^{-1}. Check the value of the characteristic impedance of the line and determine the propagation, attenuation and phase constants of the line and the signal propagation or phase velocity. If the line is insulated with a plastic dielectric determine its relative permittivity ϵ_r.

If a signal of the form $v(0,t) = 10\sin 2\pi ft$ is injected into an infinite length of the line at a frequency f of 100 MHz, calculate the signal wavelength on the line and derive expressions for the voltage and current on the line at a distance of 10 m from the input. Hence quote a value for the signal amplitude attenuation in dB per m.

Using Equation (2.20) the characteristic impedance is given by

$$\begin{aligned} Z_0 &= \sqrt{\frac{R+j\omega L}{G+j\omega C}} \\ &= \sqrt{\frac{2\times10^{-2}+j(2\pi\times10^8\times2.5\times10^{-7})}{5\times10^{-4}+j(2\pi\times10^8\times100\times10^{-12})}} \\ &= \sqrt{\frac{157.08\angle 89.993^\circ}{.06283\angle 89.544^\circ}} \\ &= 50\angle 0.2244^\circ\ \Omega \end{aligned} \tag{2.33}$$

The small phase angle results from the loss on the line due to R and G which gives a small imaginary part to Z_0. Similarly the propagation constant is also given in Equation (2.20); therefore

$$\begin{aligned}\gamma &= \sqrt{(R+j\omega L)(G+j\omega C)}\\ &= \sqrt{(157.08\angle 89.993^\circ)(0.06283\angle 89.544^\circ)}\\ &= 3.142\angle 89.768^\circ\,\Omega\\ &= 0.01273 + j3.142\,\Omega\end{aligned} \tag{2.34}$$

from which the attenuation constant α and the phase constant β can be directly determined to be

$$\left.\begin{aligned}\alpha &= 0.01273\ \text{Np m}^{-1} = 0.1106\ \text{dB m}^{-1}\\ \beta &= 3.142\ \text{rad m}^{-1}\end{aligned}\right\} \tag{2.35}$$

Note that because the amplitude of a signal propagating on a lossy line decreases exponentially, the correct unit for α is Nepers m^{-1}. To convert this to dB m^{-1}, i.e. to convert from a natural logarithmic base to the decibel notation, the value for α should be multiplied by a factor 20 $\log_{10}$ e $= 8.686$. The phase velocity is given by Equation (2.24), thus

$$v_p = \frac{2\pi f}{\beta} = \frac{2\pi \times 10^8}{3.142} = 2 \times 10^8\ \text{m s}^{-1} \tag{2.36}$$

The signal wavelength λ is easily determined to be 2 m using Equation (2.24), which should be compared to the "free-space" wavelength λ_0 of a signal at 100 MHz of 3 m using Equation (2.30). Hence from Equation (2.32) ϵ_r is found to be 2.25.

Equations (2.20) and (2.22) can be used to determine the voltage and current signals on the line 10 m away from the input. Clearly there is no reflected or backward wave because the line has infinite length and the phase of the input signal as given is zero. Therefore, neglecting the time dependency of the signal at first gives

$$\left.\begin{aligned} v(10) &= 10\ \exp(-\alpha z)\exp(-j\beta z)\\ &= 10\ \exp(-0.1273)\angle -\beta z\\ &= 8.806\angle -31.42\ \text{rad V}\\ \therefore\quad v(10,t) &= 8.806\sin(\omega t - 31.42)\ \text{V}\\ &\text{and}\\ i(10) &= \frac{8.806\angle -31.42\ \text{rad V}}{Z_0}\\ &= 0.176\angle -31.42\ \text{rad A}\\ \therefore\quad i(10,t) &= 0.176\sin(\omega t - 31.42)\ \text{A}\end{aligned}\right\} \tag{2.37}$$

Since the loss in amplitude over 1 m is given by α above, the loss over 10 m is simply 10 times this figure, i.e. 1.106 dB.

2.3.2 The Transmission Line as a Two-port Network

An alternative form of the solutions to the line equations given in Equation (2.20) can be used to derive a general equation, in matrix form, which relates the current and voltage at

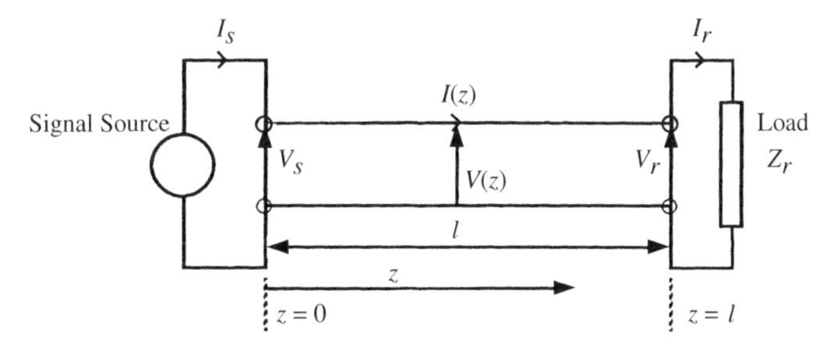

Figure 2.3 A length of transmission line as a two-port network

the input to a finite length of line to the voltage and current at the other end (output end) as if the line was acting as a two-port network as shown in Figure 2.3.

The voltage and current solutions given in Equation (2.20) are

$$\left.\begin{aligned} V(z) &= V_1 \exp(-\gamma z) + V_2 \exp(\gamma z) \\ I(z) &= \frac{1}{Z_0}[V_1 \exp(-\gamma z) - V_2 \exp(\gamma z)] \end{aligned}\right\} \tag{2.38}$$

Let V_s, I_s, V_r, and I_r be the voltages at the input to the line (sending end) and output of the line (receiving end), respectively. At the input of the line $z = 0$ therefore $V(0) = V_s$ and $I(0) = I_s$ and

$$\left.\begin{aligned} V_s &= V_1 + V_2 \\ I_s &= \frac{1}{Z_0}(V_1 - V_2) \\ \therefore \quad V_1 &= \frac{V_s + I_s Z_0}{2} \\ \text{and} \quad V_2 &= \frac{V_s - I_s Z_0}{2} \end{aligned}\right\} \tag{2.39}$$

By substitution the solutions given in Equation (2.38) become

$$\left.\begin{aligned} V(z) &= \exp(-\gamma z)\left[\frac{V_s + I_s Z_0}{2}\right] + \exp(\gamma z)\left[\frac{V_s - I_s Z_0}{2}\right] \\ I(z) &= \frac{\exp(-\gamma z)}{Z_0}\left[\frac{V_s + I_s Z_0}{2}\right] - \frac{\exp(\gamma z)}{Z_0}\left[\frac{V_s - I_s Z_0}{2}\right] \end{aligned}\right\} \tag{2.40}$$

which can be arranged into the form

$$\left.\begin{aligned} V(z) &= V_s \cosh(\gamma z) - I_s Z_0 \sinh(\gamma z) \\ I(z) &= I_s \cosh(\gamma z) - \frac{V_s}{Z_0} \sinh(\gamma z) \end{aligned}\right\} \tag{2.41}$$

These equations can be written in matrix form, i.e.

$$\begin{pmatrix} V(z) \\ I(z) \end{pmatrix} = \begin{pmatrix} \cosh(\gamma z) & -Z_0 \sinh(\gamma z) \\ -\frac{1}{Z_0}\sinh(\gamma z) & \cosh(\gamma z) \end{pmatrix} \begin{pmatrix} V_s \\ I_s \end{pmatrix} \tag{2.42}$$

If the length of the line is l then $z = l$ at the load and $V(z) = V_r$ and $I(z) = I_r$. Equation (2.42) may therefore be written as

$$\left.\begin{aligned} \begin{pmatrix} V_r \\ I_r \end{pmatrix} &= \begin{pmatrix} \cosh(\gamma l) & -Z_0 \sinh(\gamma l) \\ -\dfrac{1}{Z_0}\sinh(\gamma l) & \cosh(\gamma l) \end{pmatrix} \begin{pmatrix} V_s \\ I_s \end{pmatrix} \\ \text{or} \quad \begin{pmatrix} V_s \\ I_s \end{pmatrix} &= \begin{pmatrix} \cosh(\gamma l) & Z_0 \sinh(\gamma l) \\ \dfrac{1}{Z_0}\sinh(\gamma l) & \cosh(\gamma l) \end{pmatrix} \begin{pmatrix} V_r \\ I_r \end{pmatrix} \end{aligned}\right\} \tag{2.43}$$

2.3.3 *Impedance Relations for Terminated Lines*

From Equation (2.43) it is possible to derive an expression for the input impedance of a line terminated in an arbitrary impedance Z_r. Clearly, at the load $Z_r = V_r/I_r$, and at the input the input impedance $Z_i = V_s/I_s$. Therefore from Equation (2.43)

$$Z_i = \frac{V_s}{I_s} = \frac{V_r \cosh(\gamma l) + I_r Z_0 \sinh(\gamma l)}{\dfrac{V_r}{Z_0}\sinh(\gamma l) + I_r \cosh(\gamma l)} = Z_0 \left[\frac{\dfrac{Z_r}{Z_0} + \tanh(\gamma l)}{1 + \dfrac{Z_r}{Z_0}\tanh(\gamma l)} \right] \tag{2.44}$$

Note that if $Z_r = Z_0$, then $Z_i = Z_0$ or the input impedance of any length of transmission line which is terminated by an impedance equal to its characteristic impedance is equal to its characteristic impedance. At low frequencies or for low loss or loss-free lines α is small compared to β, therefore $\gamma \approx j\beta$. Since $\tanh(j\beta l) = j\tan(\beta l)$ Equation (2.44) is often written in the form

$$Z_i = Z_0 \left[\frac{\dfrac{Z_r}{Z_0} + j\tan(\beta l)}{1 + \dfrac{Z_r}{Z_0} j\tan(\beta l)} \right] \tag{2.45}$$

Two special cases should also be considered, namely the case when a finite length of line is short-circuited and the case when the line is left unterminated, i.e. there is an open-circuit at the load end of the line. In these two cases Equation (2.44) can be simplified as follows

$$\left.\begin{aligned} Z_r = 0 &\Rightarrow Z_i(\text{sc}) = Z_0 \tanh(\gamma l) \\ Z_r = \infty &\Rightarrow Z_i(\text{oc}) = Z_0 \coth(\gamma l) \end{aligned}\right\} \tag{2.46}$$

These two relationships are particularly important in the analysis of pulse forming line action and will be referred to extensively in later chapters. It is also useful to consider the consequence of these equations for the loss-free case when these equations can be changed to

$$\left.\begin{aligned} Z_r = 0, \alpha = 0 &\Rightarrow Z_i(\text{sc}) = Z_0 j\tan(\beta l) \\ Z_r = \infty, \alpha = 0 &\Rightarrow Z_i(\text{oc}) = -Z_0 j\cot(\beta l) \end{aligned}\right\} \tag{2.47}$$

As Z_0 normally has a purely resistive (real) value for most lines in common use, $Z_i(\text{sc})$ and $Z_i(\text{oc})$ will be purely reactive (imaginary). The magnitudes of these reactances clearly depend on Z_0 and the length of the line through either a tangent or cotangent term. The magnitudes of the reactances of both a short-circuit and open-circuit line are plotted in Figure 2.4.

The curves shown in Figure 2.4 have some interesting features. The reactances of either a short-circuited line or an open-circuited line can take any value. In other words, at a fixed

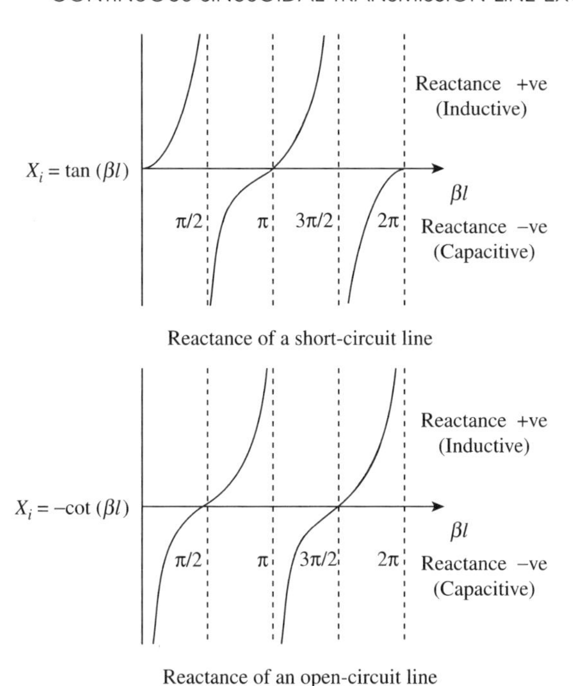

Figure 2.4 The input reactances X_i as a function of βl for short-circuited and open-circuited loss-free transmission lines

frequency, a length of short-circuited or open-circuited transmission line will behave either as a pure inductance or a pure capacitance or a combination of the two depending on length. Also, in the case of the short-circuited line, if the line is cut to a value of βl corresponding to $\pi/2$ or $3\pi/2$ (i.e. $l = \lambda/4$ or $3\lambda/4$) the input impedance of the line Z_i(sc) will be infinite. If it is cut at a value of βl of π then its input impedance Z_i(sc) will be zero. Similarly in the case of an open-circuited line, if the line is cut to a value of βl corresponding to $\pi/2$ or $3\pi/2$ (i.e. $l = \lambda/4$ or $3\lambda/4$) the input impedance of the line Z_i(sc) will be zero, or if it is cut at a value of βl of π then its input impedance Z_i(sc) will be infinite. The ability of short-circuited and open-circuited lines to behave in this way can provide a particularly useful and convenient way of matching loads to lines either using stub-matching or other reactive matching techniques (see, for example, [6]).

Two further special cases are of interest that arise from Equation (2.45) if the length of the line is either one-quarter or one-half of a wavelength long. Since $\beta = 2\pi/\lambda$ then by substitution into the equation

$$\left.\begin{aligned} l = \frac{\lambda}{4} &\Rightarrow Z_i = Z_0\left[\frac{\dfrac{Z_r}{Z_0} + j\tan(\pi/2)}{1 + \dfrac{Z_r}{Z_0}j\tan(\pi/2)}\right] = \frac{Z_0^2}{Z_r} \\ l = \frac{\lambda}{2} &\Rightarrow Z_i = Z_0\left[\frac{\dfrac{Z_r}{Z_0} + j\tan(\pi)}{1 + \dfrac{Z_r}{Z_0}j\tan(\pi)}\right] = Z_r \end{aligned}\right\} \tag{2.48}$$

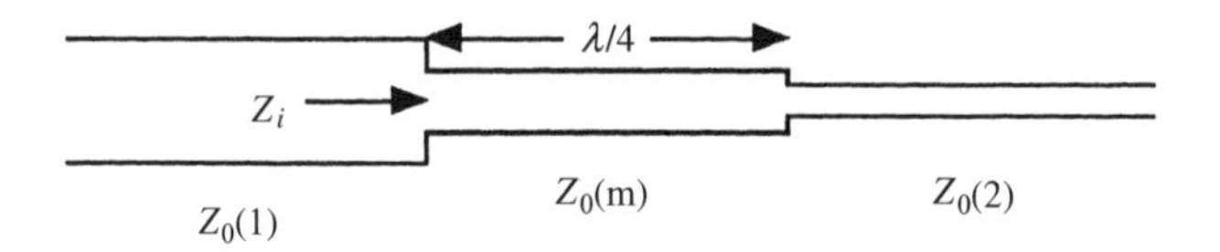

Figure 2.5 Diagram of a quarter-wave transformer

The quarter-wavelength case is of special interest as it can be used as a method of matching two lengths of transmission line whose impedances are different. This is done using the circuit shown in Figure 2.5.

If it is required to match a transmission line of characteristic impedance $Z_0(1)$ to a second line of impedance $Z_0(2)$, a transmission line one-quarter of a wavelength long with characteristic impedance $Z_0(m)$ is connected between the two lines to be matched, as shown in Figure 2.5. The matching impedance $Z_0(m)$ using Equation (2.48) for the quarter-wavelength case is given by

$$Z_0(1) = Z_i = \frac{Z_0(m)^2}{Z_0(2)}$$
$$\text{or} \quad Z_0(m) = \sqrt{Z_0(1)Z_0(2)} \tag{2.49}$$

Since the length of matching line depends on the wavelength of the signal on the line, it is clear that this matching technique can only be used at a single fixed frequency. To improve the bandwidth of operation, further quarter-wavelength sections of transmission line can be added whose impedances successively taper from the impedance of the first line to that of the line to be matched. This technique will be discussed later in this book in Chapter 6.

Example 2.2

At a fixed frequency the input impedance of a section of coaxial transmission line of length l is measured, first with the line terminated in a short circuit and then with the line terminated in an open circuit. The respective values obtained are $Z(\text{sc}) = 15 + j20\,\Omega$ and $Z(\text{oc}) = 100 - j120\,\Omega$. Calculate the characteristic impedance of the line.

How could the attenuation constant of the line be determined if its length l were known?

From Equation (2.46)

$$\begin{aligned} Z_0 &= \sqrt{Z(\text{sc})Z(\text{oc})} \\ &= \sqrt{(15 + j20)(100 - j120)} \\ &= \sqrt{(25\angle 53.13^\circ)(156.2\angle -50.19^\circ)} \\ &= 62.49\angle 1.47^\circ\Omega \\ &= 62.46 + j1.6\Omega \end{aligned} \tag{2.50}$$

Also

$$\sqrt{\frac{Z(\text{sc})}{Z(\text{oc})}} = \tanh(\gamma l) = \tanh(\alpha + j\beta)l \tag{2.51}$$

Since

$$\tanh(x) = \frac{e^x - e^{-x}}{e^x + e^{-x}} \tag{2.52}$$

Equation (2.51) can be written in the form

$$\exp(\alpha + j\beta)2l = \frac{1 + \sqrt{Z(\text{sc})/Z(\text{oc})}}{1 - \sqrt{Z(\text{sc})/Z(\text{oc})}}$$

$$\text{or} \qquad (\alpha + j\beta)l = \frac{1}{2}\ln\left[\frac{1 + \sqrt{Z(\text{sc})/Z(\text{oc})}}{1 - \sqrt{Z(\text{sc})/Z(\text{oc})}}\right] \tag{2.53}$$

As

$$\ln[A\exp(j\theta)] = \ln A + j(\theta + 2n\pi), \qquad n = 1, 2, 3\ldots \tag{2.54}$$

α can be determined to be given by

$$\alpha = \frac{1}{2l}\ln\left|\frac{1 + \sqrt{Z(\text{sc})/Z(\text{oc})}}{1 - \sqrt{Z(\text{sc})/Z(\text{oc})}}\right| \tag{2.55}$$

2.3.4 Line Reflections

If the terminating impedance at the end of a fixed length of transmission line does not equal the characteristic impedance of the line, then a discontinuity is set up at the end of the line and reflected voltage and current waves are produced which travel back to the input of the line. The line is said to be unmatched and some energy is dissipated in the terminating impedance and some is reflected back. Referring again to Figure 2.3, if the unmatched load impedance is at a distance l from the input of the line, then from Equation (2.38) the voltage and current at the load can be written as

$$\left.\begin{aligned} V(l) &= V_1\exp(-\gamma l) + V_2\exp(\gamma l) \\ I(l) &= \frac{1}{Z_0}[V_1\exp(-\gamma l) - V_2\exp(\gamma l)] \end{aligned}\right\} \tag{2.56}$$

As explained earlier, the first part of the expression for $V(l)$, i.e. $V_1\exp(-\gamma l)$, is the forward wave since its amplitude is decreasing in the positive z direction. It can also be regarded as the incident wave at the load, and as such is often given the symbol V_+. Similarly the second part of the expression, i.e. $V_2\exp(\gamma l)$, is the backward wave since its amplitude is increasing as it proceeds back towards the load. This part of the expression can be regarded as that part of the incident wave that is reflected by the load and as such is given the symbol V_-. The ratio of the reflected wave to the incident wave is known as the reflection coefficient ρ_V and is defined as

$$\rho_V = \frac{V_-}{V_+} = \frac{V_2\exp(\gamma l)}{V_1\exp(-\gamma l)} = \frac{V_2}{V_1}\exp(2\gamma l) \tag{2.57}$$

The subscript v is used to denote the fact that here the reflection coefficient ρ_V is a voltage reflection coefficient. Since ρ_V is likely to be complex it can also be written in the form

$$\rho_V = |\rho_V|\exp(j\phi) \tag{2.58}$$

where ϕ represents the change in phase angle between the incident and reflected waves. The reflection coefficient can also be expressed in terms of the characteristic impedance of the line and the load impedance. From Equation (2.56)

$$\begin{aligned} Z_r = \frac{V_r}{I_r} &= Z_0 \left[\frac{V_1 \exp(-\gamma l) + V_2 \exp(\gamma l)}{V_1 \exp(-\gamma l) - V_2 \exp(\gamma l)}\right] \\ &= Z_0 \left[\frac{1 + (V_2/V_1)\exp(2\gamma l)}{1 - (V_2/V_1)\exp(2\gamma l)}\right] \\ &= Z_0 \left(\frac{1 + \rho_V}{1 - \rho_V}\right) \end{aligned} \tag{2.59}$$

This expression can be rearranged into its more familiar form, i.e.

$$\rho_V = \frac{Z_r - Z_0}{Z_r + Z_0} \tag{2.60}$$

Note that if a similar procedure is used to determine the current reflection coefficient, the same expression results except that the sign is changed, i.e. $\rho_I = -\rho_V$. At this point it is helpful to investigate the shape of the voltage waveforms on both a short-circuit and an open-circuit length of transmission line into which a sinusoidal signal is injected at the input. Figure 2.6 gives sketches of these waveforms for the special case of a line which is exactly one wavelength long. Each sketch is a "snapshot" of the waveform at times corresponding to the phase angle of the input waveform being 0 or 90°.

In the case of the open-circuit line $Z_r = \infty$, and therefore Equation (2.60) gives $\rho_V = 1$ and $\phi = 0°$. Examination of the waveform at the end of the line ($l = \lambda$) clearly shows that there is no phase change between the incident and reflected wave and the polarity is unchanged as expected. However for a short-circuit $Z_r = 0$, and therefore Equation (2.60)

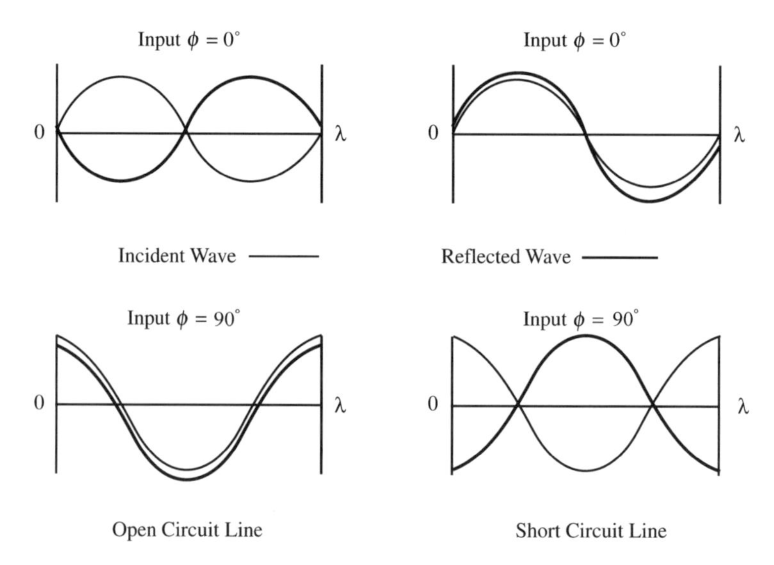

Figure 2.6 Voltage waveforms on a short circuit and open circuit transmission line one wavelength long

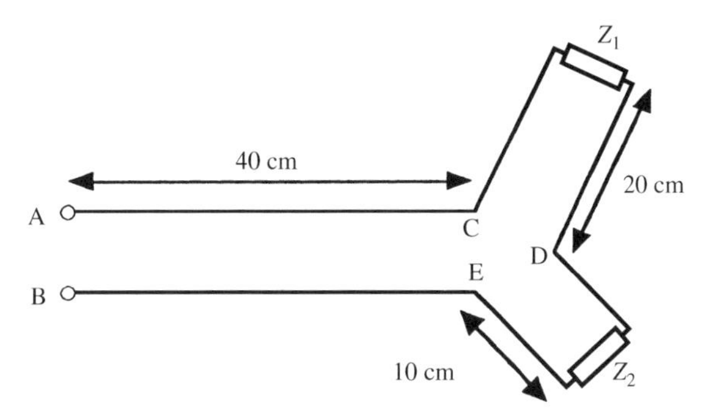

Figure 2.7 The transmission line system for Example 2.3

gives $\rho_V = -1$ and $\phi = 180°$. Again examination of the waveforms for the short-circuit line clearly show the required phase shift and change in polarity.

Example 2.3

Calculate the value of the load Z_1 and hence the impedance at the input terminals AB of the loss-free air-insulated transmission line system shown in Figure 2.7. Load Z_1 has a reflection coefficient ρ_{1V} of $0.5 + j0.5$; load Z_2 is purely reactive and has a value $j25\,\Omega$. All three transmission lines have a characteristic impedance of $50\,\Omega$ and the system is operated at a frequency of 375 MHz.

At a frequency of 375 MHz, the line wavelength on an air-insulated line will be $\lambda = c/3.75 \times 10^8 = 80$ cm. From Equation (2.59)

$$Z_1 = Z_0\left(\frac{1+\rho_{1V}}{1-\rho_{1V}}\right) = 50\frac{(1.5+j0.5)}{(0.5-j0.5)} = 50 + j100\,\Omega \tag{2.61}$$

The length of the line from Z_1 to the terminals CD is 20 cm, i.e. $\lambda/4$. Thus this section of line acts as a quarter-wave transformer and the input impedance at terminals CD is given by

$$Z_i(\mathrm{CD}) = \frac{Z_0^2}{Z_1} = \frac{2500}{50+\mathrm{j}100} = 10 - \mathrm{j}20 \tag{2.62}$$

The input impedance at terminals DE looking into the 10 cm length of line terminated by Z_2 is given by

$$Z_i(\mathrm{DE}) = Z_0\left[\frac{Z_2 + jZ_0\tan(\beta l)}{Z_0 + jZ_2\tan(\beta l)}\right]$$

$$\text{where}\quad \beta l = \frac{2\pi l}{\lambda} = \frac{2\pi 10}{80} = \frac{\pi}{4} \quad\text{and}\quad \tan(\beta l) = 1$$

$$\therefore\quad Z_i(\mathrm{DE}) = 50\left[\frac{j25 + j50}{50 - 25}\right] = j150\ \Omega \tag{2.63}$$

The input impedance looking into terminals CE must be equal to the sum of the input impedances at terminals CD and DE, i.e. $10 + j130\ \Omega$. Since the length of line from AB to CE is $\lambda/2$, the input impedance looking into the input at terminals AB must be equal to that at terminals CE, i.e. $10 + j130\ \Omega$.

2.4 TRANSIENT TRANSMISSION LINE RESPONSE

Having investigated the electrical characteristics of transmission lines subject to continuous sinusoidal excitation, their behaviour when subjected to the input of transient electrical signals can now be analysed. This analysis is of fundamental importance to the understanding of the use of transmission lines in pulse forming which is covered in Chapter 3. In order to carry out this analysis, it is necessary to return to the generalised expressions for voltage and current on a transmission line, expressed in terms of the complex frequency parameter **s** and the position coordinate z as given above. It is important to note that some of the relationships and concepts derived in the last section can be extended to cover the analysis of transient response problems.

2.4.1 Transient Response of the Infinite Line

Here the response of an infinite length of transmission line is analysed when subjected to a transient input signal of the form $v(0, t)$ or $\mathbf{V}(0, \mathbf{s})$ in the complex frequency domain. It is assumed that the transient signal is injected into the input of the line from a generator with internal impedance $Z_g(\mathbf{s})$, as shown in Figure 2.8. Clearly if the line is lossy the current $i(z, t)$ and voltage $v(z, t)$ on the line must tend to zero and there will be no reflected or backwards signals. Therefore from Equations (2.14) and (2.17) the voltage and current on the line are given by

$$\left.\begin{aligned}\mathbf{V}(z, \mathbf{s}) &= \mathbf{V}_1(\mathbf{s})\exp(-\gamma(\mathbf{s})z)\\ \mathbf{I}(z, \mathbf{s}) &= \frac{1}{\mathbf{Z}_0(\mathbf{s})}[\mathbf{V}_1(\mathbf{s})\exp(-\gamma(\mathbf{s})z)]\end{aligned}\right\} \tag{2.64}$$

At the input of the line $(z = 0)$ $\mathbf{V}(0, \mathbf{s}) = \mathbf{V}_1(\mathbf{s})$, an expression for $\mathbf{V}_1(\mathbf{s})$ can be determined from the circuit in Figure 2.8 to be

$$\mathbf{V}_1(\mathbf{s}) = \frac{\mathbf{Z}_0(\mathbf{s})}{\mathbf{Z}_0(\mathbf{s}) + \mathbf{Z}_g(\mathbf{s})}\mathbf{V}_g(\mathbf{s}) \tag{2.65}$$

Therefore the voltage on the line $\mathbf{V}(z, \mathbf{s})$, expressed as a function of **s** is given by

$$\mathbf{V}(z, \mathbf{s}) = \frac{\mathbf{Z}_0(\mathbf{s})}{\mathbf{Z}_0(\mathbf{s}) + \mathbf{Z}_g(\mathbf{s})}\mathbf{V}_g(\mathbf{s})\exp(-\gamma(\mathbf{s})z) \tag{2.66}$$

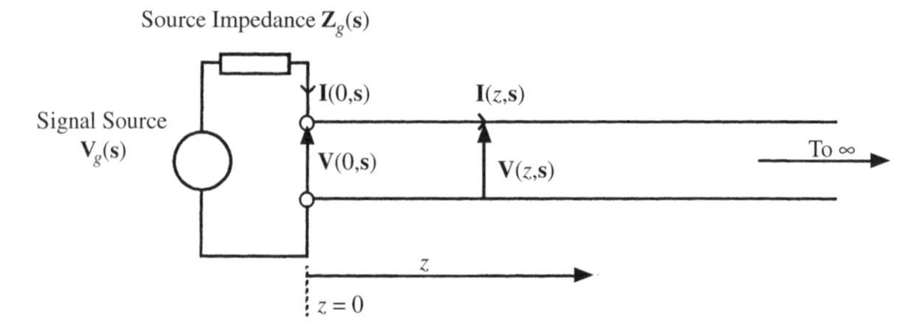

Figure 2.8 Circuit for the analysis of the transient behaviour of an infinitely long transmission line

In order to invert this expression to determine the voltage and current on the line as functions of time, it is easiest to consider the case of the loss-free line where R and $G = 0$. In this case using Equations (2.12) and (2.29) the propagation constant $\gamma(\mathbf{s})$ is given by

$$\gamma(\mathbf{s}) = \mathbf{s}\sqrt{LC} = \frac{\mathbf{s}}{v_p} \tag{2.67}$$

where v_p is the propagation velocity of a signal or transient on the line and is in the same units of length s^{-1} as the units of length over which the line parameters L and C are defined. Also from Equation (2.17) it is clear that the line impedance Z_0 will now be real. If it is assumed that the generator also has a real impedance Z_g then Equation (2.66) becomes

$$\mathbf{V}(z, \mathbf{s}) = \frac{Z_0}{Z_0 + Z_g}\mathbf{V}_g(\mathbf{s})\exp(-\mathbf{s}z/v_p) \tag{2.68}$$

This expression can be inverted to find an expression for the voltage on the line at any position co-ordinate z using the shifting property given by Equation (1.79). This gives

$$\left.\begin{aligned} v(z, t) &= \frac{Z_0}{Z_0 + Z_g} v_g(t - z/v_p)u(t - z/v_p) \\ \text{where} \quad v_g(t) &= \mathscr{L}^{-1}\{\mathbf{V}(0, \mathbf{s})\} \end{aligned}\right\} \tag{2.69}$$

This equation basically states that the voltage $v(z, t)$ at any point on the line will be the same as that produced by the generator but is delayed by a time equal to z/v_p, i.e. the time taken for the signal to propagate to the position given by z.

2.4.2 Transient Response of Lossy Transmission Lines

In the last section the treatment of the propagation of a transient signal was simplified by assuming the line was loss-free. However, in reality of course, transmission lines are not loss-free. The two main sources of loss are dielectric loss in the dielectric medium which insulates the lines and resistive loss in the conductors (usually copper). A full treatment of the transient response of a line when these two sources of loss are included (by including terms for both R and G) becomes difficult, the problem being exacerbated because, in both cases, the loss is frequency-dependent. The problem is an important one since it deals with the way in which pulses are distorted as they propagate on lossy lines. The standard method of dealing with this difficult problem is to consider the relative sizes of these two sources of loss at frequencies in which the loss and consequent distortion are likely to be significant. Expressions for loss in terms of frequency-dependent attenuation constants are available [8], and for a coaxial line they take the form

$$\left.\begin{aligned} \alpha_d &= (0.0903 p_f\sqrt{\varepsilon_r}) f \,\text{dBm}^{-1} \qquad \text{dielectric loss} \\ \alpha_r &= \left[\frac{3.617 \times 10^{-4}}{Z_0}\left(\frac{1}{a} + \frac{1}{b}\right)\right]\sqrt{f}\,\text{dBm}^{-1} \qquad \text{resistive loss} \end{aligned}\right\} \tag{2.70}$$

where p_f is the power factor of the dielectric, ϵ_r is the relative permittivity or dielectric constant of the dielectric, f is the frequency in MHz, and a and b are the diameters of the inner and outer conductors, respectively, in metres. The expression for the resistive attenuation constant depends on $\sqrt{f}$ because the main source of loss is due to the skin effect which is also dependent on $\sqrt{f}$. To proceed further it is necessary to consider the relative size of each loss contribution to the total attenuation constant. Referring to [8] it can be shown that in typical coaxial lines resistive losses dominate below frequencies of 1 GHz and therefore the total attenuation constant can be expressed in the form

$$\alpha = K\sqrt{f} \tag{2.71}$$

where K is a constant which is determined by the physical size and characteristic impedance of the coaxial line through Equation (2.70). Consideration of the way in which the skin effect will affect the physical parameters of a transmission line, leads to the conclusion that not only will the resistance of the conductors increase with frequency but the inductance will also be frequency-dependent. A full analysis of the way in which the impedance of a round wire will change as a result of the skin effect [4,9] suggests that the wire should be represented as having a skin impedance $\mathbf{Z}_{sk}(\mathbf{s})$ given by

$$\left.\begin{aligned} \mathbf{Z}_{sk}(\mathbf{s}) &= R_{sk}\sqrt{\mathbf{s}} \\ \text{where} \qquad R_{sk} &= \frac{1}{2\pi r_0}\sqrt{\frac{\mu}{\sigma}} \end{aligned}\right\} \tag{2.72}$$

and where r_0 is the radius of the round wire, μ its permeability and σ its conductivity. The dependence on $\sqrt{\mathbf{s}}$ is not surprising given the dependence on $\sqrt{f}$ of the attenuation constant α. Clearly this impedance can be used to replace the line resistance parameter R in the investigation of the transient behaviour of a coaxial transmission line. Thus the propagation constant of the line, at high frequencies, can be approximated to

$$\begin{aligned} \gamma(\mathbf{s}) &= \sqrt{(R + \mathbf{s}L)(\mathbf{s}C)} \\ &= \sqrt{(\mathbf{Z}_{sk}(\mathbf{s}) + \mathbf{s}L)(\mathbf{s}C)} \\ &= \sqrt{(\mathbf{s}^2 LC + \mathbf{s}^{3/2} CR_{sk})} \end{aligned} \tag{2.73}$$

This expression can be further approximated by use of the binomial expansion to give

$$\begin{aligned} \gamma(\mathbf{s}) &\approx \mathbf{s}\sqrt{LC} + \sqrt{\mathbf{s}}\frac{R_{sk}}{2Z_0} + \cdots \\ \text{where} \qquad Z_0 &= \sqrt{\frac{L}{C}} \end{aligned} \tag{2.74}$$

Equation (2.66) can be rewritten as

$$\begin{aligned} \mathbf{V}(z, \mathbf{s}) &= \frac{Z_0}{Z_0 + Z_g}\mathbf{V}_g(\mathbf{s})\exp\left[-\left(\mathbf{s}\sqrt{LC} + \sqrt{\mathbf{s}}\frac{R_{sk}}{2Z_0}\right)z\right] \\ &= \frac{Z_0}{Z_0 + Z_g}\mathbf{V}_g(\mathbf{s})\exp(-\mathbf{s}z/v_p)\exp\left(-\sqrt{\mathbf{s}}\frac{R_{sk}}{2Z_0}z\right) \end{aligned} \tag{2.75}$$

To find the response of the line to a unit step input $\mathbf{V}_g(\mathbf{s})$ is replaced by $1/\mathbf{s}$ and the inverse transform must be taken. By reference to Equation (2.69), it is easy to see that the first exponential term in $\mathbf{s}$ simply gives the delay of the step to the position of interest on the line given by the z coordinate. The second term, however, describes the way in which the step is distorted after propagation. By reference to the table of Laplace transforms at the rear of this book, this term can be inverted as follows

$$
\begin{aligned}
v(z,t) &= \frac{Z_0}{Z_0+Z_g}\mathscr{L}^{-1}\left[\frac{1}{\mathbf{s}}\exp\left(-\sqrt{\mathbf{s}}\frac{R_{sk}}{2Z_0}z\right)\right] \\
&= \frac{Z_0}{Z_0+Z_g}\left[1-\operatorname{erf}\left(\frac{zR_{sk}}{4Z_0\sqrt{t}}\right)\right]
\end{aligned} \tag{2.76}
$$

The function given is the complementary error function and can be found listed in comprehensive books of mathematical tables. It is important to remember that this function starts after a delay given by the time for the unit step to propagate to the position on the line of interest, i.e. after a delay of z/v_p. In order to understand the way in which this function distorts the step leading to a loss in rise-time, a normalised plot of the function against time is given in Figure 2.9.

Although the time taken for the amplitude of the pulse to reach 50% of full amplitude, after the start of the rising edge of the distorted step, may be quite short, it takes 29 times as long to reach 90% of full amplitude. This can cause quite serious loss of pulse rise-time and distortion if fast pulses are to be propagated over long distances. The only way to minimise this problem is to use large-diameter coaxial cables in which the copper loss due to the skin effect can be reduced.

2.4.3 Transient Response of Terminated Lines

The behaviour of a finite length of transmission line driven from a source with impedance $Z_g(\mathbf{s})$ and terminated in a load $Z_r(\mathbf{s})$ can be deduced using the basic circuit shown in Figure 2.10. The general solutions for both the current and voltage on the line are given by Equations (2.14) and (2.16), i.e.

$$
\left.
\begin{aligned}
\mathbf{V}(z,\mathbf{s}) &= \mathbf{V}_1(\mathbf{s})\exp(-\gamma(\mathbf{s})z) + \mathbf{V}_2(\mathbf{s})\exp(\gamma(\mathbf{s})z) \\
\mathbf{I}(z,\mathbf{s}) &= \frac{1}{\mathbf{Z}_0(\mathbf{s})}[\mathbf{V}_1(\mathbf{s})\exp(-\gamma(\mathbf{s})z) - \mathbf{V}_2(\mathbf{s})\exp(\gamma(\mathbf{s})z)]
\end{aligned}
\right\} \tag{2.77}
$$

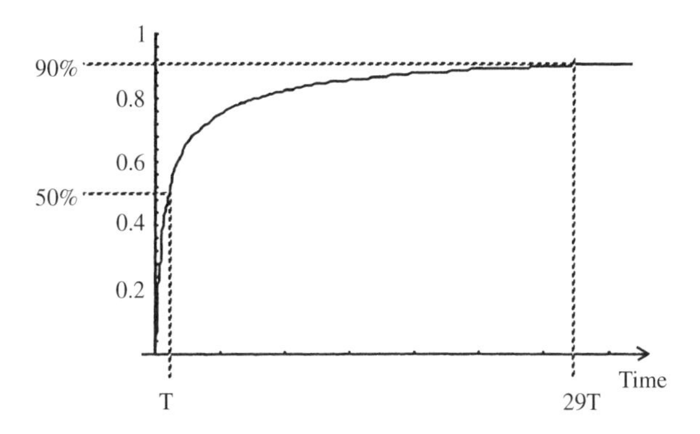

Figure 2.9 A normalised plot of Equation (2.76) showing the serious degradation in rise-time of a pulse caused by skin-effect losses in transmission lines

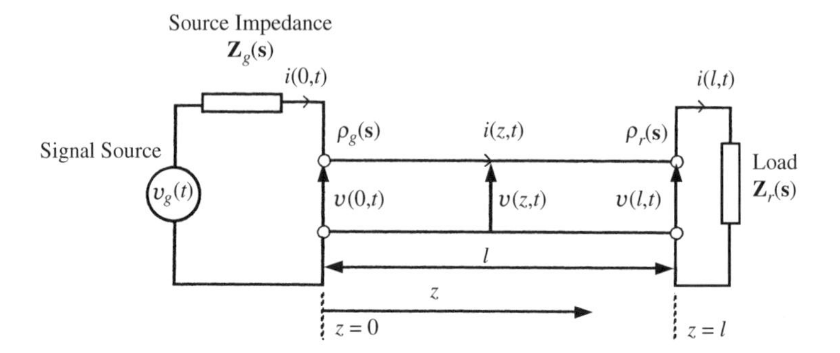

Figure 2.10 Circuit representation of the terminated line problem

As explained above, the first term in the first two expressions is interpreted as a forward travelling signal or transient and the second term as a backward travelling signal. Thus, using the nomenclature used earlier to distinguish forward and backward travelling signals, one can write, for example,

$$\left.\begin{aligned} \mathbf{V}_+(z,\mathbf{s}) &= \mathbf{V}_1(\mathbf{s})\exp(-\gamma(\mathbf{s})z) \\ \mathbf{V}_-(z,\mathbf{s}) &= \mathbf{V}_2(\mathbf{s})\exp(+\gamma(\mathbf{s})z) \\ \mathbf{V}(z,\mathbf{s}) &= \mathbf{V}_+(z,\mathbf{s}) + \mathbf{V}_-(z,\mathbf{s}) \end{aligned}\right\} \tag{2.78}$$

Similar expressions can written for the current on the line. As before, the relationship between the forward and backward travelling signals at the load $Z_\mathbf{r}(\mathbf{s})$ can be determined from the condition

$$\mathbf{V}(l,\mathbf{s}) = \mathbf{I}(l,\mathbf{s})\mathbf{Z}_r(\mathbf{s}) \tag{2.79}$$

Substitution from Equation (2.78) gives

$$\mathbf{V}_+(l,\mathbf{s}) + \mathbf{V}_-(l,\mathbf{s}) = \frac{\mathbf{Z}_r(\mathbf{s})}{\mathbf{Z}_0(\mathbf{s})}[\mathbf{V}_+(l,\mathbf{s}) - \mathbf{V}_-(l,\mathbf{s})]$$

$$\text{or} \quad \frac{\mathbf{V}_-(l,\mathbf{s})}{\mathbf{V}_+(l,\mathbf{s})} = \frac{\mathbf{Z}_r(\mathbf{s}) - \mathbf{Z}_0(\mathbf{s})}{\mathbf{Z}_r(\mathbf{s}) + \mathbf{Z}_0(\mathbf{s})} = \rho_r(\mathbf{s}) \tag{2.80}$$

Here a generalised voltage reflection coefficient $\rho_r(\mathbf{s})$ has been defined which is in a similar form to that given in Equation (2.60), yet will enable the reflection of transient signals at a non-matched load to be dealt with. As noted before, the current reflection coefficient at the load takes exactly the same form, in terms of the load and line impedances, with the exception that its sign is changed. In the case where the load and line impedances are equal, $\rho_r(\mathbf{s})$ is zero and there is no reflected signal. Again the line is said to be matched and, in effect, the line behaves in a similar way to the infinite line for transient signals propagating at a velocity $v_p(= 1/\sqrt{LC}$ for a loss-free line) from the generator to the load.

If the impedance of the generator $\mathbf{Z}_\mathrm{g}(\mathbf{s})$ is not equal to the impedance of the line $\mathbf{Z}_0(\mathbf{s})$ then any reflected signal from the load which propagates back towards the input of the line will also be reflected and will then propagate back again towards the load. Thus in an analogous way to the determination of the reflection coefficient at the load, a reflection coefficient at the input $\rho_g(\mathbf{s})$ can be defined as

$$\rho_g(\mathbf{s}) = \frac{\mathbf{Z}_g(\mathbf{s}) - \mathbf{Z}_0(\mathbf{s})}{\mathbf{Z}_g(\mathbf{s}) + \mathbf{Z}_0(\mathbf{s})} \tag{2.81}$$

A generalised expression for the voltage $\mathbf{V}(z, \mathbf{s})$ on the line, in the case where the line and load impedances are not equal, can be determined by starting from Equation (2.77) and by applying boundary conditions at both the input and load ends of the line. Therefore at the input to the line

$$\left.\begin{aligned} \mathbf{V}(0, \mathbf{s}) &= \mathbf{V}_g(\mathbf{s}) - \mathbf{I}(0, \mathbf{s})\mathbf{Z}_g(\mathbf{s}) = \mathbf{V}_1(\mathbf{s}) + \mathbf{V}_2(\mathbf{s}) \\ \mathbf{I}(0, \mathbf{s}) &= \frac{\mathbf{V}_1(\mathbf{s}) - \mathbf{V}_2(\mathbf{s})}{Z_0(\mathbf{s})} \end{aligned}\right\} \tag{2.82}$$

Eliminating $\mathbf{I}(0, \mathbf{s})$ from these two equations gives

$$\mathbf{V}_g(\mathbf{s}) \frac{\mathbf{Z}_0(\mathbf{s})}{\mathbf{Z}_0(\mathbf{s}) + \mathbf{Z}_g(\mathbf{s})} = a\mathbf{V}_g(\mathbf{s}) = \mathbf{V}_1(\mathbf{s}) - \rho_g(\mathbf{s})\mathbf{V}_2(\mathbf{s})$$

$$\text{where} \qquad a = \frac{\mathbf{Z}_0(\mathbf{s})}{\mathbf{Z}_0(\mathbf{s}) + \mathbf{Z}_g(\mathbf{s})} \tag{2.83}$$

where a can be considered to be a form of attenuation constant. Substitution of Equations (2.77) into Equation (2.79) gives

$$[\mathbf{V}_1(\mathbf{s})\exp(-\gamma(\mathbf{s})l) - \mathbf{V}_2(\mathbf{s})\exp(+\gamma(\mathbf{s})l)] \frac{\mathbf{Z}_r(\mathbf{s})}{\mathbf{Z}_0(\mathbf{s})} = [\mathbf{V}_1(\mathbf{s})\exp(-\gamma(\mathbf{s})l) + \mathbf{V}_2(\mathbf{s})\exp(+\gamma(\mathbf{s})l)]$$

$$\text{or} \quad \rho_r(\mathbf{s})\mathbf{V}_1(\mathbf{s})\exp(-\gamma(\mathbf{s})l) - \mathbf{V}_2(\mathbf{s})\exp(+\gamma(\mathbf{s})l) = 0 \tag{2.84}$$

Solving these two equations for the coefficients $\mathbf{V}_1(\mathbf{s})$ and $\mathbf{V}_2(\mathbf{s})$ gives

$$\begin{aligned} \mathbf{V}_1(\mathbf{s}) &= a\mathbf{V}_g(\mathbf{s})\left[1 + \frac{\rho_g(\mathbf{s})\rho_r(\mathbf{s})}{\exp(2\gamma(\mathbf{s})l) - \rho_g(\mathbf{s})\rho_r(\mathbf{s})}\right] \\ \mathbf{V}_2(\mathbf{s}) &= \frac{a\mathbf{V}_g(\mathbf{s})\rho_r(\mathbf{s})}{\exp(2\gamma(\mathbf{s})l) - \rho_g(\mathbf{s})\rho_r(\mathbf{s})} \end{aligned} \tag{2.85}$$

Substitution of these two equations into the expression for the voltage on the line (Equation (2.77)), and rearranging, results in

$$\mathbf{V}(z, \mathbf{s}) = a\mathbf{V}_g(\mathbf{s})\left[\frac{\exp(-\gamma(\mathbf{s})z) + \rho_r(\mathbf{s})\exp(\gamma(\mathbf{s})(z - 2l))}{1 - \rho_g(\mathbf{s})\rho_r(\mathbf{s})\exp(-2\gamma(\mathbf{s})l)}\right] \tag{2.86}$$

This equation can be further modified for the special case where the line is loss-free. In this case $\gamma(\mathbf{s})$ is given by Equation (2.67). The time delay per unit length of line (δ') can be determined from the propagation velocity on the line, therefore the propagation constant can be written as

$$\gamma(\mathbf{s}) = \frac{\mathbf{s}}{v_p} = \mathbf{s}\delta'$$

$$\text{where} \qquad \delta' = \sqrt{LC} \tag{2.87}$$

Therefore by substitution Equation (2.86) becomes

$$\mathbf{V}(z,\mathbf{s}) = a\mathbf{V}_g(\mathbf{s})\left[\frac{\exp(-\mathbf{s}\delta' z) + \rho_r(\mathbf{s})\exp(\mathbf{s}\delta'(z-2l))}{1-\rho_g(\mathbf{s})\rho_r(\mathbf{s})\exp(-2\mathbf{s}\delta' l)}\right] \tag{2.88}$$

At first sight, this equation appears to be rather complicated and certainly not easy to invert so that the behaviour of transient signals on transmission lines can be investigated. However, in practice this is not the case and the equation provides a very elegant way of representing the way in which transient signals are propagated and reflected through multiple reflections at both the input and load ends of the line. For example, if the voltage at the input of the line is required, as a function of time, putting $z = 0$ into Equation (2.88) gives

$$\mathbf{V}(0,\mathbf{s}) = a\mathbf{V}_g(\mathbf{s})\left[\frac{1 + \rho_r(\mathbf{s})\exp(-2\mathbf{s}\delta' l)}{1-\rho_g(\mathbf{s})\rho_r(\mathbf{s})\exp(-2\mathbf{s}\delta' l)}\right] \tag{2.89}$$

If it is now assumed that the load and generator impedances are both resistive quantities, i.e. the reflection coefficients no longer depend on $\mathbf{s}$, then by expanding the denominator in Equation (2.89) using a binomial expansion

$$\mathbf{V}(0,\mathbf{s}) = a\mathbf{V}_g(\mathbf{s})[1+\rho_r\exp(-2\mathbf{s}\delta' l)]\left[\begin{array}{c} 1 + \rho_g\rho_r\exp(-2\mathbf{s}\delta' l) + (\rho_g\rho_r)^2\exp(-4\mathbf{s}\delta' l) \\ +(\rho_g\rho_r)^3\exp(-6\mathbf{s}\delta' l) + \cdots \end{array}\right] \tag{2.90}$$

It is now a simple matter to find the inverse transform, as each exponential term, by reference to the Laplace transform table, is recognised as a delay term; the delay time being in units of $2\delta' l$, i.e. after expansion

$$\begin{aligned} v(0,t) &= av_g(t) + a\rho_r(1+\rho_g)v_g(t-2\delta' l)u(t-2\delta' l) \\ &+ a\rho_g\rho_r^2(1+\rho_g)v_g(t-4\delta' l)u(t-4\delta' l) \\ &+ a\rho_g^2\rho_r^3(1+\rho_g)v_g(t-6\delta' l)u(t-6\delta' l)\ldots \\ &+ a(\rho_g\rho_r)^{n-1}\rho_r(1+\rho_g)v_g(t-2n\delta' l)u(t-2n\delta' l) \end{aligned} \tag{2.91}$$

Thus the voltage at the input of the line is equal to $av_g(t)$ for a time $t = 2\delta' l$ after which a component $a\rho_r(1+\rho_g)v_g(t-2\delta' l)$ is added for a time t given by $2\delta' l \leq t \leq 4\delta' l$. After this time period a further component $a\rho_g\rho_r^2(1+\rho_g)v_g(t-4\delta' l)$ is added and so on.

It is instructive to examine some special cases to illustrate just how useful the above mathematical method can prove to be. This is done for the cases where the load $\mathbf{Z}_r(\mathbf{s})$ is a short circuit, an open circuit, and is matched and the input to the line is a unit step or a rectangular pulse. In each case it is assumed that the generator is matched to the line such that $a = 1/2$ and $\rho_g = 0$. Therefore Equation (2.89) becomes

$$\mathbf{V}(0,\mathbf{s}) = \frac{\mathbf{V}_g(\mathbf{s})}{2}[1 + \rho_r(\mathbf{s})\exp(-2\mathbf{s}\delta' l)] \tag{2.92}$$

Clearly there is now only one reflection at the load and just one delayed component to be added to $v(0,t)$ after a time $t = 2\delta' l$. Looking at the three load conditions, the reflection coefficient for each will be given by Equation (2.80) and therefore Equation (2.92) becomes

$$\left.\begin{aligned}
\mathbf{V}(0,\mathbf{s}) &= \frac{\mathbf{V}_g(\mathbf{s})}{2}[1-\exp(2\mathbf{s}\delta' l)], \ \mathbf{Z}_r(\mathbf{s}) = 0 && \text{(short circuit)}\\
\mathbf{V}(0,\mathbf{s}) &= \frac{\mathbf{V}_g(\mathbf{s})}{2}[1+\exp(2\mathbf{s}\delta' l)], \ \mathbf{Z}_r(\mathbf{s}) = \infty && \text{(open circuit)}\\
\mathbf{V}(0,\mathbf{s}) &= \frac{\mathbf{V}_g(\mathbf{s})}{2}, \qquad \mathbf{Z}_r(\mathbf{s}) = \mathbf{Z}_0(\mathbf{s}) && \text{(matched load)}
\end{aligned}\right\} \tag{2.93}$$

The inverse transforms of these equations are

$$\begin{aligned}
v(0,t) &= \left[\frac{v_g(t) - v_g(t-2\delta' l)u(t-2\delta' l)}{2}\right] && \text{(short circuit)}\\
v(0,t) &= \left[\frac{v_g(t) + v_g(t-2\delta' l)u(t-2\delta' l)}{2}\right] && \text{(open circuit)}\\
v(0,t) &= \frac{v_g(t)}{2} && \text{(matched load)}
\end{aligned} \tag{2.94}$$

The resulting pulse shapes for the voltage at the input of the line are now straightforward to determine for the cases of an unit step function and a unit rectangular pulse. In order to illustrate the resulting response of the transmission line, as a function of time, a series of plots of the potential along a length of transmission line l, at various times after the start of the input, are given in Figure 2.11. It is a useful exercise to check that the plots given match the voltage profiles given by Equation (2.94) for the voltage at the input of the line for the two different inputs.

2.4.4 Input Impedance of Terminated Lines for Transient Signals

The input impedance $\mathbf{Z}_i(\mathbf{s})$ presented by a terminated line to a transient input signal can be determined by reference to the circuit given in Figure 2.10. By inspection the input voltage to the line $\mathbf{V}(0,\mathbf{s})$ is related to the generator voltage $\mathbf{V}_g(\mathbf{s})$ by the equation

$$\mathbf{V}(0,\mathbf{s}) = \frac{\mathbf{Z}_i(\mathbf{s})}{\mathbf{Z}_i(\mathbf{s}) + \mathbf{Z}_g(\mathbf{s})}\mathbf{V}_g(\mathbf{s}) \tag{2.95}$$

Using Equations (2.89), (2.81) and (2.83) the input impedance $\mathbf{Z}_i(\mathrm{s})$ can be derived to be

$$\mathbf{Z}_i(\mathbf{s}) = \mathbf{Z}_0(\mathbf{s})\frac{1+\rho_r(\mathbf{s})\exp(-2\mathbf{s}\delta' l)}{1-\rho_r(\mathbf{s})\exp(-2\mathbf{s}\delta' l)} \tag{2.96}$$

In the special cases of a short circuit and open circuit, the load $\rho_{\mathrm{r}}(\mathbf{s})$ takes the values -1 and $+1$, respectively; therefore Equation (2.96) can be simplified to

$$\left.\begin{aligned}
\mathbf{Z}_i(\mathbf{s}) &= \mathbf{Z}_0(\mathbf{s})\frac{1-\exp(-2\mathbf{s}\delta.l)}{1+\exp(-2\mathbf{s}\delta.l)} = \mathbf{Z}_0(\mathbf{s})\tanh(-\mathbf{s}\delta' l) && \text{(short circuit)}\\
\mathbf{Z}_i(\mathbf{s}) &= \mathbf{Z}_0(\mathbf{s})\frac{1+\exp(-2\mathbf{s}\delta.l)}{1-\exp(-2\mathbf{s}\delta.l)} = \mathbf{Z}_0(\mathbf{s})\coth(-\mathbf{s}\delta' l) && \text{(open circuit)}
\end{aligned}\right\} \tag{2.97}$$

These equations are identical to those derived for continuously excited lines (Equation (2.46)) if the substitution for the propagation constant $\gamma(\mathbf{s})$ given in Equation (2.87) is used. Both equations are extremely useful in developing the theory of pulse forming lines and will be used extensively in later chapters.

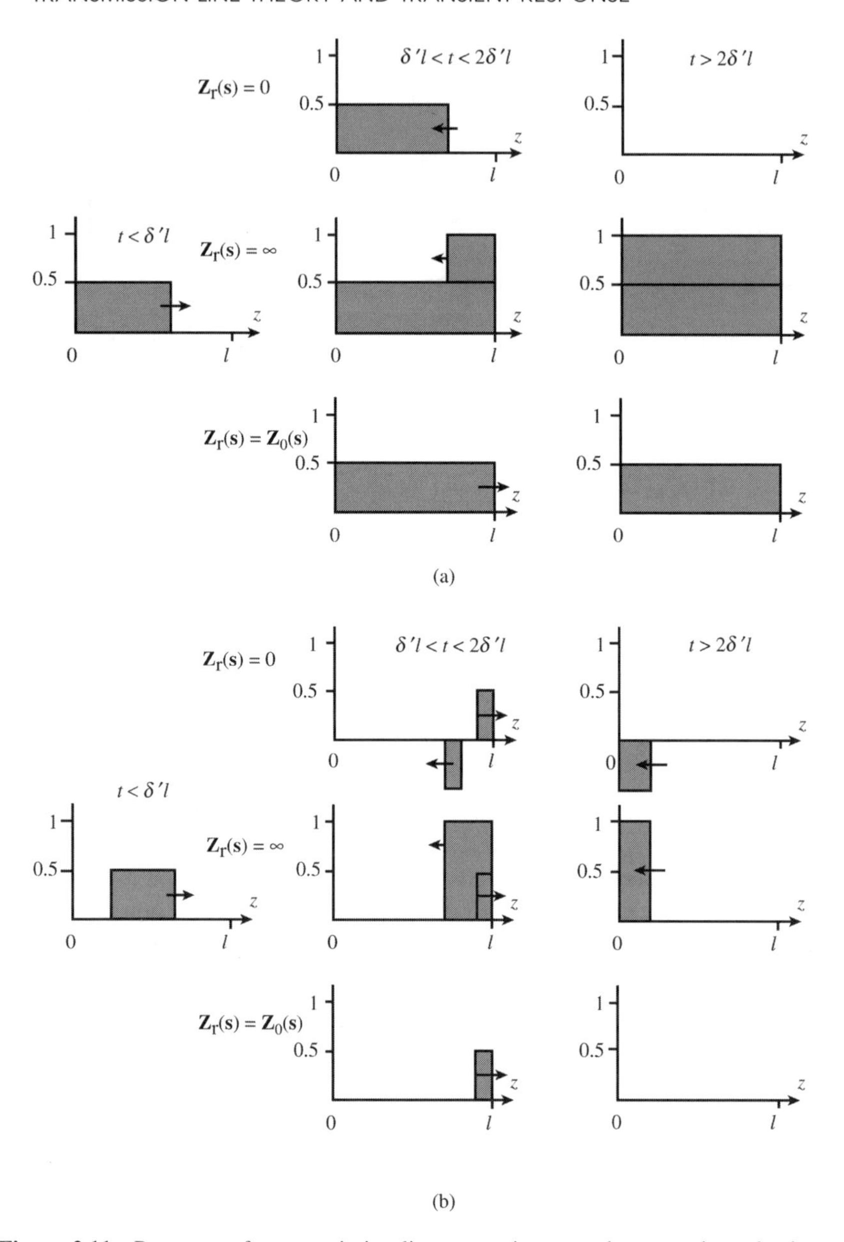

Figure 2.11 Response of a transmission line to a unit step and rectangular pulse input

2.4.5 Reflections on Lines with Reactive Terminations

It may be the case that a line, impedance Z_0, is terminated not in a pure resistance but in a load which is partly resistive and partly capacitive or partly resistive and partly inductive. For example if the load is a resistor R_r in parallel with an capacitor C_r then the load impedance $\mathbf{Z}_r(\mathbf{s})$ will be given by

$$\mathbf{Z}_r(\mathbf{s}) = \frac{R_r}{\mathbf{s}C_rR_r + 1} \tag{2.98}$$

Substituting this expression into Equation (2.80), the reflection coefficient of the termination is found to be

$$\rho_r(\mathbf{s}) = \frac{-(\mathbf{s} - a)}{(\mathbf{s} + b)}$$

$$\text{where} \qquad a = \frac{R_r - Z_0}{C_rR_rZ_0}$$

$$b = \frac{R_r + Z_0}{C_rR_rZ_0} \tag{2.99}$$

When the load is purely capacitive, i.e. $R_r = \infty$, then

$$a = b = \frac{1}{C_rZ_0} \tag{2.100}$$

If the input to the line is a voltage step of amplitude V from a matched generator then from Equation (2.99) the reflected step from the load $\mathbf{V}_-(\mathbf{s})$ will be given by

$$\mathbf{V}_-(\mathbf{s}) = \frac{V}{2\mathbf{s}} \frac{-(\mathbf{s} - a)}{(\mathbf{s} + b)} \tag{2.101}$$

Taking the inverse transform of this expression, after some rearrangement, gives

$$v_-(t) = \frac{V}{2}\left[\frac{a}{b} - \left(\frac{a+b}{b}\right)\exp(-bt')\right]$$

$$\text{where} \qquad t' = t - \delta' l \tag{2.102}$$

Adding this signal to the input step then gives an expression for the voltage signal on the load, which is

$$v_r(t) = \frac{V}{2}\left[\frac{a+b}{b}\right][1 - \exp(-bt')] \tag{2.103}$$

Noting that

$$\frac{a+b}{b} = \frac{2R_r}{(R_r + Z_0)} \tag{2.104}$$

the voltage waveforms at the input and output of the line are then as shown in Figure 2.12. The final voltage V_f on the line is therefore

$$V_f = \frac{VR_r}{(R_r + Z_0)} \tag{2.105}$$

Thus the capacitive part of the load has resulted in an output voltage waveform which has a rise-time with time constant τ_C

$$\tau_C = \frac{R_rC_rZ_0}{(R_r + Z_0)} \tag{2.105}$$

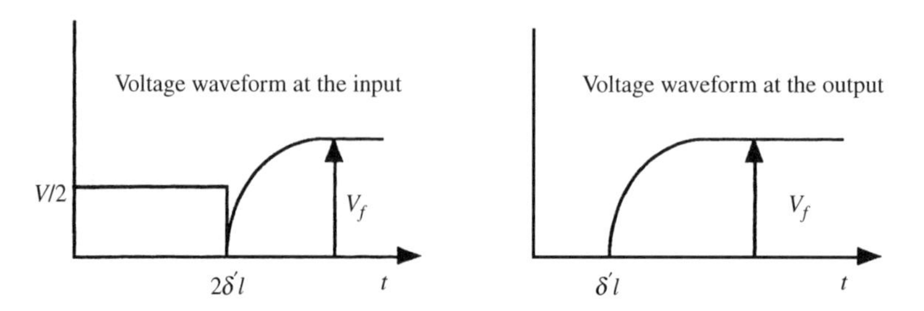

Figure 2.12 Input and output waveforms for a line terminated in a partly capacitive load

A very similar analysis can be carried out for a load which comprises an inductor in series with a resistor. In this case the voltage waveforms at the input and output of the line are as shown in Figure 2.13.

The waveforms overshoot the final voltage and settle with a time constant τ_L

$$\tau_L = \frac{L_r}{R_r + Z_0} \tag{2.106}$$

where L_r is the inductive part of the load.

2.4.6 Reflection Charts or Lattice Diagrams

Although the Laplace transform method of analysing reflections on a transmission line is mathematically elegant, the expressions that result from an analysis can be rather complex and long. An alternative method of analysing the effect of reflections on unmatched transmission lines is to use a reflection chart or lattice diagram. This technique will be used extensively later in the book because of the simplicity it offers.

The method can be best described by reference to the general diagram of a transmission line shown in Figure 2.10. In this case it will be assumed that both the generator and load are unmatched to the line, and for the method to be usable the impedances of the generator, the line and the load must be purely resistive. Thus there will be voltage reflection coefficients at the input and output of the line ρ_0 and ρ_r given by

$$\rho_0 = \frac{R_g - Z_0}{R_g + Z_0} \quad \text{and} \quad \rho_r = \frac{R_r - Z_0}{R_r + Z_0} \tag{2.107}$$

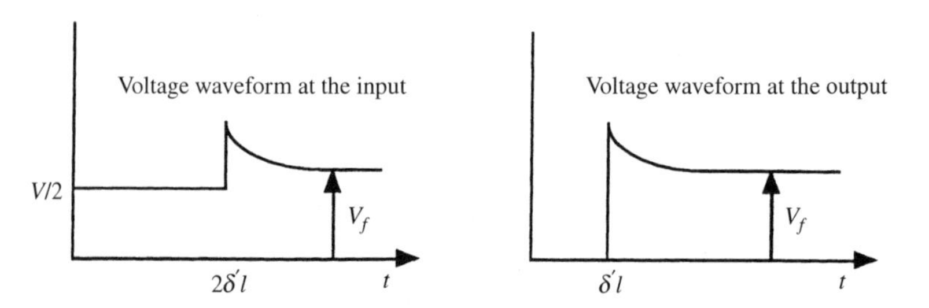

Figure 2.13 Input and output waveforms for a line terminated in a partly inductive load

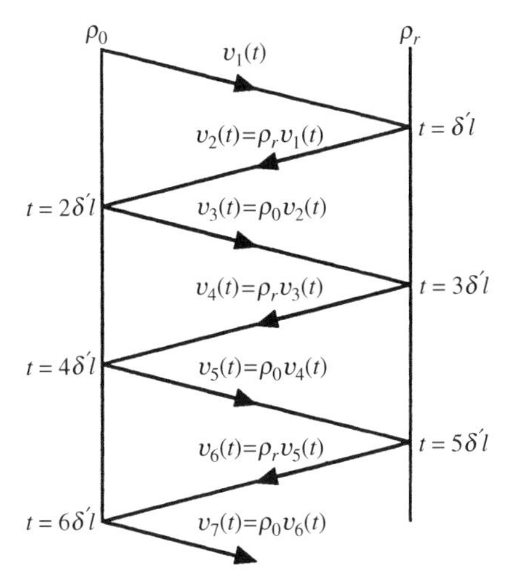

Figure 2.14 An example of reflection chart or lattice diagram

where R_r and R_g are the resistive impedances of the load and generator, respectively.

If a voltage signal $v_1(t)$ is injected on to the line, then after a time $\delta' l$ it will arrive at the load and be reflected with a reflection coefficient ρ_r. It will then be reflected, and the amplitude $v_2(t)$ of the reflected signal and the voltage on the load $v_r(t)$ are

$$v_2(t) = \rho_r v_1(t) \quad \text{and} \quad v_r(t) = (1 + \rho_r) v_1(t) \tag{2.108}$$

The reflected signal now returns to the generator end of the line where it is reflected with a reflection coefficient ρ_0. The amplitude of the reflected signal $v_3(t)$ and the voltage at the input of the line $v_i(t)$ are given by

$$v_3(t) = \rho_0 v_2(t) \quad \text{and} \quad v_i(t) = (1 + \rho_0) v_2(t) \tag{2.109}$$

This process repeats and a lattice diagram can be drawn of the repeated reflections, as shown in Figure 2.14. Notice that to determine the voltage on the load or at the input to the line the

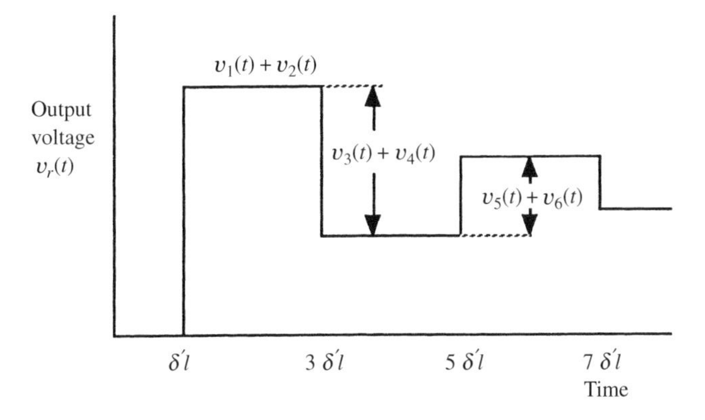

Figure 2.15 Output waveform for a transmission line based on the lattice diagram given in Figure 2.14

potentials as marked on the diagram are added successively with a delay between successive terms of $2\delta' l$. This process is best illustrated by the hypothetical output waveform shown in Figure 2.15. This waveform has been drawn on the assumptions that the load impedance is greater than the line impedance and the generator impedance is less than the line impedance.

REFERENCES

[1] Dunlop J. and Smith D. G. "Telecommunications Engineering". Van Nostrand Reinhold (1984) ISBN 0-442-30585-0
[2] Chipman R. A. "Transmission Lines". Shaum's Outline Series, McGraw-Hill (1969)
[3] Connor F. R. "Wave Transmission". Edward Arnold (1972) ISBN 0-7131-3278-7
[4] Matick R. E. "Transmission Lines for Digital and Communications Networks". McGraw Hill (1969)
[5] Zepler E. E. and Nichols K. G. "Transients in Electronic Engineering" Chapman and Hall (1971) ISBN 412-10130-0
[6] Rizzi P. A. "Microwave Engineering Passive Circuits". Prentice-Hall International (1988) ISBN 0-13-581711-0
[7] Lewis I. A. D. and Wells F. H. in "Millimicrosecond Pulse Techniques", Pergamon Press (1953) 109–111
[8] Dreher T. "Cabling fast pulses? Don't trip on the steps". The Electronic Engineer (1969) 71–75
[9] Ramo S., Whinnery J. R. and Van Duzer T. "Fields and Waves in Communications Electronics". John Wiley and Sons (1965)

3

Pulse-forming Lines

3.1 INTRODUCTION

There are numerous applications in both physics and electrical engineering for short (~ 10 ns $< t_p <$ 100 μs) electrical pulses. These applications often require that the pulses have a "good" square shape, i.e. they have fast rise t_r and fall t_f times (t_r, $t_f \ll t_p$). Although there are many ways for generating such pulses, the pulse-forming line is one of the simplest techniques and can be used even at extremely high pulsed power levels. There are many variants of the basic pulse-forming method which relies on the use of a single length of transmission line and a fast switch. Several of these methods will be discussed in this chapter after the operation of the simple pulse-forming line has been described. In building most of these circuits, practical considerations such as switch speed and current rating, switch and load inductance, and transmission line power handling must be carefully assessed to ensure successful pulse generation.

3.2 THE SINGLE PULSE-FORMING LINE

Figure 3.1 gives a basic circuit diagram of a pulse generator based on a single pulse-forming line. A length l of transmission line, not necessarily coaxial, as shown, is charged to a potential V by a DC power supply, usually via a resistor or inductor whose impedance is high enough to limit the charging current to the maximum corresponding to the output power specification of the supply. If a capacitor charging supply is used, such an impedance may not be necessary provided that the switch recovery time is at least as fast as the response time of the supply. This is important if a switch is used, such as a spark gap, that can "latch" into a permanently on-state being fed by the current generated by the charging supply. It is also important that the output impedance of the supply, or the supply and charging impedance, are large in comparison to the characteristic impedance of the transmission line Z_o.

Once the line has been charged the switch is closed and the line is allowed to discharge into the load Z_l. It is assumed in this description that the impedance of the load Z_l is purely resistive. Because the stored energy source for the pulse is a transmission line, rather than a simple capacitor, a rectangular pulse is generated in the load whose duration is equal to twice the transit time that a voltage step would take to propagate along the line from one end to the other. The propagation velocity of such a step v_p was given in Equation (2.31), i.e.

$$v_p = \frac{c}{\sqrt{\varepsilon_r}} \tag{3.1}$$

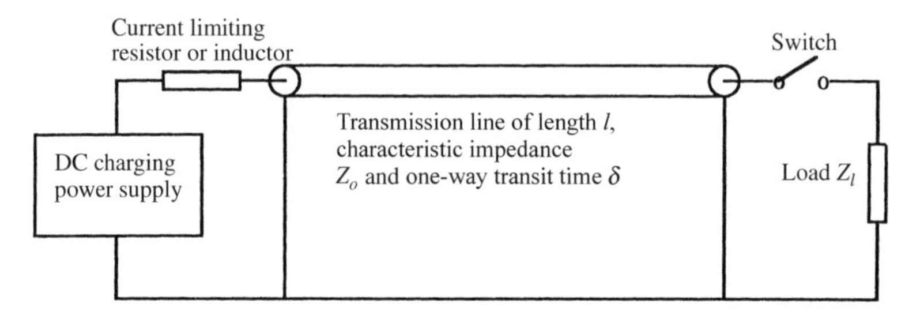

Figure 3.1 The basic pulse-forming line

assuming that the relative permeability of the line is unity. As before, c is the velocity of light and ε_r is the relative permittivity of the dielectric material which insulates the line. The duration of the pulse t_p that is generated is simply given by

$$t_p = \frac{2l}{v_p} = 2\delta \tag{3.2}$$

It is important that the closing and opening times of the switch are much shorter than the duration of the pulse to be generated, as these characteristic times dictate, in part, the rise and fall times of the pulse that is generated. It is also important for short duration pulses that the stray inductance L_s of the part of the circuit which connects the output end of the line to the load is minimised. If this is not the case the rise time of the pulse will be limited by this stray inductance with a time constant τ_r given by

$$\tau_r = \frac{L_s}{Z_o + Z_l} \tag{3.3}$$

In order to understand the way in which a rectangular pulse is generated, when a charged length of transmission line is discharged into a load, a number of methods can be used. Perhaps the easiest, for illustration purposes, is to draw a series of voltage/time diagrams which show the potential on the line at various times after the switch is closed. Such a series of diagrams is given in Figure 3.2 in which the impedances of the load and line are matched, i.e. $Z_o = Z_l$.

On closure of the switch, the voltage on the load rises from zero to a value determined by the initial charge potential V, the characteristic impedance of the line Z_o and the load impedance Z_l, i.e. the voltage on the load is given by

$$V_l = V\frac{Z_l}{Z_l + Z_o} \tag{3.4}$$

In the case where the impedances of the line and load are equal then clearly the voltage on the load will be $V/2$. Simultaneously a voltage step V_s is propagated away from the load towards the charging end of the line. This step has an amplitude V_s given by

$$V_s = V\left(\frac{Z_l}{Z_l + Z_o} - 1\right) = V\left(\frac{-Z_o}{Z_l + Z_o}\right) \tag{3.5}$$

Again, in the case of the line and load impedances being equal this step will simply be $-V/2$. When this step reaches the charging end of the line after a time l/v_p, the step is then

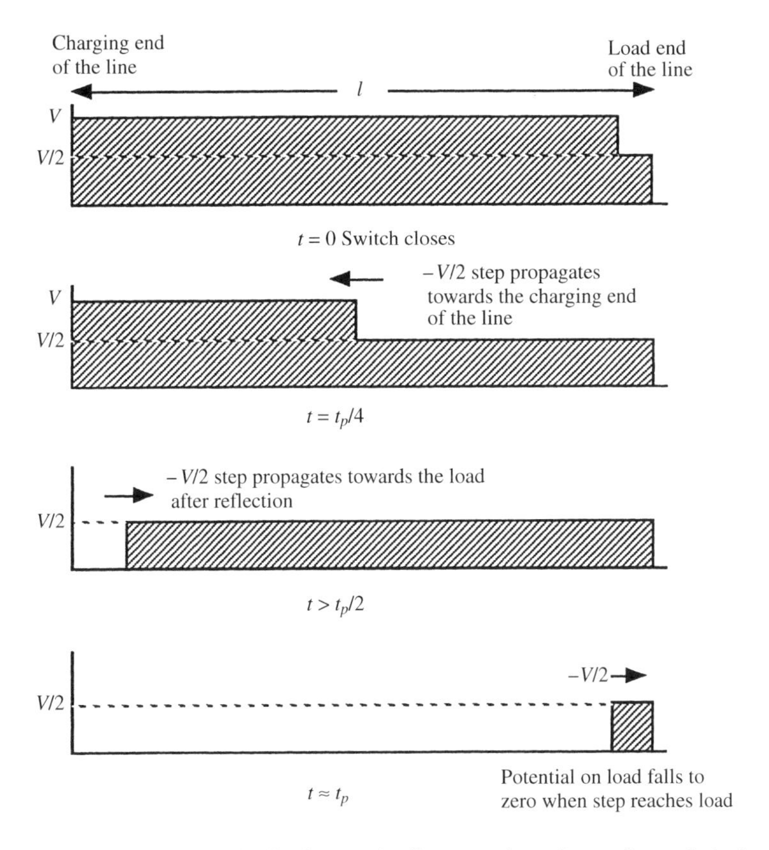

Figure 3.2 The potential distribution on the line at various times after switch closure

reflected (see Chapter 2) with a reflection coefficient $\rho(\mathbf{s}) \approx +1$ because the output impedance of the charging supply or the charging supply plus the charging impedance is much greater than the impedance of the line. Thus the potential at the charging end of the line falls to zero and a voltage step of amplitude V_s propagates back towards the load, altering as it goes the potential on the line. Again, if the line and load impedances are matched, this step has an amplitude $-V/2$ and, as it propagates, it completely discharges the line, i.e. the line potential is reduced to zero. When the step finally reaches the load, the line is completely discharged, the potential on the load falls to zero and the pulse terminates. If the impedances of the load and line are not matched, further reflections of the propagating step occur at the load and the resulting output pulse has a more complicated shape, rather than that of a single rectangular pulse.

Commonly, pulse-forming line generators are built using standard coaxial cable, but unfortunately this is usually supplied in fixed impedances, the most common being $50\,\Omega$. This limitation can to some degree be overcome by using equal lengths of line which are either connected in parallel to build lower output impedance generators or connected in series to build higher output impedance generators. Alternatively parallel plate type lines can be made fairly easily to provide an exact match to loads which have unusual impedances.

3.2.1 Lattice Diagram Representation of Pulse-forming Action using a Single Transmission Line

Pulse-forming action by the simple circuit given in Figure 3.1 can also be described using a lattice diagram (see Chapter 2) or by carrying out a Laplace transform type of analysis which results in a complete mathematical description of the pulse as a function of time. The lattice diagram approach can be quite useful, particularly in accounting for the reflections and pulse shape that result when the line and load impedances are not matched. The lattice diagram analysis proceeds by first considering the reflection coefficients at either end of the line, as shown in Figure 3.3.

On closure of the switch, the potential on the load is given as before by Equation (3.4), which is modified for convenience to

$$V_l = V\frac{Z_l}{Z_l + Z_o} = \alpha V \tag{3.6}$$

The lattice diagram can be drawn as before by considering the reflection coefficients at either end of the line and noting the one-way transit time along the line δ which separates the various reflections at either end of the line.

The lattice diagram is constructed as described in Chapter 2 and is shown in Figure 3.4. As before, the potential on the load, as a function of time, is determined by adding the reflected voltage components at the load sequentially to the initial potential on the load given by Equation (3.6). Thus the potential on the load is given by

$$\left.\begin{aligned}
&V_l = \alpha V, \quad 0 < t < 2\delta \\
&V_l = \alpha V + (\alpha - 1)\gamma V, \quad 2\delta < t < 4\delta \qquad \text{where} \quad \gamma = \beta + 1 \\
&V_l = \alpha V + (\alpha - 1)\gamma V + \beta(\alpha - 1)\gamma V, \qquad 4\delta < t < 6\delta \\
&V_l = \alpha V + (\alpha - 1)\gamma V + \beta(\alpha - 1)\gamma V + \beta^2(\alpha - 1)\gamma V, \qquad 6\delta < t < 8\delta \\
&\quad \text{and so on} \ldots
\end{aligned}\right\} \tag{3.7}$$

Alternatively the potential on the load can be written as

$$V_l = V\left[\alpha + (\alpha - 1)\gamma\left(1 + \beta + \beta^2 + \beta^3 + \cdots\right)\right] \tag{3.8}$$

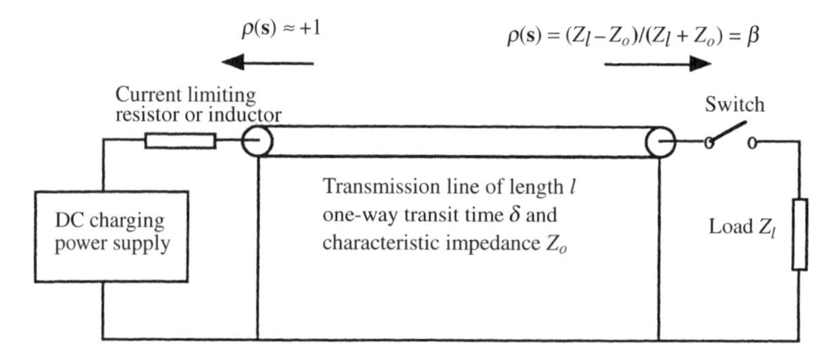

Figure 3.3 Circuit for the lattice diagram analysis of pulse-forming action

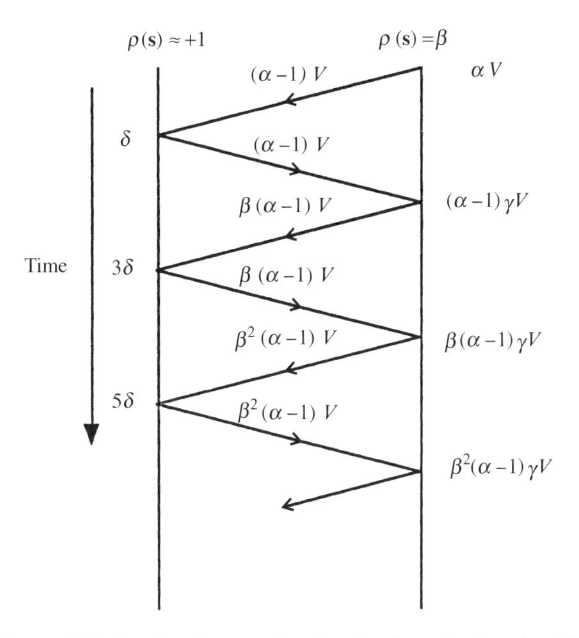

Figure 3.4 Lattice diagram for the simple pulse-forming line

where each successive term in the square brackets is separated in time by twice the transit time δ of the transmission line.

It is useful to consider what the shape of the waveforms will be in the cases where the load impedance is greater than, equal to or less than that of the transmission line impedance. It is assumed in this analysis that the load and line impedances are purely resistive. If the line and load impedances are equal then clearly the reflection coefficient ρ $(=\beta)$ at the load will be zero and a single pulse of amplitude $V/2$ will be generated at the load.

If the load impedance is greater than that of the line, a potential that is larger than $V/2$ will be initially generated at the load, and a voltage step which is smaller than $-V/2$ is propagated down the line towards the charging end. This step returns and is reflected at the load with a reflection coefficient that is positive. The resulting voltage that is added to the load potential is consequently negative and insufficient to reduce the load potential to zero. This process continues and the potential on the load takes the form of a "staircase" type waveform.

Conversely if the load impedance is less than that of the line, a potential that is smaller than $V/2$ will be initially generated at the load, and a voltage step which is larger than $-V/2$ is propagated down the line towards the charging end. This step returns and is reflected at the load with a reflection coefficient that is now negative. The resulting voltage that is added to the load potential is consequently negative and sufficient to reduce the load potential to a negative value. This process continues and the potential on the load reverses on successive load reflections. Typical waveforms for these three cases are shown in Figure 3.5.

3.3 PULSE-FORMING USING THE BLUMLEIN PULSE-FORMING LINE

An important disadvantage of the simple pulse-forming line described is that the pulse generated into a matched load is equal to half the voltage to which the transmission line is

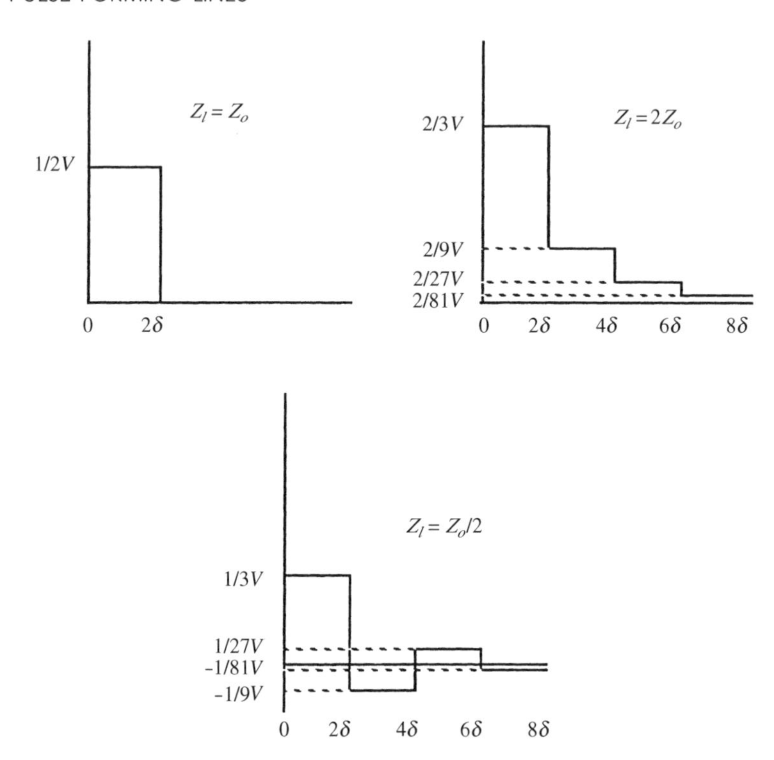

Figure 3.5 Typical waveforms from the pulse-forming line under matched and unmatched conditions

charged. This can be a serious disadvantage in pulse-forming lines which are used to generate pulses at high voltages, as the power supply used to charge the line must have an output potential which is twice that of the required pulse amplitude.

This problem can be avoided using the Blumlein pulse forming line invented by A. D. Blumlein. In this case two equal lengths of transmission line are used together with a switch to construct the generator, as shown in Figure 3.6. Again, analysis of the behaviour of this generator can be described using a series of voltage time diagrams which show the potential on the two lines during pulse-forming action. Alternatively, as with the simple

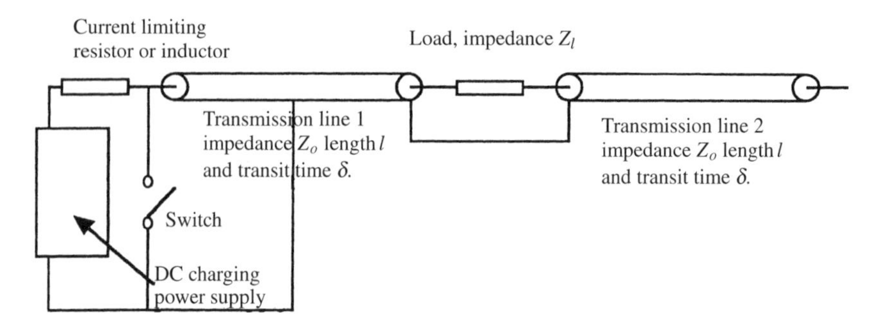

Figure 3.6 Circuit diagram of the basic Blumlein pulse-forming line

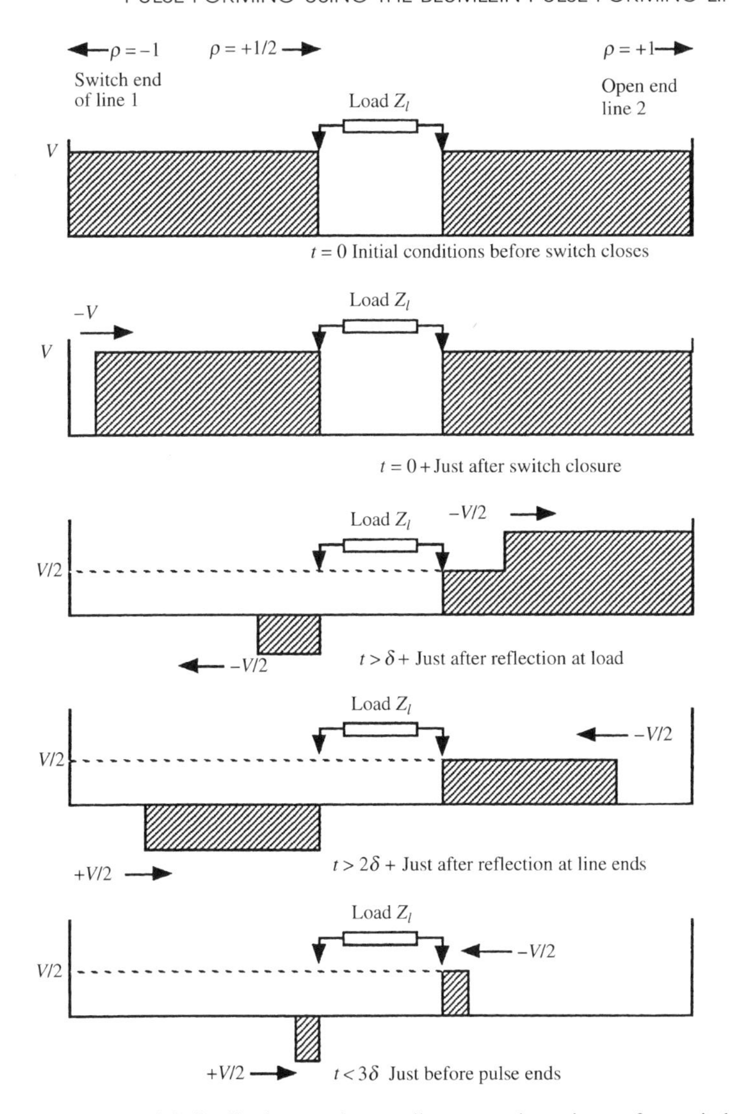

Figure 3.7 The potential distribution on the two lines at various times after switch closure

pulse-forming line, the output waveforms can be deduced either by the use of a lattice diagram or by a mathematical analysis using Laplace transforms. All three techniques will be described.

Operation of the Blumlein pulse forming line using a series of voltage time diagrams is shown in Figure 3.7. After switch closure, which initiates pulse-forming action, the end of line 1 next to the switch is effectively shorted; thus the reflection coefficient at this end of line 1 is -1. At the open end of line 2 the reflection coefficient is $+1$. At the junction of line 1 and the load the reflection coefficient can be calculated by the observation that line 1 is effectively loaded by the combination of the load in series with the impedance of line 2. This can be seen by reference to Figure 3.6. For correct operation of the Blumlein pulse-forming line the load impedance must be equal to twice the characteristic impedance of the two lines. Thus line 1 is loaded by an impedance of $3Z_o$ and the reflection coefficient at the load is given by

$$\rho(\mathbf{s}) = \rho = \frac{3Z_o - Z_o}{3Z_o + Z_o} = \frac{1}{2} \tag{3.9}$$

These reflection coefficients are marked on Figure 3.7. After the switch has closed the potential at the end of line 1 is reduced to zero and a step of amplitude $-V$ is launched along this line towards the load. As it propagates it discharges the line. When it reaches the load, after a transit time delay δ the step is partly transmitted and partly reflected with a reflection coefficient of $+1/2$ as previously noted. Thus a step of $-V/2$ is launched back into line 1 propagating towards the switch. The transmitted step will have an amplitude which is equal to the sum of the incident step $V_+ = -V$ and the reflected step $V_- = -V/2$, i.e. $-3V/2$. The simple diagram given in Figure 3.8 helps to explain how this step is divided between the load and line 2. By consideration of what amounts to a simple voltage divider it can be seen that a step of amplitude $-V/2$ is launched into line 2 and the potential across the load is V.

This second step propagates along line 2, discharging the line to a potential of $V/2$, and then reflects at the open end of this line with a coefficient of $+1$ to fully discharge the line as the $-V/2$ step then propagates back to the load. The step in line 1 first charges the line to a potential of $-V/2$ and then reflects with a coefficient of -1 so that it also discharges line 1 as it propagates back to the load. When the two steps arrive at the load they effectively annihilate each other and the pulse on the load terminates. The pulse duration is clearly equal to 2δ, where δ can be calculated as shown above.

3.3.1 Lattice Diagram Representation of Pulse-forming Action using a Blumlein Pulse-forming Line

The lattice diagram is drawn for the general case, as for the simple pulse-forming line, where the load impedance is not matched, i.e. $Z_l \neq 2Z_o$. As before it is assumed in this analysis that the load and line impedances are purely resistive. In this case the reflection coefficient at the load is given by

$$\rho(\mathbf{s}) = \rho = \frac{(Z_l + Z_o) - Z_o}{(Z_l + 2Z_o)} = \frac{Z_l}{Z_l + 2Z_o} \tag{3.10}$$

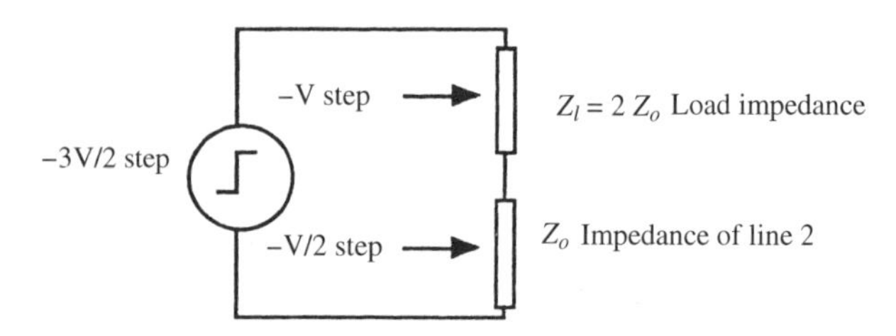

Figure 3.8 Division of the voltage step between line 2 and the load

The reflected step at the load thus has an amplitude

$$V_{-} = -\rho V = -V\frac{Z_l}{Z_l + 2Z_o} \tag{3.11}$$

The step V_T transmitted on to the load and second line has an amplitude given by

$$V_T = V_{+} + V_{-} = -V\left(1 + \frac{Z_l}{Z_l + 2Z_o}\right) = -2V\frac{(Z_l + Z_o)}{(Z_l + 2Z_o)} \tag{3.12}$$

This step is divided in the same way that has been described using Figure 3.8, and therefore the fraction of the step that is impressed on the load Z_l is given by

$$V_l = -2V\frac{Z_l}{(Z_l + Z_o)}\frac{(Z_l + Z_o)}{(Z_l + 2Z_o)} = -2V\frac{Z_l}{(Z_l + 2Z_o)} = -\alpha V$$

$$\text{where} \qquad \alpha = \frac{2Z_l}{(Z_l + 2Z_o)} \tag{3.13}$$

The fraction of the step propagated on to the second line V_{2T} is given by

$$V_{2T} = \frac{Z_o}{(Z_l + 2Z_o)}(-2V)) = -\beta V$$

$$\text{where} \qquad \beta = \frac{2Z_o}{(Z_l + 2Z_o)} \tag{3.14}$$

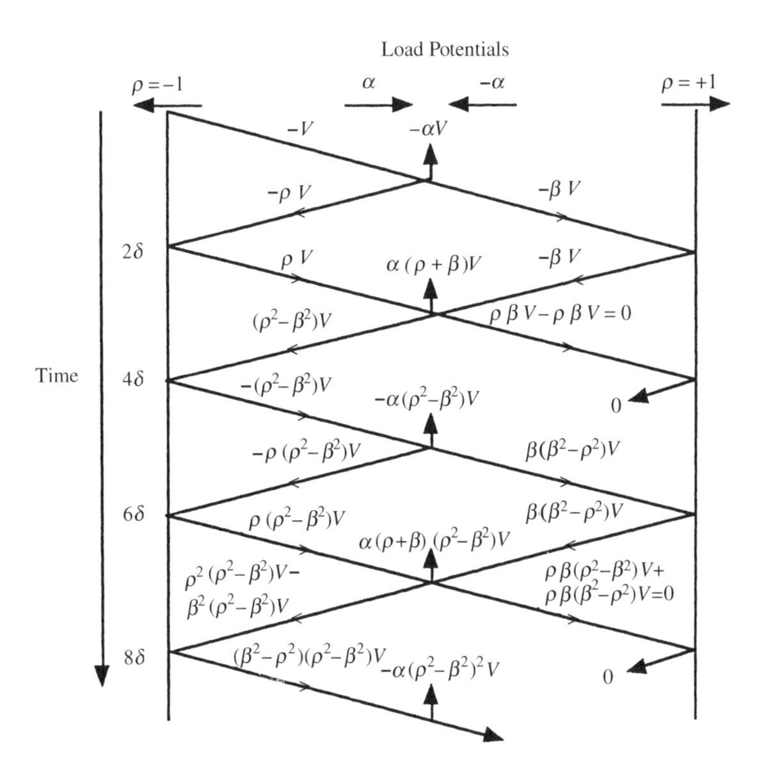

Figure 3.9 Lattice diagram for the Blumlein pulse-forming line

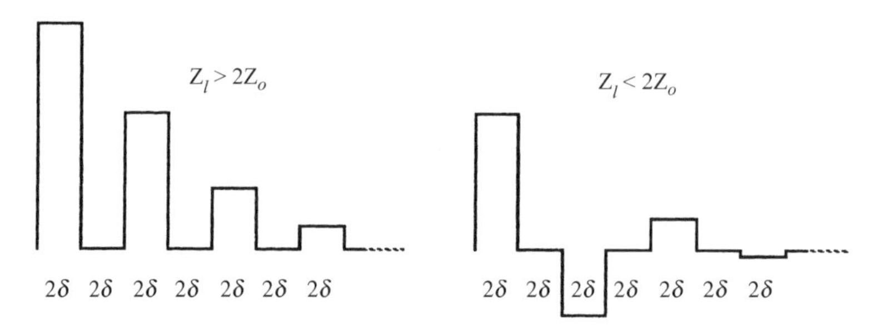

Figure 3.10 Typical waveforms generated by the Blumlein pulse-forming line into unmatched loads

From these relationships the lattice diagram can be drawn, as before, and is shown in Figure 3.9. Note from the diagram how steps propagate from left to right and vice versa, leading, in some cases, to multiple component steps. As these steps propagate across the load, the fraction which is deposited at the load changes sign depending on the direction of propagation. It is also interesting to note that in two cases the two components making up the propagating step are equal but of opposite sign, so they cancel each other.

As described on pages 68 and 69, the potential on the load is determined by adding, sequentially, the fractions of the propagating steps which are impressed on the load to the initial potential impressed on the load $-\alpha V$. Thus the potential on the load is given by

$$V_l = -\alpha V\left[1 - (\rho + \beta) + (\rho^2 - \beta^2) - (\rho + \beta)(\rho^2 - \beta^2) + (\rho^2 - \beta^2)^2 \ldots\right] \tag{3.15}$$

where each successive term in the square brackets is delayed by twice the transit time δ of the two lines. From Equations (3.10) and (3.14) it can be seen that

$$\rho + \beta = 1 \tag{3.16}$$

Equation 3.15 can therefore be written as

$$V_l = -\alpha V\left[1 - 1 + (\rho - \beta) - (\rho - \beta) + (\rho - \beta)^2 - (\rho - \beta)^2 \ldots\right] \tag{3.17}$$

It is clear from this equation that, for any unmatched load, the output waveform will comprise a series of rectangular pulses of width 2δ which are interspersed with spaces, also of width 2δ, where the potential on the load is zero. Sketches of typical waveforms for the cases where $Z_l > 2Z_o$ and $Z_l < 2Z_o$ are shown in Figure 3.10.

3.4 THE LAPLACE TRANSFORM ANALYSIS OF PULSE-FORMING ACTION BY TRANSMISSION LINES

3.4.1 Pulse-forming by the Simple Pulse-forming Line

The sheer power and range of applicability of the Laplace transform method for analysing transient behaviour in electrical circuits is elegantly demonstrated in this section. The analysis begins by reference back to Figure 3.3, which gives the circuit for the basic pulse-forming line. In this analysis, it is again assumed that the impedance of the current limiting resistor or inductor is very large during pulse-forming action. Thus the line can be regarded as a length l of open circuit transmission line which is charged and then discharged into a load. It will be again assumed that the load and line impedances are resistive or non-reactive. The input impedance of an open circuit length of transmission line is given by Equation (2.46), i.e.

$$Z_i(\text{oc}) = Z_o \coth(\gamma l) \tag{3.18}$$

For a loss-free line the attenuation constant α is zero and thus

$$\gamma = j\beta \tag{3.19}$$

From Equations (2.23) and (2.24)

$$\beta = \frac{2\pi}{\lambda} = \frac{\omega}{v_p}, \quad \Rightarrow \beta l = \omega\delta \tag{3.20}$$

Therefore Equation (3.18) can be written as

$$\begin{aligned} \mathbf{Z}_i(\mathbf{s}) &= Z_o \coth(\mathbf{s}\delta), \\ \mathbf{s} = j\omega \quad &\because \quad \sigma = 0, \qquad \text{(the line is loss-free)} \end{aligned} \tag{3.21}$$

Note that this equation is different to Equation (2.97) as δ here is the transit time of the whole line, whereas in Equation (2.97) δ' is defined as the delay per unit length of line. Equation (3.21) can then be used to produce a revised version of the circuit shown in Figure 3.3 for the transient analysis of the basic pulse forming line. The revised circuit is shown in Figure 3.11.

In the Laplace equivalent circuit the charged line is represented as an impedance (a given by Equation (3.21)) in series with a step voltage source of amplitude V. It is assumed that the switch has just been closed, i.e. the analysis proceeds from the time $t = 0_+$. The analysis can now be seen to be simply an exercise to determine $\mathbf{I(s)}$ and hence $i(t)$ and $v(t)$ on the load Z_l. The current in the load is determined to be

$$\mathbf{I(s)} = \frac{V}{\mathbf{s}[Z_l + Z_o \coth(\mathbf{s}\delta)]} \tag{3.22}$$

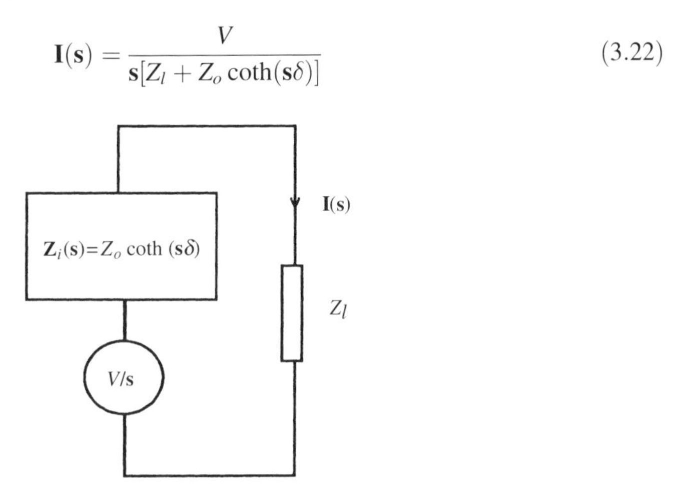

Figure 3.11 The Laplace transformed equivalent circuit of the basic pulse-forming line

From the identity

$$\coth(x) = \frac{1 + \exp(-2x)}{1 - \exp(-2x)} \tag{3.23}$$

Equation (3.22) can be written as

$$\mathbf{I(s)} = \frac{V}{\mathbf{s}(Z_l + Z_o)} \frac{1 - \exp(-2\mathbf{s}\delta)}{1 + \left(\frac{Z_o - Z_l}{Z_o + Z_l}\right)\exp(-2\mathbf{s}\delta)} \tag{3.24}$$

Using the binomial expansion

$$(1 + x)^n = 1 + nx + \frac{n(n-1)x^2}{2!} + \frac{n(n-1)(n-2)}{3!} + \cdots, \tag{3.25}$$

Equation (3.24) becomes

$$\mathbf{I(s)} = \frac{V[1 - \exp(-2\mathbf{s}\delta)]}{\mathbf{s}(Z_l + Z_o)} \begin{bmatrix} 1 - \left(\frac{Z_o - Z_l}{Z_o + Z_l}\right)\exp(-2\mathbf{s}\delta) \\ + \left(\frac{Z_o - Z_l}{Z_o + Z_l}\right)^2 \exp(-4\mathbf{s}\delta) \\ - \left(\frac{Z_o - Z_l}{Z_o + Z_l}\right)^3 \exp(-6\mathbf{s}\delta) \\ + \cdots \end{bmatrix} \tag{3.26}$$

Taking the inverse transform

$$i(t) = \frac{V}{(Z_o + Z_l)} \begin{Bmatrix} 1 - u(t - 2\delta) - \left(\frac{Z_o - Z_l}{Z_o + Z_l}\right)[u(t - 2\delta) - u(t - 4\delta)] \\ + \left(\frac{Z_o - Z_l}{Z_o + Z_l}\right)^2 [u(t - 4\delta) - u(t - 6\delta)] \\ - \left(\frac{Z_o - Z_l}{Z_o + Z_l}\right)^3 [u(t - 6\delta) - u(t - 8\delta)] \\ + \ldots \end{Bmatrix}$$

$$\text{where} \quad u(\Delta t) = 1 \quad \text{for } \Delta t > 0, \quad u(\Delta t) = 0 \quad \text{for } \Delta t < 0$$
$$\text{and} \quad \Delta t = (t - n\delta), \quad n = 2, 4, 6, \cdots \tag{3.27}$$

The voltage $v(t)$ can be easily derived from this equation and the pulse shapes for the cases where $Z_o < Z_l$ and $Z_o > Z_l$ correspond, as should be the case, to those shown in Figure 3.5. Note also that if $Z_o = Z_l$, i.e. the line is matched to the load, $v(t)$ can derived from Equation (3.27) to be

$$v(t) = \frac{V}{2}[1 - u(t - 2\delta)] \tag{3.28}$$

Clearly $v(t)$ is a single rectangular pulse of amplitude $V/2$ and duration 2δ, as expected.

3.4.2 Pulse-forming by the Blumlein Pulse-forming Line

A similar method can be used to carry out the analysis of the transient behaviour of the Blumlein pulse-forming line. It will again be assumed that the impedances of the lines and the load are resistive or non-reactive. As before, the analysis starts by deducing a Laplace equivalent circuit of the Blumlein pulse-forming line circuit. In this case there are now two lines. Referring to the circuit given in Figure 3.6, it can be seen that line 2 is an open circuit length of transmission line and therefore can be represented as an impedance given by Equation (3.21). The behaviour of line 1 in the circuit is more complex. It cannot be represented as a simple length of charged transmission line shorted at the switch end, as it initially propagates a step of amplitude $-V$ along the line towards the load after switch closure. Thus the action of switch closure can be described as the initiation of this step from a generator with zero output impedance (which is equivalent to the short circuit produced by the switch). This line must therefore be treated as a transmission line using the matrix of Equation (2.43) given in Chapter 2. Thus the Laplace equivalent circuit for the Blumlein circuit is as given in Figure 3.12.

In this circuit both lines are represented as being charged, at the start of pulse-forming action ($t = 0_+$) by the two step generators, of amplitude V, on the right of the diagram. The polarities of the two generators are marked, so that it is possible to see that they in fact cancel each other and their presence can be ignored. The problem therefore is to determine the current flowing in the load $\mathbf{I}(\mathbf{s})$. It is therefore clear that the matrix of Equation (2.43) must be used to establish the relationship between the input voltage to line 1, represented as a negative step generator of amplitude V, and the output current from line 1 which is equal to $\mathbf{I}(\mathbf{s})$. From Equation (2.43) it can be deduced that this relationship is given by

$$V_s = V_r \cosh(\gamma l) + I_r Z_o \sinh(\gamma l)$$

$$\therefore \quad \frac{V_s}{I_r} = Z_r \cosh(\gamma l) + Z_o \sinh(\gamma l)$$

$$\text{where} \quad V_s = -\frac{V}{\mathbf{s}} \quad \text{and} \quad I_r = \mathbf{I}(\mathbf{s}) \tag{3.29}$$

As above it is assumed that the lines are loss-free and therefore

$$\gamma l = \mathbf{s}\delta \tag{3.30}$$

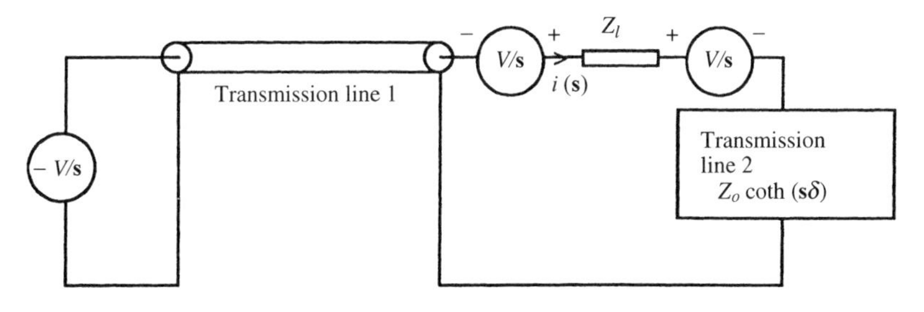

Figure 3.12 The Laplace transformed equivalent circuit of the Blumlein pulse-forming line

The load Z_r comprises the load on the Blumlein (Z_l) in series with the impedance of line 2. Therefore

$$Z_r = Z_l + Z_o \coth(\mathbf{s}\delta) \tag{3.31}$$

The potential on the load $\mathbf{V}_l$ ($\mathbf{s}$) is given by

$$\begin{aligned}\mathbf{V}_l(\mathbf{s}) = \mathbf{I}(\mathbf{s})Z_l &= -\frac{VZ_l}{\mathbf{s}} \frac{1}{(Z_l + Z_o \coth(\mathbf{s}\delta))\cosh(\mathbf{s}\delta) + Z_o \sinh(\mathbf{s}\delta)} \\ &= -\frac{VZ_l}{\mathbf{s}} \frac{\sinh(\mathbf{s}\delta)}{Z_l \cosh(\mathbf{s}\delta)\sinh(\mathbf{s}\delta) + Z_o + 2Z_o \sinh^2(\mathbf{s}\delta)}\end{aligned} \tag{3.32}$$

Converting the hyperbolic functions into their exponential identities gives

$$\begin{aligned}\mathbf{V}_l(\mathbf{s}) &= \frac{-VZ_l}{\mathbf{s}} \frac{\dfrac{\exp(\mathbf{s}\delta) - \exp(-\mathbf{s}\delta)}{2}}{\dfrac{Z_l(\exp(\mathbf{s}\delta) + \exp(-\mathbf{s}\delta))(\exp(\mathbf{s}\delta) - \exp(-\mathbf{s}\delta))}{4} + Z_o\left(1 + \dfrac{(\exp(\mathbf{s}\delta) - \exp(-\mathbf{s}\delta))^2}{2}\right)} \\ &= \frac{-VZ_l}{\mathbf{s}(Z_l + 2Z_o)} \exp(-\mathbf{s}\delta) \frac{(1 - \exp(-2\mathbf{s}\delta))}{\left[1 - \left(\dfrac{Z_l - 2Z_o}{Z_l + 2Z_o}\right)\exp(-4\mathbf{s}\delta)\right]}\end{aligned} \tag{3.33}$$

This expression, as with the basic pulse-forming line, can be rewritten using the binomial expansion (Equation (3.25)) to give

$$\mathbf{V}(\mathbf{s}) = \frac{-VZ_l}{\mathbf{s}(Z_l + 2Z_o)} \exp(-\mathbf{s}\delta)(1 - \exp(-2\mathbf{s}\delta)) \begin{bmatrix} 1 + \left(\dfrac{Z_l - 2Z_o}{Z_l + 2Z_o}\right)\exp(-4\mathbf{s}\delta) \\ + \left(\dfrac{Z_l - 2Z_o}{Z_l + 2Z_o}\right)^2 \exp(-8\mathbf{s}\delta) \\ + \left(\dfrac{Z_l - 2Z_o}{Z_l + 2Z_o}\right)^3 \exp(-12\mathbf{s}\delta) \\ + \cdots \end{bmatrix} \tag{3.34}$$

Taking the inverse transform gives

$$v(t) = \frac{-VZ_l u(t - \delta)}{(Z_l + 2Z_o)} \begin{bmatrix} 1 - u(t - 3\delta) + \left(\dfrac{Z_l - 2Z_o}{Z_l + 2Z_o}\right)[u(t - 5\delta) - u(t - 7\delta)] \\ + \left(\dfrac{Z_l - 2Z_o}{Z_l + 2Z_o}\right)^2 [u(t - 9\delta) - u(t - 11\delta)] \\ + \left(\dfrac{Z_l - 2Z_o}{Z_l + 2Z_o}\right)^3 [u(t - 13\delta) - u(t - 15\delta)] \\ + \cdots \end{bmatrix}$$

$$\text{where} \quad u\,(\Delta t) = 1 \quad \text{for } \Delta t > 0, \quad u(\Delta t) = 0 \quad \text{for } \Delta t < 0$$
$$\text{and} \quad \Delta t = (t - n\delta), \quad n = 1, 3, 5, \ldots \tag{3.35}$$

From Equations (3.10), (3.14) and (3.16)

$$(\rho - \beta) = \left(\frac{Z_l - 2Z_o}{Z_l + 2Z_o}\right) \tag{3.36}$$

Also using Equation (3.13), Equation (3.35) can be written as

$$v(t) = -\alpha V u(t-\delta)\begin{bmatrix} 1 - u(t-3\delta) + (\rho-\beta)[u(t-5\delta) - u(t-7\delta)] \\ +(\rho-\beta)^2[u(t-9\delta) - u(t-11\delta)] \\ +(\rho-\beta)^3[u(t-13\delta) - u(t-15\delta)] \\ +\cdots \end{bmatrix} \tag{3.37}$$

Equation (3.37) is a more mathematically rigorous way of writing Equation (3.17) and it includes the delay between switch closure and the start of the pulse and the delay between successive terms.

3.5 SOME OTHER PULSE-FORMING LINE VARIANTS

3.5.1 The Stacked Blumlein Pulse-forming Line Generator

Figure 3.13 gives a simplified circuit for a two-stage stacked Blumlein pulse-forming line generator. It is easier to explain the operation of the circuit by reference to a two-stage device although the stacking technique can be used with many stages. The importance of this circuit is that the voltage of the output pulse is greater than the voltage to which the circuit is initially charged [2]. For the two-stage device shown, the output pulse is twice that of the initial potential to which the circuit is charged when operated into a matched load of $4Z_o$. However it is quite possible to build stacked circuits where the voltage of the output pulse is many times that of the initial potential.

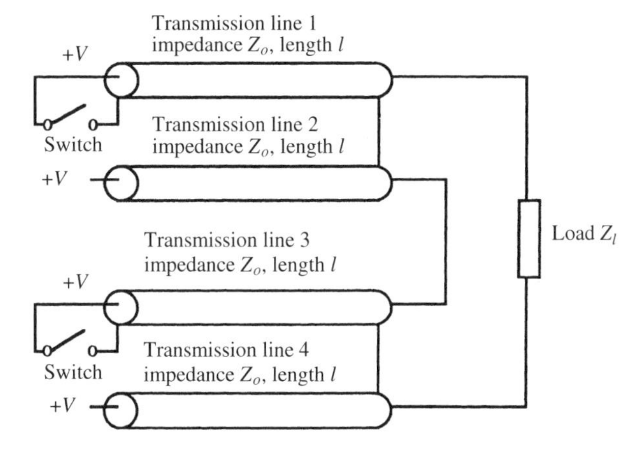

Figure 3.13 A simplified circuit diagram of a two-stage stacked Blumlein pulse-forming line

By reference to Figure 3.6 it can be seen that the two-stage circuit is effectively two Blumlein pulse-forming lines connected in series. Therefore the load needed to match the circuit is twice that of the match for an individual Blumlein pulse-forming line, i.e. $4Z_o$, where Z_o is the characteristic impedance of an individual transmission line. Each of the four lines has the same length and consequently the same transit time δ. The central conductor of each line is charged to a positive potential $+V$ and lines 1 and 3 are shorted simultaneously to initiate pulse-forming action using the switches shown.

To explain the operation of the circuit a series of voltage time diagrams is given in Figure 3.14.

In Figure 3.14 the voltage time diagrams for lines 2 and 4 are inverted to those of lines 1 and 3 to properly account for the way in which all four lines are connected in the circuit. The potential on the lines and its direction are marked on the right hand side of each set of diagrams so that the potentials can be added to give the total potential on the load. Before switch closure it can be seen that the direction of the potential on lines 2 and 4 are reversed

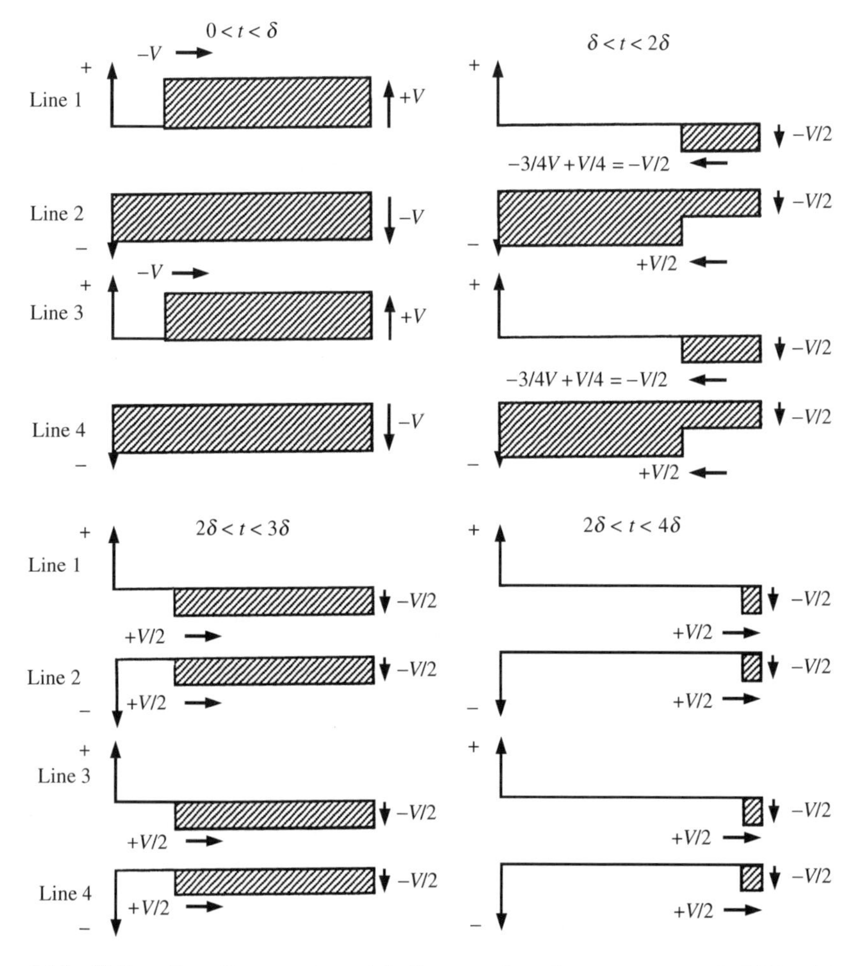

Figure 3.14 Voltage time diagrams to explain the operation of a two-stage stacked Blumlein pulse-forming line

to that on lines 1 and 3. Consequently the potentials cancel and the potential on the load is zero. To initiate pulse-forming action the switches are simultaneously closed, shorting the ends of lines 1 and 3. This causes a voltage step of amplitude $-V$ to be launched into lines 1 and 3 which propagates towards the load end of the circuit. Careful examination of the circuit reveals that lines 1 and 3 are terminated by the load in series with the remaining three lines, i.e. lines 1 and 3 are terminated in an impedance of $7Z_o$. Therefore the reflection coefficient at the end of lines 1 and 3 is given by

$$\rho = \frac{7Z_o - Z_o}{7Z_o + Z_o} = \frac{3}{4} \tag{3.38}$$

Once again it is assumed that the load and line impedances are resistive. When the step on line 1 reaches the load end of the line, a step of amplitude $-3V/4$ is reflected back into the line and a step of amplitude $-7V/4$ is impressed on the load and the other three lines. As in the analysis of the operation of the basic Blumlein pulse-forming line, this step is divided between the load and the other three lines such that a step of $-V$ is placed on the load and individual steps of $-V/4$ are launched back into the other three lines. Similarly line 3 also impresses a step of amplitude $-V$ on the load and launches individual steps of amplitude $-V/4$ back into the other three lines. The total potential on the load is therefore $-2V$, steps of amplitude $+V/2$ are launched into lines 2 and 4 (note the inversion of the lines hence the polarity reversal) and steps of total amplitude $-V/2$ are launched back into lines 1 and 3. These steps reflect at the start of the lines either with a reflection coefficient of -1 for the shorted, switched lines 1 and 3 or with a coefficient of $+1$ for the open circuit lines 2 and 4. In both cases the propagation of the steps discharges all four lines after a round trip of transit time 2δ which ends the pulse.

In practice this circuit can be greatly simplified by using a single switch and by joining the open ends of lines 1 and 3, as shown in Figure 3.15. Transmission line 3 which now has a length of $2l$ in the figure is equivalent to linking lines 2 and 4 in Figure 3.14. In the figure it will be noted that the connection between the lines at the place where load would be connected has been left open. This is because 4 modes of operation of this revised circuit is possible. If points A and B and C and D are connected, then the load can be placed either

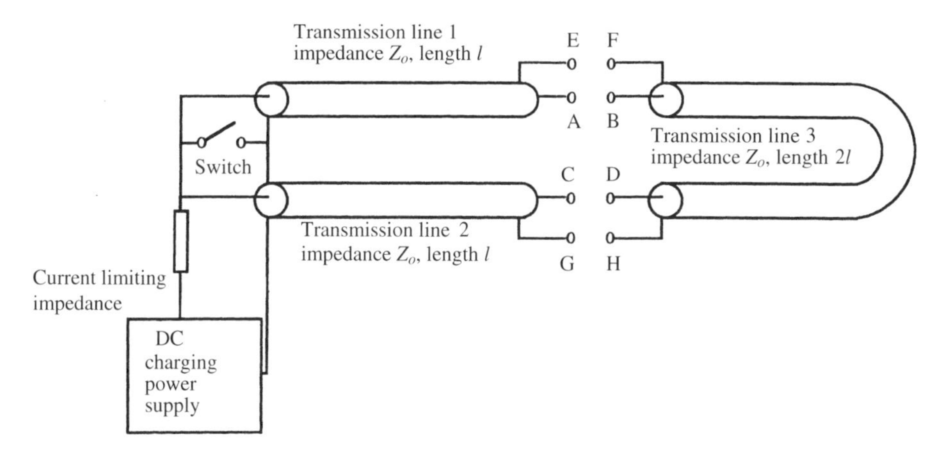

Figure 3.15 A more practical realisation of the two-stage Blumlein pulse-forming line

across points E to H with points F and G connected or across points F and G with points E and H connected. This change simply reverses the polarity of the pulse generated in the load which is connected to the outer conductors of the lines. It can also be connected across the inner conductors of the lines if points E and F and G and H are connected then the load can be placed either across points A to D with points B and C connected or across points B and C with points A and D connected.

In making these various different connections, examination of the circuits will show that the connections made lead to short circuit paths being set up, which can lead to currents flowing in the wrong part of the circuit and cause serious distortion of the output wave-form. This problem is usually solved by winding the lines that make up the stacked Blumlein pulse-forming line inductively (see [2] for more detail). Although this does not affect the normal propagation mode of steps in the lines, the inductance created makes the short circuit paths inductive. Provided the inductance is large enough, its impedance, during the pulse, will be also be large and prevent significant short circuit currents flowing. This problem is dealt with in more detail in Chapter 6.

3.5.2 *The Darlington Circuits*

The Blumlein pulse-forming line is, in fact, the simplest member of a whole family of circuits known as the Darlington circuits, named after S. Darlington who invented them. These circuits are of interest because they not only generate rectangular pulses, but they also produce them at potentials which can be many multiples of the potential to which the circuit is charged initially, i.e. there is pulse-forming action and voltage gain in the same circuit. The circuits can be constructed using transmission lines or pulse-forming networks in the form of *LC* ladders, as will be discussed in Chapter 4. A typical Darlington circuit is shown in Figure 3.16. The circuit comprises a number of transmission lines of constant length but of different characteristic impedances. The circuit has $(n-1)$ lines connected in series "ladder fashion", as shown, and one open ended line (the nth) connected in series with the load. The lines are charged as usual from a DC charging supply via a current limiting resistor or impedance, and the first line in the series is shorted via a switch that initiates pulse generation.

The impedances of the lines increase successively from left to right in the circuit shown. A single rectangular pulse will be generated in the load with a duration that is equal to twice the transit time of any of the lines, provided that the impedance relationship for the lines

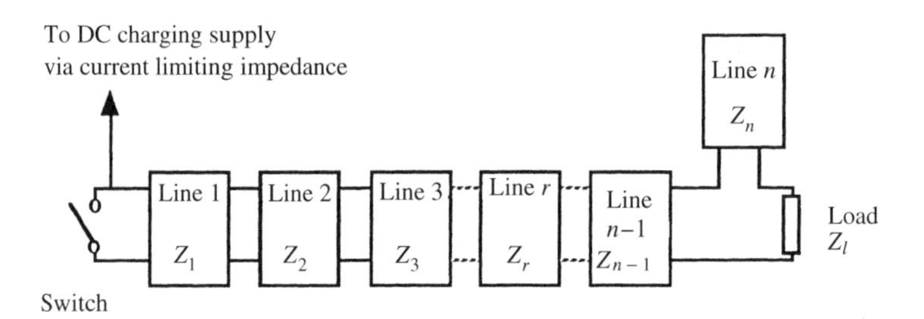

Figure 3.16 Basic configuration of a Darlington circuit

obeys the following relationships

$$\left.\begin{aligned} Z_r &= Z_l\left[\frac{r(r+1)}{n^2}\right] \\ &\text{where } n \text{ is number of lines and} \\ Z_n &= \frac{Z_l}{n} \end{aligned}\right\} \tag{3.39}$$

The amplitude of the pulse produced on the load is given by

$$V_l = \frac{nV}{2} \tag{3.40}$$

To understand the operation of the circuit, a series of voltage time diagrams is drawn in Figure 3.17 for a Darlington circuit with $n = 3$. If it is assumed that the impedance of the first line is Z_o then, using Equation (3.39), the impedance of the next line will be $3Z_o$ and the impedance of the third line will be $3Z_o/2$. The matched load will be $9Z_o/2$. It should be noted that as with all Darlington generators, the matched load impedance is equal to the sum of the nth (3rd) and $(n-1)$th (2nd) lines. In Figure 3.17 the circuit is drawn in a slightly

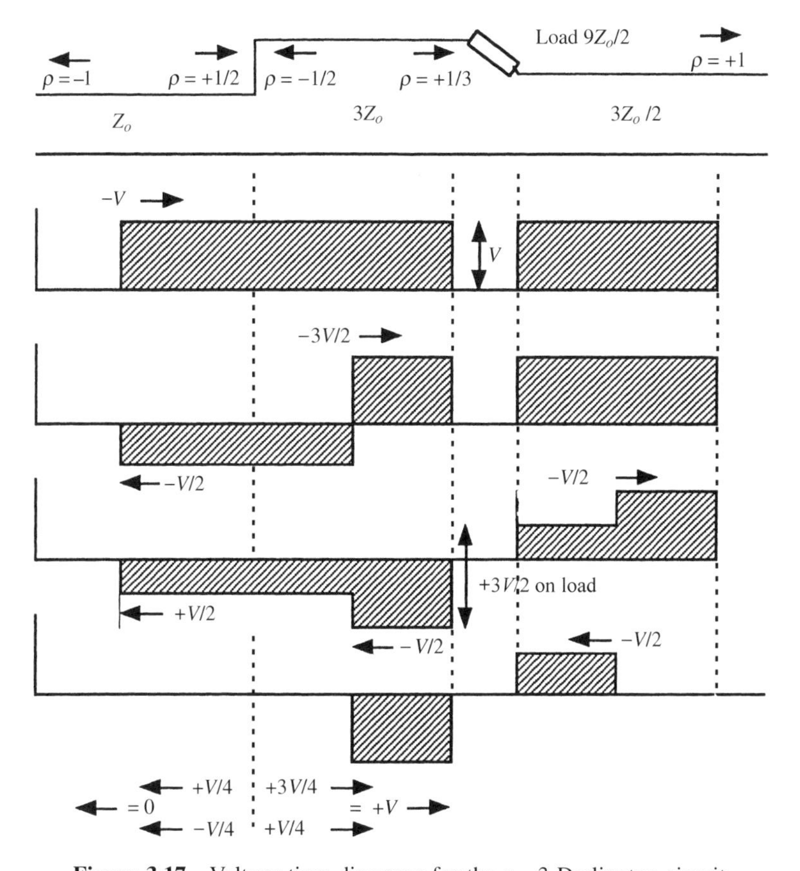

Figure 3.17 Voltage time diagrams for the $n = 3$ Darlington circuit

different way so as to facilitate the set of voltage time diagrams. As with the other pulse-forming line circuits described in this chapter, voltage reflection coefficients are marked at the junctions between lines, at the load, and at the ends of the 1st and 3rd lines.

The lines are charged to a potential V, as shown, from a DC charging power supply via a current limiting impedance as usual. As with the Blumlein circuit, pulse-forming action is initiated by the closure of the switch. A $-V$ step then propagates to the right and is partly transmitted and partly reflected when it reaches the first junction. The transmitted step reaches the load where a potential of $3V/2$ is impressed on the load, and a $-V/2$ step is transmitted into the line on the right. This step then reflects and returns to the load completely discharging this line. In the meantime, two steps propagate on the two lines to the left and eventually combine to form a step of $+V$ which propagates back to the load, again discharging the two lines on the left of the diagram.

There are several interesting features of the Darlington circuits. For larger numbers of stages n it is found that the impedances of successive lines moving from the left to the right of the circuits are proportional to the successive sums of the natural integers, i.e. 1, $1+2=3$, $1+2+3=6$, etc., and the $n=2$ Darlington circuit is simply a Blumlein pulse-forming line. Pulse-forming networks (see the next chapter) are usually used instead of lines because of the need to have lines with very non-standard impedances. Alternatively strip lines can be constructed to provide the odd impedances required. The output impedance of the $(n-1)$th section is given by

$$Z_{n-1} = Z_l \left[\frac{n-1}{n} \right] \tag{3.41}$$

As previously noted, for matched conditions the sum of the impedances of the nth and the $(n-1)$th lines must equal that of the load, i.e.

$$Z_{n-1} + Z_n = Z_l \left[\frac{n-1}{n} \right] + Z_l \left[\frac{1}{n} \right] = Z_l \tag{3.42}$$

For an n section circuit the value of the load impedance Z_l in terms of the impedance of the first line Z_1 is given by

$$Z_l = \frac{n^2 Z_1}{2} \tag{3.43}$$

Finally the operation of a $n=5$ Darlington circuit is given in Figure 3.18.

3.5.3 Further Darlington-like Pulse-forming Lines

There are many other circuits [3] similar to the Darlington circuits which can form pulses either at higher potentials than the initial charge potential of the circuit or which can be used to generate high current pulses, i.e. they act in an analogous way to a current transformer. In most cases these circuits require different sections to be charged to different potentials and may require the use of more than one switch. As with the Darlington circuits these generators are made up of a number of transmission lines of differing impedances but of equal lengths. Two examples will be given and their operation described by the use of

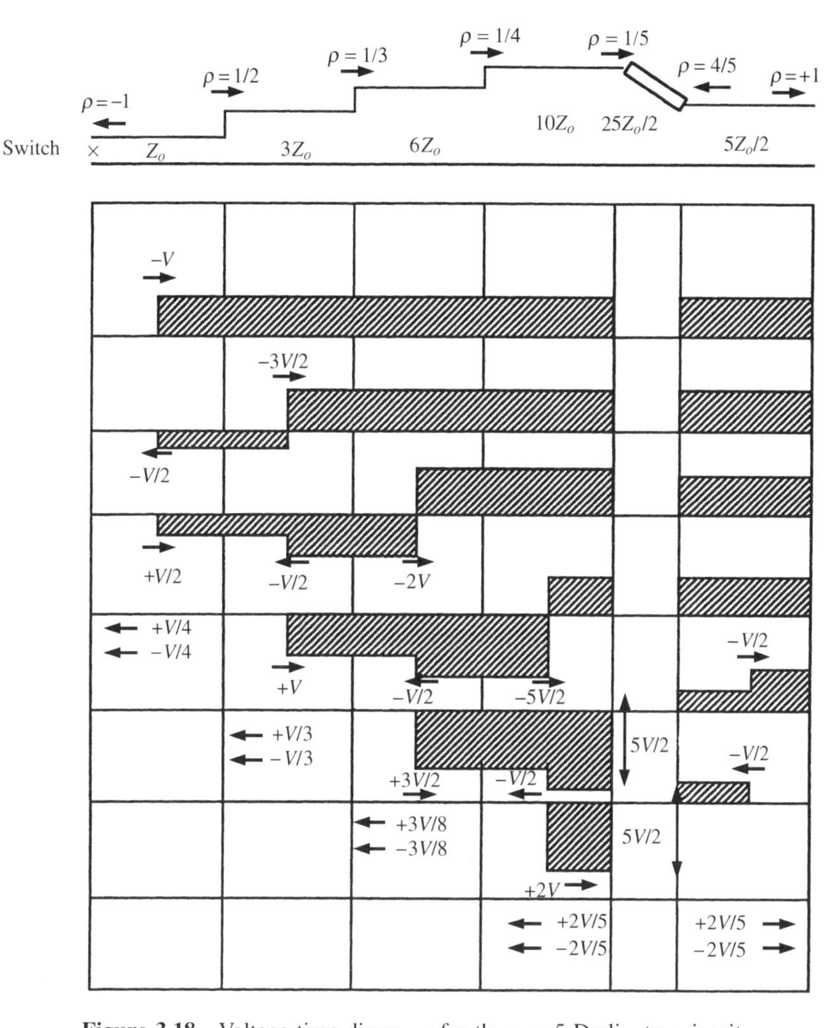

Figure 3.18 Voltage time diagrams for the $n = 5$ Darlington circuit

voltage time diagrams. In Figure 3.19 a circuit is shown which acts like a pulse-forming current transformer, where the impedance transformation from the primary line to the load is 5 : 1 and the matched pulse-forming line current is increased by a factor of 2.5. In Figure 3.20 a circuit is given where the output pulse voltage is increased by 50% from the potential on the primary line and the output impedance is increased by a factor of 6. It is interesting to note that in this circuit a number of the lines are charged to a negative potential yet the output pulse is positive.

In order to understand, in more detail, the way in which these circuits operate, it is necessary to consider the way in which power flows across the junction between two transmission lines of differing characteristic impedances which are pre-charged to a DC potential [3]. This is represented by the diagram given in Figure 3.21.

Referring to Figure 3.21, consider the voltage and current that flows across the pre-charged junction when a step of amplitude V propagates from line 1 to line 2. Assume that

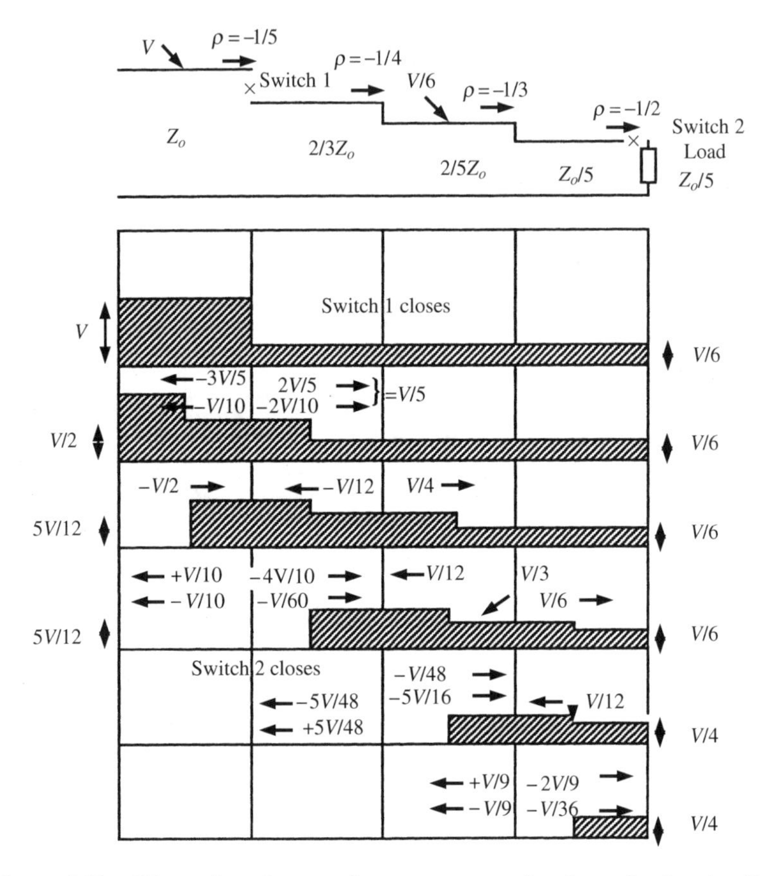

Figure 3.19 Voltage time diagrams for a current transforming pulse-forming line

two lines or junction are pre-charged to a potential αV. The current that flows across the junction after the step has reached the junction is given by

$$I = \frac{2V}{(Z_1 + Z_2)} \tag{3.44}$$

The corresponding total potential at the junction V_J is given by

$$\begin{aligned} V_J &= \alpha V + V_+ + V_- = \alpha V + V_+ + \rho V_+ \\ &= \alpha V + V_+ + \frac{(Z_2 - Z_1)}{(Z_2 + Z_1)} V_+ \\ &= \alpha V + 2V \frac{Z_2}{(Z_1 + Z_2)} \end{aligned} \tag{3.45}$$

where V_+ and V_- are the incident and reflected steps at the junction. The power flowing across the junction P can be calculated simply by multiplying the total junction potential by the current as given in Equations (3.44) and (3.45). It is then interesting to find out what the relationship between Z_1 and Z_2 should be in order to maximise the power flow across the

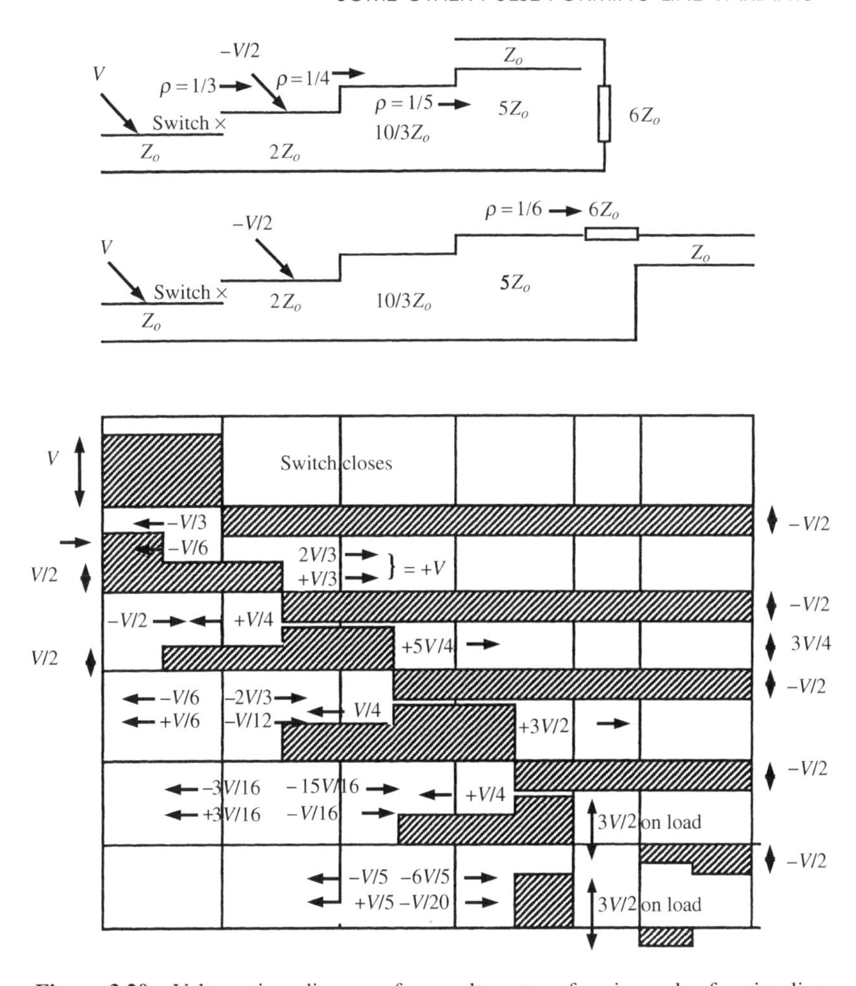

Figure 3.20 Voltage time diagrams for a voltage transforming pulse-forming line

junction. This can be done by differentiating P with respect to Z_2 and setting the differential to zero. If this is carried out then it is found that the condition

$$Z_2 = Z_1 \frac{(2-\alpha)}{(2+\alpha)} \tag{3.46}$$

will maximise the power flow. Clearly if the lines are not pre-charged then α is zero and the expected matched condition $Z_1 = Z_2$ results. The maximum power flow that results if Equation (3.46) is satisfied is given by

$$P_{\max} = \left(1+\frac{\alpha}{2}\right)^2 \frac{V^2}{Z_1}$$

$$= (V+\alpha V)\frac{V}{Z_1} + \frac{(\alpha V)^2}{4Z_1} \tag{3.47}$$

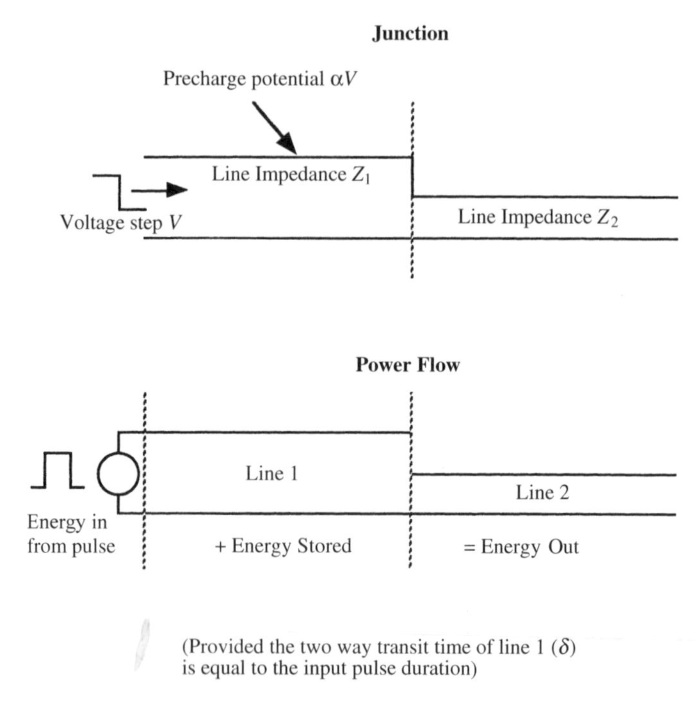

Figure 3.21 Power flow through the pre-charged junction

Examination of the rearranged form of Equation (3.47), given in the second line, shows that the first term is the power in the step as it propagates on line 1. The second term is the maximum power that the pre-charged line 1 could deliver if switched on to a matched load. If the step is replaced by a pulse with a duration which is equal to twice the transit time of line 1, as shown in Figure 3.21, then all of the energy in such a pulse plus all the energy stored in the pre-charged line will be propagated into line 2, i.e. no energy will be reflected. It is on this principle that the two circuits given in Figures 3.19 and 3.20 are based.

Referring again to the circuits given in Figures 3.19 and 3.20, it can be seen that the circuits comprise a primary pulse-forming line, charged to a potential V which is switched into a series of pre-charged lines. It is therefore also necessary to consider what the relationship should be between the impedances of this line and the line into which it is switched such that all the energy in the line is discharged into the succeeding pre-charged lines. This line behaves in a way which is similar to the single pulse-forming line described above. Therefore a step with an amplitude which is equal to half the initial potential on the line must be propagated back into the line in order to fully discharge the line after a double transit.

Referring to Figure 3.22, it can be seen that when the switch is closed a back-propagating step will be launched into line 1 due to the mismatch in impedances because $Z_1 \neq Z_2$, and an additional step is propagated into line 1 due to the discharge of line 2 into it. The sum of the amplitudes of these two steps must be equal to $-V_1/2$ if line 1 is to be fully discharged, in a double transit, in an identical way to the operation of the single pulse-forming line described earlier. The back step propagated into line 1 due to the mismatch

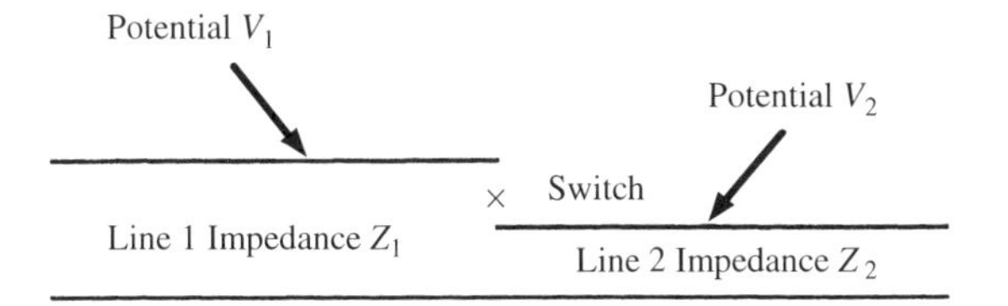

Figure 3.22 Primary pulse-forming line discharge circuit

is given by

$$V_{s1} = V_1\left(\frac{Z_2}{Z_1+Z_2} - 1\right) = V_1\left(\frac{-Z_1}{Z_1+Z_2}\right) \quad (3.48)$$

The step propagated into line 1 from line 2 is given by

$$V_{s2} = V_2\left(\frac{Z_1}{Z_1+Z_2}\right) \quad (3.49)$$

Thus the total amplitude of the step propagated into line 1 when the switch is closed is given by

$$V_{s1} + V_{s2} = \frac{Z_1}{Z_1+Z_2}(V_2 - V_1) = \frac{Z_1}{Z_1+Z_2}V_1(\alpha - 1) \quad \text{if} \quad V_2 = \alpha V_1 \quad (3.50)$$

This step must have an amplitude equal to $-V_1/2$ for line 1 to completely discharge in a double transit. Therefore

$$\frac{-V_1}{2} = \frac{Z_1}{Z_1+Z_2}V_1(\alpha - 1) \quad \text{i.e.} \quad Z_2 = Z_1(1 - 2\alpha) \quad (3.51)$$

Equations (3.46) and (3.51) now provide the basis for the design of this family of pulse-forming circuits.

It is interesting to check the operation of these equations by reference to the circuit given in Figure 3.19. The primary pulse-forming line is charged to a potential V and the successive lines are pre-charged to a potential of $V/6$. Thus $\alpha = 1/6$. Using Equation (3.51), it is found that if the primary line has an impedance of Z_o then the line into which it is switched must have an impedance of $2/3Z_o$, as shown in the circuit. Examination of the second of the voltage time diagrams in Figure 3.19 reveals that a step of $V/3$ propagates from the second to the third line. Since the pre-charge voltage is $V/6$, α in Equation (3.46) is $1/2$. As the impedance of the line on which this step is propagating is $2/3Z_o$, then using Equation (3.46) it is found that the impedance of the third line must be $2/5Z_o$. It is left as an exercise for the reader to check that the impedances of the other lines in Figure 3.19 and that the impedances of the lines in Figure 3.20 are correct.

3.5.4 The Self-matching Pulse-forming Line

All of the pulse-forming lines described so far must be discharged into a matched load in order to generate a single rectangular pulse. However there are circumstances in which it is desirable to produce a single rectangular pulse in a load which cannot be matched to the output impedance of a particular pulse-forming line. The pulse-forming line described by Ishii and Yamada [4] can supply a single rectangular pulse into a load regardless of its resistance. The basic circuit of this pulse generator is given in figure 3.23. In the Figure the circuit is also drawn in a modified way in order to facilitate the description of the operation of the circuit.

This relatively simple circuit comprises a length of transmission line of impedance Z_o, the outer conductor of which is charged to a potential $2V$ by a DC charging power supply which is, as usual, current-limited with a resistor or inductor. At one end of the line the inner conductor is connected to a terminating resistor R and at the other it is connected to a load Z_l, where $Z_l \neq Z_0$. Pulse-forming action is initiated by closure of the switch shown. Closure of the switch causes the outer conductor of the line to be grounded and consequently the inner conductor is raised to a potential of $-2V$. In the equivalent circuit the switch is replaced by two switches and it is assumed that the closure of these two switches is simultaneous to describe the operation of the circuit. When these switches are closed, a step of amplitude $-2V(\alpha-1)$ is launched into the line at the load end of the line and a step of amplitude $-2V(\beta-1)$ is launched into the line at the end of the line connected to the

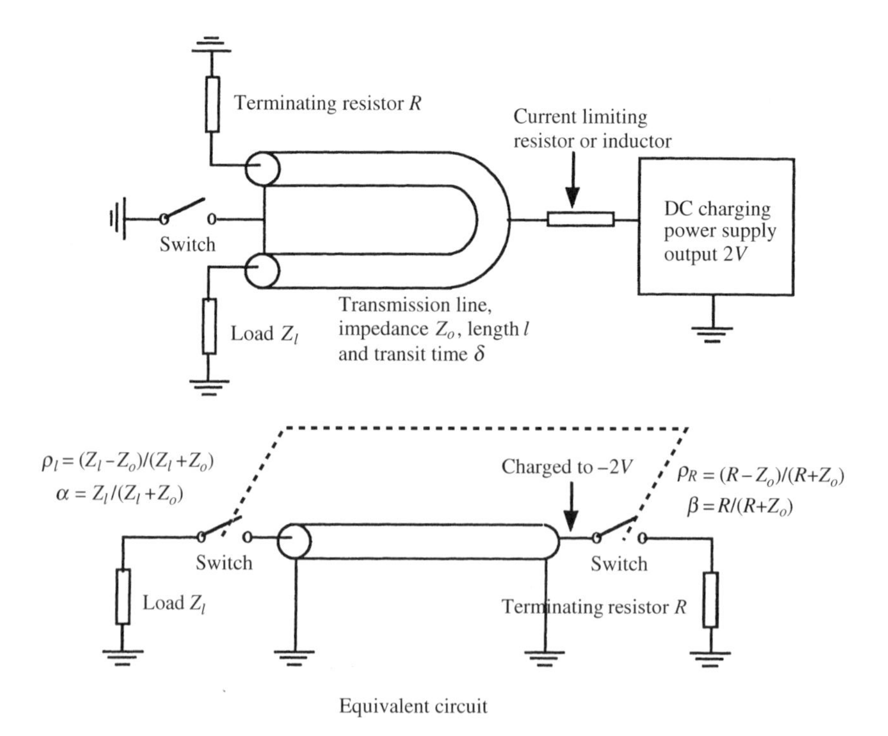

Figure 3.23 Circuit diagram of the self-matching pulse-forming line and its equivalent circuit

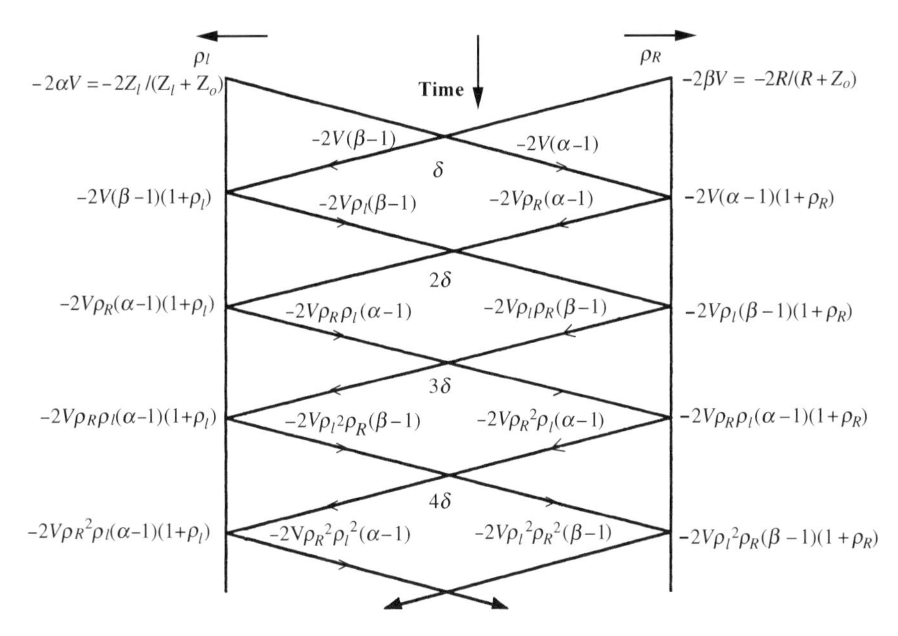

Figure 3.24 Lattice diagram for the self-matching pulse-forming line

resistor. α and β are defined in the diagram. The operation of the circuit can be best described with the aid of the lattice diagram given in Figure 3.24.

From this diagram the voltage on the load can be found to be

$$v_l(t) = -2V \begin{bmatrix} \alpha + (\beta-1)(1+\rho_l)u(t-\delta) \\ +\rho_R(\alpha-1)(1+\rho_l)u(t-2\delta) \\ +\rho_R\rho_l(\beta-1)(1+\rho_l)u(t-3\delta) \\ +\rho_R^2\rho_l(\alpha-1)(1+\rho_l)u(t-4\delta) \\ +\cdots \end{bmatrix} \tag{3.52}$$

where $u(\Delta t) = 1 \quad \text{for } \Delta t > 0, \quad u(\Delta t) = 0 \quad \text{for } \Delta t < 0$
and $\Delta t = (t - n\delta), \quad n = 1, 2, 3, \ldots$

The Laplace transform of this voltage can be written as

$$\mathbf{V}_l(\mathbf{s}) = -\frac{4V\alpha}{\mathbf{s}} \begin{bmatrix} \dfrac{1}{2} + [(\beta-1)\exp(-\mathbf{s}\delta) + \rho_R(\alpha-1)\exp(-2\mathbf{s}\delta)] \\ \times \left[1 + \sum_1^\infty (\rho_l\rho_R\exp(-2\mathbf{s}\delta))^n\right] \end{bmatrix}$$

$$\text{since} \quad 2\alpha = (1+\rho_l) \tag{3.53}$$

If the value of the terminating resistor is made equal to the impedance of the line, i.e. $R = Z_o$, then $\rho_R = 0$ and $\beta = 1/2$. Equation (3.53) then becomes

$$\mathbf{V}_l(\mathbf{s}) = \frac{-2V\alpha}{\mathbf{s}}[1 - \exp(-\mathbf{s}\delta)] \tag{3.54}$$

which is recognised as a single rectangular pulse with duration equal to a single transit time of the transmission line independently of the value of the load impedance Z_l, i.e. the pulse-forming line is self-matching.

Furthermore, if the load impedance is infinite, $\rho_l = +1$ and the potential on the load will be $-2V$. Since the line impedance is equal to the terminating resistor, a step of amplitude V will be launched into the cable at the resistor end of the line and will travel towards the load. When this step reaches the load it will reflect and deposit a potential of $2V$ on the load terminating the pulse. The reflected step then returns to the terminating resistor, where $\rho_R = 0$ and is fully absorbed. Thus the duration of the pulse on the load is δ and the duration of the pulse on the terminating resistor is 2δ. This is always the case, independent of the value of the load impedance.

3.5.5 The Bi-directional or Zero Integral Pulse-forming Line

Some applications, for example linear induction accelerators [5,6,7], require the generation of a bi-directional pulse in which the polarity of the pulse changes sign in the middle of the pulse, causing the net integral of the pulse to be zero. Such a pulse can be generated using the quite simple circuit shown in Figure 3.25.

The operation of this circuit is best described by the use of a series of voltage time diagrams as given in Figure 3.26. For correct operation the impedance of the load must be matched to that of the cable. When the switch closes the outer conductor of the transmission line is grounded, forcing the potential of the inner conductor to be raised to $-V$. At the switched end of the line this causes a step of amplitude $+V$ to be launched into the cable. At the other end of the line, which is connected to the load, a step of $+V/2$ is launched into the line because the load is matched to the impedance of the line and the voltage on the load is $-V/2$. The two propagating steps pass through each other and when the $+V$ step reaches the load the polarity of the potential on the load simply changes to $+V/2$. The reflection coefficient at the switched end of the line is -1, thus the $+V/2$ step reverses polarity when it reaches the switch and returns to the load, terminating the pulse. Therefore the pulse on the load is $-V/2$ for one transit of the line and $+V/2$ for a further transit of the line. The pulse has a duration 2δ and changes polarity in the middle of the pulse after a time δ.

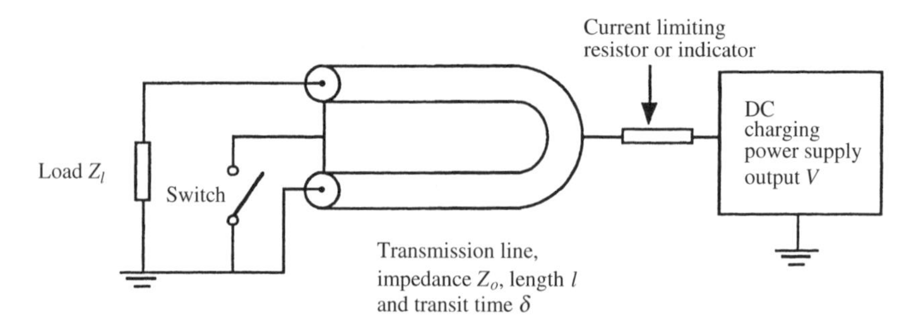

Figure 3.25 The bi-directional pulse-forming line

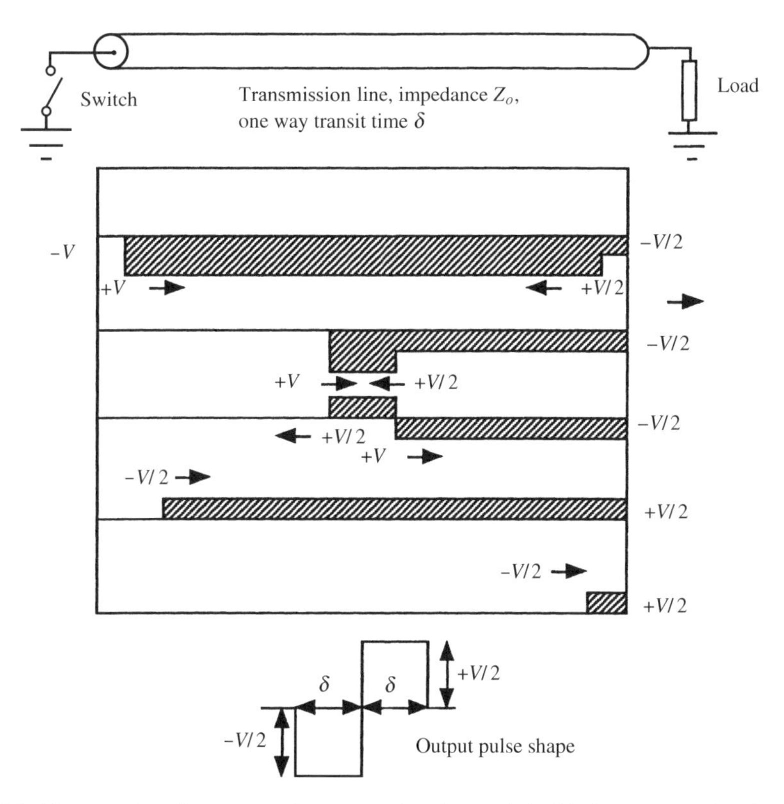

Figure 3.26 Voltage time diagrams and output pulse shape of the bi-directional pulse-forming line

3.5.6 A Pseudo-repetitive Pulse-forming Line

This rather ingenious pulse-forming line system [8] can generate a repetitive stream of rectangular pulses and can be used to generate a synthetic RF pulse if the pulse stream is suitably filtered with a low pass filter. The circuit, together with pulse waveforms at various points in the circuit, is given in Figure 3.28. However, to understand the operation of the circuit, it is helpful to first consider the operation of the rather simpler circuit given in Figure 3.27.

The circuit shown in Figure 3.27 comprises a transmission line connected to a matched load which is also connected half-way along its length to a second short circuited length of line or stub. The second line has an impedance which is half that of the first line and a length and one-way transit time which differs from the first line. A step generator is connected to the input of the first line and, to simplify the description of the operation of the circuit, it is given an amplitude of $2V$. It is also assumed that the output impedance of the generator is matched to that of the first line, hence a step of amplitude V is launched into the line at its input at point A. After a time δ_1 this step reaches point B in the circuit where the reflection coefficient is given by

$$\rho = \frac{[(Z_o \,||\, Z_o/2) - Z_0]}{[(Z_o \,||\, Z_o/2) + Z_0]} = \frac{[(Z_o/3 - Z_0]}{[(Z_o/3 + Z_0]} = -\frac{1}{2} \tag{3.55}$$

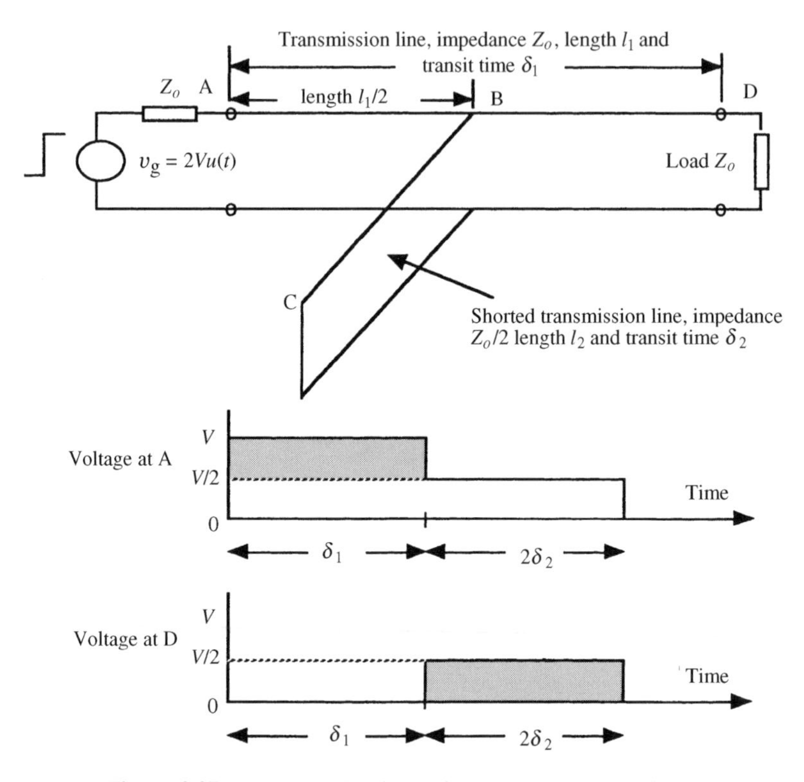

Figure 3.27 Example circuit and its response to a step input

Thus a step of amplitude $(1+\rho)V$or $V/2$ is propagated into the short-circuited length of line towards point C and also along the first line towards point D. A $-V/2$ step is also reflected back to the generator. Since the load and generator are matched to the impedance of the first line, the steps arriving at point A and point D are completely absorbed without further reflections. The step arriving at C is reflected with a reflection coefficient of -1 because of the short circuit. Thus a step of amplitude $-V/2$ propagates back towards point B. This step is not reflected at point B as the second line is matched to the impedance of the two halves of line 1 which are effectively in parallel. Thus the $-V/2$ step is propagated towards the input of the line at point A and the load at point D. When this step reaches point D it adds to the existing potential $V/2$. Hence a rectangular pulse is generated on the load with duration $2\delta_2$ and amplitude $V/2$ after a delay of δ_1.

A circuit diagram of the repetitive pulse-forming line is shown in Figure 3.28. It can be seen that it consists of a cascade of the example circuit shown in Figure 3.27. The circuit consists of a main primary transmission line of impedance Z_o with four secondary lines connected at equal distances along the line. The first two of these secondary lines are short-circuited to form stubs and have an impedance which is half that of the primary line. The next two are twice and four times the length of the first two and are both open-circuit. Again they have an impedance equal to half that of the primary line. Injection of a step into the input of the primary line gives a rectangular pulse at point A of amplitude $V/2$ and duration $2\delta_2$. When this pulse is fed into the next section of the circuit, a delayed and attenuated replica of the pulse is propagated to point B. After a further delay of $2\delta_2$ seconds, a similar

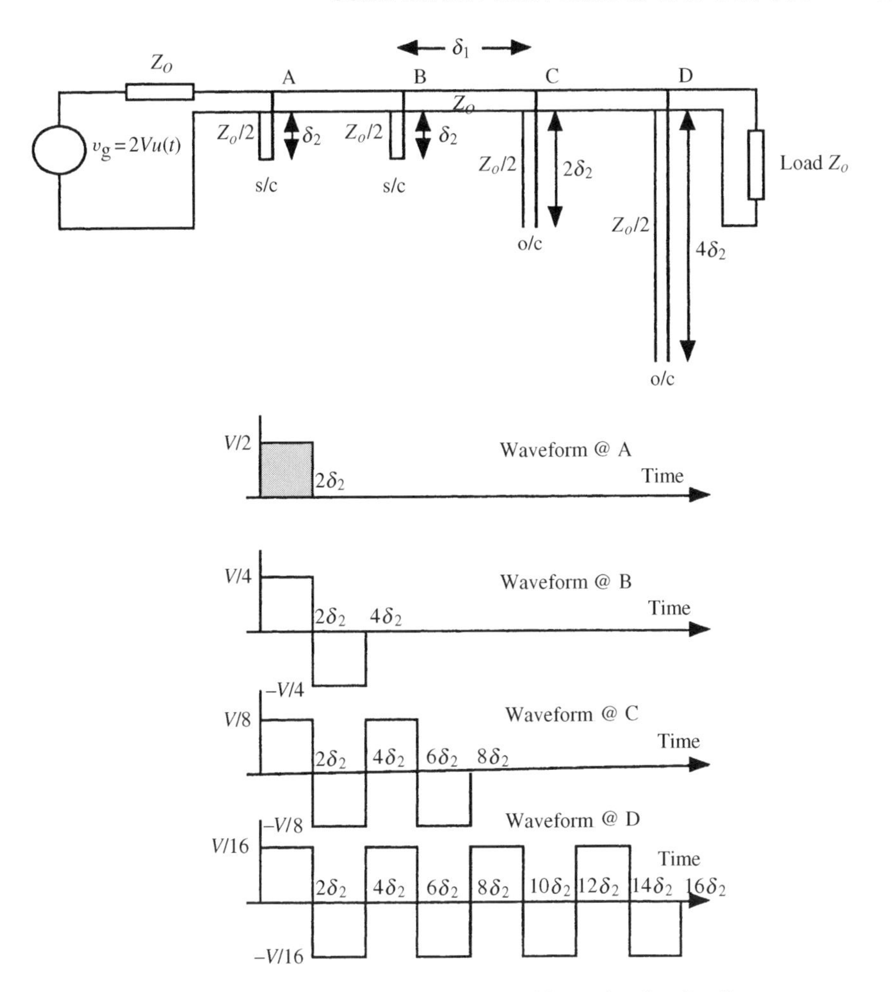

Figure 3.28 Circuit diagram of the repetitive pulse-forming line

but inverted pulse joins the first pulse at point B, giving a single cycle square wave whose amplitude is $\pm V/4$. This square wave propagates into the third section of the circuit and reaches point C, having been reduced in amplitude again by a half. Part of the wave also propagates in the third open circuit transmission line. This wave reflects at the open circuit without change of sign and joins the first wave at C, after a delay of $4\delta_2$ seconds. Therefore, at C the waveform is now two cycles long with an amplitude of $V/8$. Finally this waveform is fed into the final, fourth, section of the circuit and, as in the preceding section, is doubled in length and halved in amplitude to give a waveform which is four cycles long with an amplitude of $V/16$.

For a circuit with n secondary lines the number of cycles N at the nth section is given by

$$N = 2^{n-2} \tag{3.56}$$

and the amplitude of the output waveform $v_o(t)$ is given by

$$v_o(t) = 2^{-n} \tag{3.57}$$

The overall pulse duration T and fundamental frequency of the waveform f_o are

$$\left.\begin{aligned} T &= N4\delta_2 = 2^n\delta_2 \\ f_o &= \frac{1}{4\delta_2} \end{aligned}\right\} \tag{3.58}$$

Clearly, if this waveform is passed through a low pass filter whose cut-off frequency is approximately $2f_o$ then a sinusoidal RF pulse will be generated at a frequency f_o.

3.5.7 Current-fed Pulse-forming Lines

In all of the pulse-forming line circuits described so far the lines have been charged from a DC voltage power supply, such that the energy released in the pulse(s) is originally stored as electrostatic energy in the capacitance of the lines. As an alternative, energy can be stored in the lines magnetically by feeding or charging the lines from a DC current source [9,10]. The energy that is subsequently released in the pulse is stored in the inductance of the lines rather than the capacitance. This mode of operation is analogous to inductive storage (see Example 1.12) and will be discussed further in the next chapter and Chapter 7. Inductive energy storage is an attractive option for passive pulse generators, in particular large pulsed power systems, because it can be more compact than capacitive energy storage. The main problem is the need to use an opening switch rather than a closing switch to initiate the release of energy from the inductive store. At relatively low potentials this does not present a problem as there are a number of suitable semiconductor switches that will work in this type of circuit. However at high voltages and powers, despite many years of research effort, no ideal switch has been designed that will operate both reliably and efficiently.

In [9,10] circuits are given for the current-fed analogue or dual of the single pulse-forming line and also the Blumlein pulse-forming line. A circuit for the line, together with a series of current and voltage time diagrams, is given in Figure 3.29. It is assumed that a circulating current $I(0)$ has been established in the line with the opening switch in its closed position. Clearly, at both the switch and the shorted end of the line the voltage reflection coefficients are -1, whereas at the load it is zero. Before the switch is opened the circulating current can be considered as two current steps I_+ and I_-, of equal amplitude, which are propagating in opposite directions along the line, as shown in Figure 3.29. These current steps establish voltage steps which also propagate as shown. Note that the sum of these two steps must be zero since, as the line is shorted circuited at both ends and the steps propagate around the line, they reverse polarity as they encounter the short circuits. When the switch is opened the forward propagating voltage step V_+ is fed to the load and the backward step V_- starts to propagate towards the shorted end of the line, leaving a net potential on the line of V_+. When the backward step reaches the shorted end of the line it reflects with a coefficient -1, reverses polarity and sets up a second forward propagating step with an amplitude V_+ which follows the original forward propagating step. Notice that this step terminates as the original step circulation pattern is interrupted by the opening of the switch. Thus a pulse is generated in a matched load with an amplitude $Z_oI(0)/2$ and a duration equal to twice the transit time along the line.

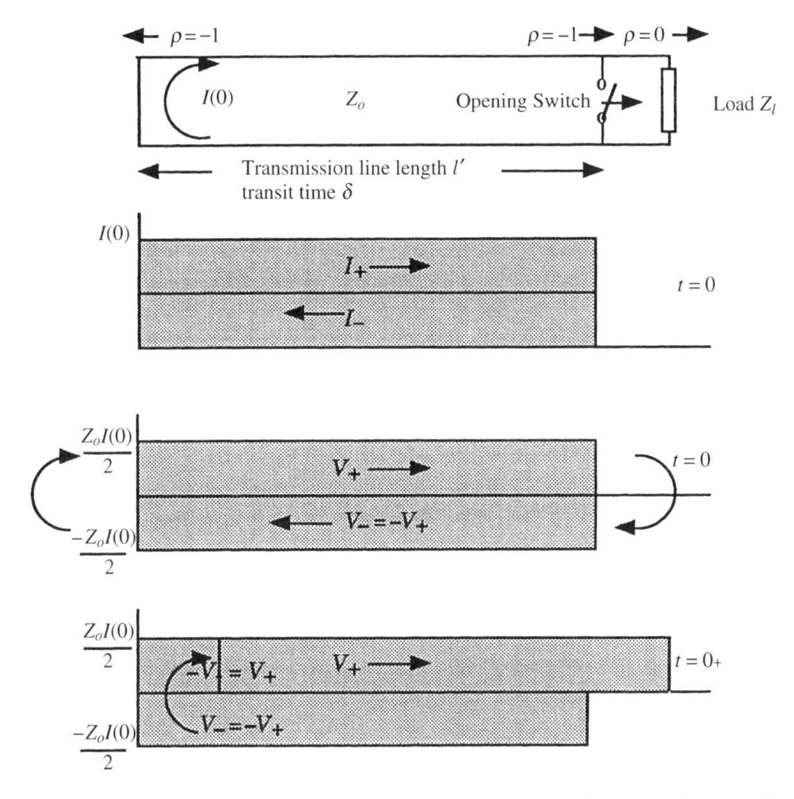

Figure 3.29 Operation and circuit diagram of the current-fed pulse-forming line

In the case of loads which are not matched to the pulse-forming line the waveforms that are produced can be determined using the circuit shown in Figure 3.30. Since the line is an inductive store and is short-circuited at its far end, the equivalent circuit is based on the equivalent circuit of an inductor with an initial current flowing through it (see Figure 1.8).

From the circuit shown in Figure 3.30

$$\begin{aligned}\frac{I(0)}{\mathbf{s}} &= \mathbf{V}(\mathbf{s})\left(Y_i(\mathbf{s}) + \frac{1}{Z_l}\right)\\ &= \mathbf{V}(\mathbf{s})\left(\frac{1}{Z_0\tanh(\mathbf{s}\delta)} + \frac{1}{Z_l}\right)\\ &= \mathbf{V}(\mathbf{s})\left(\frac{Z_l + Z_0\tanh(\mathbf{s}\delta)}{Z_l Z_0\tanh(\mathbf{s}\delta)}\right)\end{aligned} \tag{3.59}$$

$$\begin{aligned}\mathbf{V}(\mathbf{s}) &= \frac{I(0)}{\mathbf{s}}\frac{Z_l Z_0\tanh(\mathbf{s}\delta)}{Z_l + Z_0\tanh(\mathbf{s}\delta)}\\ &= \frac{I(0)}{\mathbf{s}}\frac{Z_l Z_0}{Z_l\coth(\mathbf{s}\delta) + Z_0}\end{aligned} \tag{3.60}$$

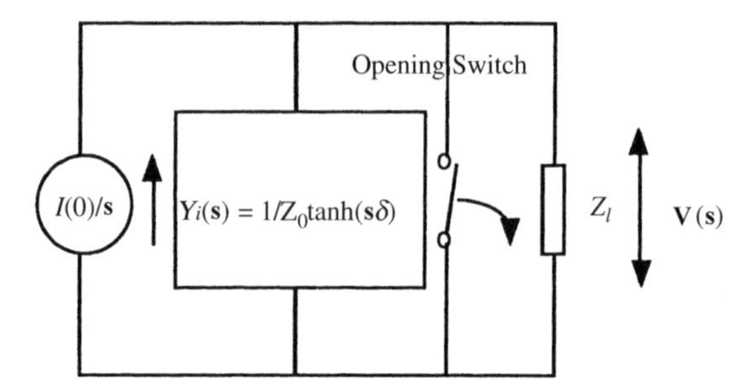

Figure 3.30 Laplace equivalent circuit of a current-fed pulse-forming line

Using the identity

$$\coth(x) = \frac{1 + \exp(-2x)}{1 - \exp(-2x)} \tag{3.61}$$

Equation (3.60) can be written as

$$\mathbf{V}(\mathbf{s}) = \frac{I(0)Z_l Z_0}{\mathbf{s}(Z_l + Z_0)} \frac{1 - \exp(-2\mathbf{s}\delta)}{1 + \left(\frac{Z_l - Z_0}{Z_l + Z_0}\right)\exp(-2\mathbf{s}\delta)} \tag{3.61}$$

Using the binomial expansion

$$(1 + x)^n = 1 + nx + \frac{n(n-1)x^2}{2!} + \frac{n(n-1)(n-2)}{3!} + \cdots, \tag{3.63}$$

Equation (3.62) becomes

$$\mathbf{V}(\mathbf{s}) = \frac{I(0)Z_l Z_0[1 - \exp(-2\mathbf{s}\delta)]}{\mathbf{s}(Z_l + Z_o)} \left[\begin{array}{l} 1 - \left(\frac{Z_l - Z_0}{Z_l + Z_0}\right)\exp(-2\mathbf{s}\delta) \\ + \left(\frac{Z_l - Z_0}{Z_l + Z_0}\right)^2 \exp(-4\mathbf{s}\delta) \\ - \left(\frac{Z_l - Z_0}{Z_l + Z_0}\right)^3 \exp(-6\mathbf{s}\delta) \\ + \cdots \end{array} \right] \tag{3.64}$$

Taking the inverse transform

$$v(t) = \frac{I(0)Z_l Z_0}{(Z_o + Z_l)} \left\{ \begin{array}{l} 1 - u(t - 2\delta) - \left(\frac{Z_l - Z_0}{Z_l + Z_0}\right)[u(t - 2\delta) - u(t - 4\delta)] \\ + \left(\frac{Z_l - Z_0}{Z_l + Z_0}\right)^2 [u(t - 4\delta) - u(t - 6\delta)] \\ - \left(\frac{Z_l - Z_0}{Z_l + Z_0}\right)^3 [u(t - 6\delta) - u(t - 8\delta)] \\ + \cdots \end{array} \right\} \tag{3.65}$$

where $u(\Delta t) = 1$ for $\Delta t > 0$, $u(\Delta t) = 0$ for $\Delta t < 0$

and $\Delta t = (t - n\delta)$, $n = 2, 4, 6, \ldots$

The shapes of the voltage pulses $v(t)$ for the cases where $Z_o < Z_l$ and $Z_o > Z_l$ correspond, as should be the case, to those shown in Figure 3.5. Typical waveforms are shown in Figure 3.31. Note also that if $Z_o = Z_l$, i.e. the line is matched to the load, $v(t)$ can derived from Equation (3.64) to be

$$v(t) = \frac{I(0)Z_0}{2}[1 - u(t - 2\delta)] \tag{3.66}$$

Clearly $v(t)$ is a single rectangular pulse of amplitude $V/2$ and duration 2δ as expected.

Finally it is easy to see that the circuit of the current-fed pulse-forming line given in Figure 3.29 would be non-realizable as there is no way to establish the initial circulating current $I(0)$. Two experimental realizable circuits are shown in Figure 3.32 for the single current-fed pulse-forming line and the current-fed or dual of the Blumlein pulse-forming line.

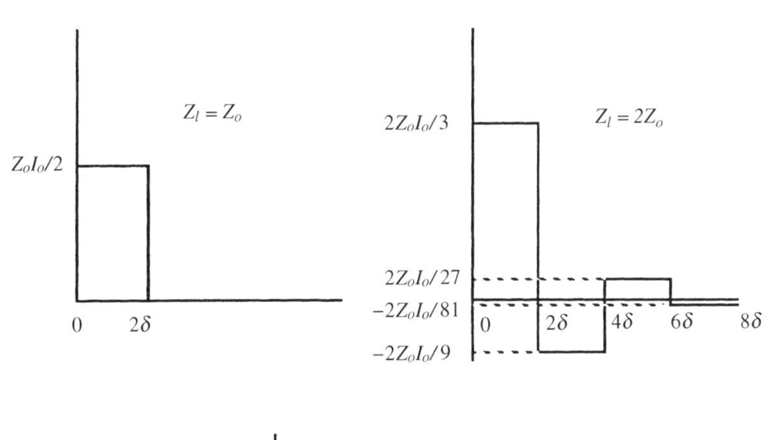

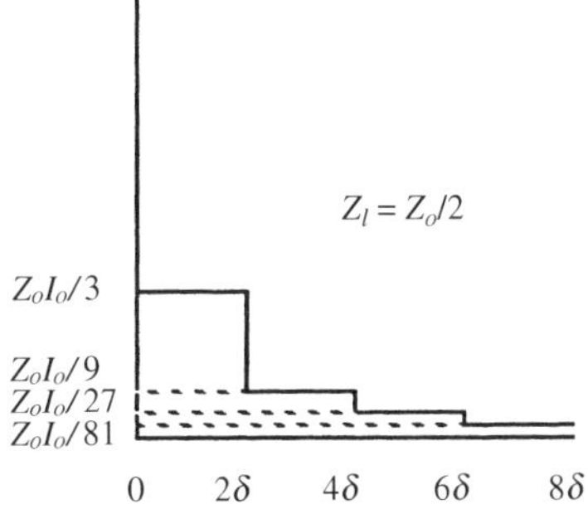

Figure 3.31 Typical waveforms from the current-fed pulse-forming line under matched and unmatched conditions

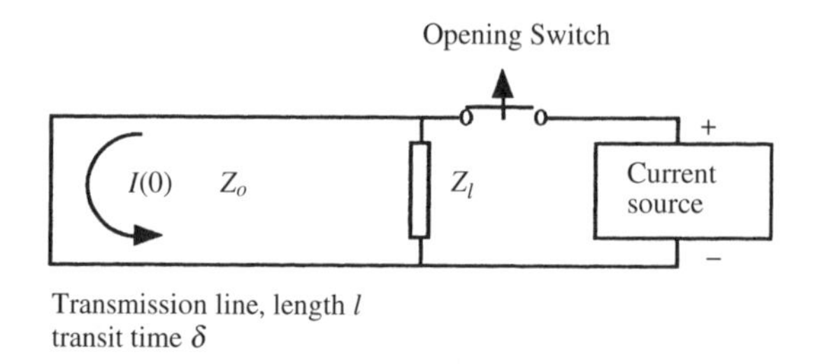

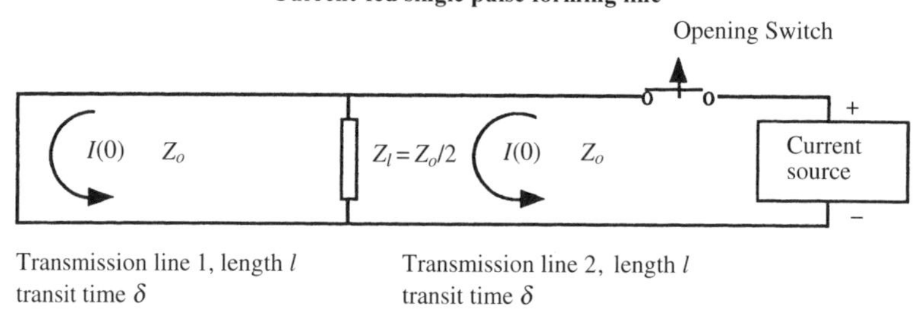

Figure 3.32 Realizable forms of the current-fed single pulse-forming line and Blumlein pulse-forming line

In these circuits the initial circulating current is supplied from a current source and the switch is moved to a position which is in series with this source. Note that the matched load for the current-fed Blumlein pulse-forming line is $Z_o/2$.

REFERENCES

[1] Blumlein A. D. UK Patent 589127 (1941)

[2] Somerville I. C., MacGregor S. J. and Farish O. "An Efficient Stacked-Blumlein HV Pulse Generator". *Meas. Sci. Technol.* (1990), **1**, 865–868

[3] Smith I. D. "Principles of the Design of Lossless Tapered Transmission Line Transformers". Proc. 7th IEEE Pulsed Power Conference (1989), 103–107

[4] Ishii M. and Yamada H. "Self-Matched High-voltage Rectangular Wave Pulse Generator". *Rev. Sci. Instrum.* (1985), **56**(11), 2116–2118

[5] Eccleshall D. and Temperley J. K. "Transfer of Energy from Charged Transmission Lines with Applications to Pulsed High-current Accelerators". *J. Appl. Phys.* (1978), **49**(7), 3649–3655

[6] Smith I. "Linear Induction Accelerators made from Pulse-line Cavities with External Pulse Injection". *Rev. Sci. Instrum.* (1979), **50**(6), 714–718

[7] Hotta E. and Abe M. "Bidirectional Pulsers with High Efficiency for Linear Induction Accelerators". *Electrical Engineering in Japan* (1984), **104**(4), 41–53

[8] Ross G. F. "The Synthetic Generation of Phase-coherent Microwave Signals for Transient Behaviour Measurements". IEEE Trans. on Microwave Theory and Techniques (1965), 704–705

[9] Rhee M. J., Fine, T. A. and Kung C. C. "Basic circuits for Inductive-energy Pulsed, Power Systems". *J. Appl. Phys.* (1990), **67**(9), 4333–4337

[10] Rhee M. J. and Ding B. N. "Repetitive Square Pulse Generation by Inductive Pulse-forming Lines with a Field Effect Transistor as an Opening Switch". *Rev. Sci. Instrum.* (1993), **64**(4), 1665–1666

4

Pulse-forming Networks

4.1 INTRODUCTION

In the previous chapter, the use of lengths of transmission line to generate various types of rectangular pulse was described. Unfortunately standard transmission line, which is commercially available, usually coaxial line, is only supplied in standard impedances, 50 Ω being the most common. It is of course possible to obtain other impedances by making pulse-forming lines from stripline type lines, but this can be both expensive and difficult if the lines are to be used at relatively high potentials. A further disadvantage is the speed of propagation of electromagnetic waves along transmission lines. From Equation (3.1) it can be seen that the propagation velocity depends on the relative permittivity or dielectric constant of the material used to insulate the two conductors which make up the line. The material that is commonly used is some type of polymer plastic such as polypropylene, and the dielectric constant tends to be quite low ($\varepsilon_r = 2\text{–}3$). Thus the propagation velocity on the line is around 2×10^8 metres per second. In terms of transit time, frequently referred to in the last chapter, this is 20 cm per nanosecond. Therefore to generate a 1 μs long pulse using the single pulse-forming line described above would require 100 m of transmission line. Clearly, for longer pulses the lengths of line that are required become impractical unless stripline is used which is insulated with materials with high dielectric constants. Again, the practical consequences and cost of building such lines generally make this option unattractive.

The alternative approach is to build a simulated line using a ladder network of inductors and capacitors. In Chapter 2, the propagation characteristics of a transmission line were examined by approximating the line to a ladder network comprising an inductance, a capacitance, a resistance and a conductance, which are all defined for a given unit length. It is therefore not surprising that an artificial transmission line can be constructed from a ladder of inductors and capacitors. It is, of course, not necessary or desirable to add resistance to simulate the resistance of the conductors and the conductance between the conductors, as this would simply add unnecessary loss and cause pulse distortion. Such an artificial line is sometimes known as a delay line because of its ability to achieve substantial propagation delays along relatively short networks.

Usually the energy that is released from the line, in the generated rectangular pulse, is stored in the capacitors which make up the ladder. Such a network is known as a voltage-fed network. However it is also possible to store the energy as magnetic energy in the inductors which make up the ladder. In this case the network is called a current-fed network.

Before looking at the use of such ladder networks to generate rectangular pulses, the impedance and propagation characteristics of such networks will be examined. As in the case of pulse-forming lines there are several different types of network that can be used, and the analysis and synthesis methods for generating such networks will be described.

4.2 LC LADDER NETWORKS

4.2.1 The Impedance Characteristics of an LC Ladder Network

Figure 4.1 gives a basic circuit of a generalised ladder network in which the series elements are impedances Z_1 and the parallel elements are impedances Z_2. These impedances are kept constant at every stage of the ladder.

To find the input impedance of the ladder, a voltage source such as a battery with potential V is switched into the ladder and the input current i_1 is determined. It is assumed that the battery has zero internal impedance. The network comprises n sections and, using Kirchhoff's voltage law, the currents flowing in some abitrary intermediate kth mesh must obey the relationship

$$\begin{aligned}(i_k - i_{k-1})Z_2 + i_k Z_1 + (i_k - i_{k+1})Z_2 &= 0\\ i_{k-1}Z_2 - i_k(Z_1 + 2Z_2) + i_{k+1}Z_2 &= 0\end{aligned} \tag{4.1}$$

This is a linear constant-coefficient difference equation of the form

$$\left(a\mathbf{E}^2 - b\mathbf{E} + c\right)i_k = 0 \tag{4.2}$$

where $\mathbf{E}$ is the shift operator, $a = Z_2$, $b = Z_1 + 2Z_2$ and $c = Z_2$ [1]. This equation has a solution of general form [2]

$$\left.\begin{aligned} &i_k = A\exp(k\gamma) + B\exp(-k\gamma)\\ &\text{where } \gamma \text{ is given by}\\ &\cosh\gamma = \left(1 + \frac{Z_1}{2Z_2}\right) \quad \text{if} \quad \frac{Z_1}{4Z_2} > 0\\ &\cos\gamma = \left(1 + \frac{Z_1}{2Z_2}\right) \quad \text{if} \quad 0 > \frac{Z_1}{4Z_2} > -1\\ &-\cosh\gamma = \left(1 + \frac{Z_1}{2Z_2}\right) \quad \text{if} \quad \frac{Z_1}{4Z_2} < -1 \end{aligned}\right\} \tag{4.3}$$

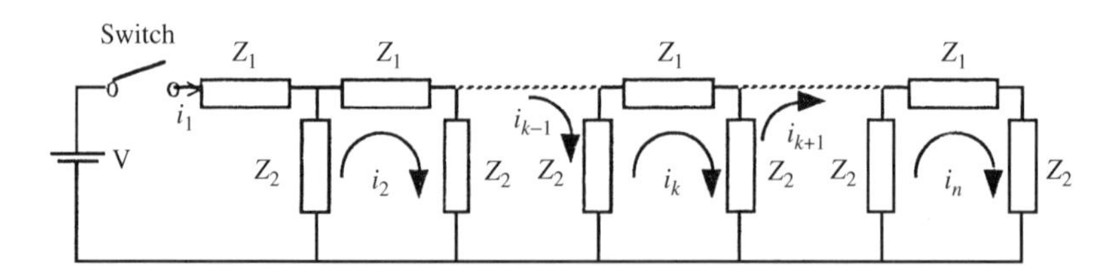

Figure 4.1 Generalised n-section ladder network

A and B are coefficients which can be determined by Kirchhoff's voltage law to the first and last (nth) mesh. From the first mesh

$$\begin{aligned} V &= i_1 Z_1 + (i_1 - i_2) Z_2 \\ &= i_1 (Z_1 + Z_2) - i_2 Z_2 \end{aligned} \tag{4.4}$$

From the last mesh

$$\left.\begin{aligned} (i_n - i_{n-1}) Z_2 + i_n Z_1 + i_n Z_2 = 0 \\ i_n (Z_1 + 2Z_2) - i_{n-1} Z_2 = 0 \end{aligned}\right\} \tag{4.5}$$

Use of Equation (4.3) with $k = 1,2$ gives

$$\left.\begin{aligned} i_1 &= A \exp(\gamma) + B \exp(-\gamma) \\ i_2 &= A \exp(2\gamma) + B \exp(-2\gamma) \end{aligned}\right\} \tag{4.6}$$

Substituting Equations (4.6) into Equation (4.4) gives

$$V = [A \exp(\gamma) + B \exp(-\gamma)](Z_1 + Z_2) - [A \exp(2\gamma) + B \exp(-2\gamma)] Z_2 \tag{4.7}$$

It is assumed that a wave propagating on the ladder would be attenuated as it propagates. In reality this is likely to be the case for most practical ladders due to loss in the reactive components making up the impedances Z_1 and Z_2. For this reason the first equation for γ, given in Equation (4.3), is used [1], i.e.

$$\left.\begin{aligned} \cosh \gamma = \frac{\exp(\gamma) + \exp(-\gamma)}{2} &= 1 + \frac{Z_1}{2Z_2} \\ \text{or} \qquad \exp(\gamma) + \exp(-\gamma) &= 2 + \frac{Z_1}{Z_2} \end{aligned}\right\} \tag{4.8}$$

Equation (4.7) can therefore be rearranged into the form

$$\frac{V}{Z_2} = A + B - A \exp(\gamma) - B \exp(-\gamma) \tag{4.9}$$

Similarly using Equation (4.3) with $k = n$, $n - 1$ gives

$$\left.\begin{aligned} i_n &= A \exp(n\gamma) + B \exp(-n\gamma) \\ i_{n-1} &= A \exp((n-1)\gamma) + B \exp(-(n-1)\gamma) \end{aligned}\right\} \tag{4.10}$$

Substituting these equations into Equation (4.5) gives

$$\begin{aligned} (A \exp(n\gamma) + B \exp(-n\gamma))(Z_1 + 2Z_2) - (A \exp((n-1)\gamma) \\ + B \exp(-(n-1)\gamma)) Z_2 = 0 \\ \text{or} \qquad A \exp((n+1)\gamma) = -B \exp(-(n+1)\gamma) \end{aligned} \tag{4.11}$$

Equations (4.9) and (4.11) can now be solved simultaneously to obtain expressions for the constants A and B.

$$\left.\begin{aligned} A &= \frac{\dfrac{V}{Z_2}\exp(-(n+1)\gamma)}{2\sinh(n\gamma) - 2\sinh((n+1)\gamma)} \\ B &= \frac{-\dfrac{V}{Z_2}\exp(+(n+1)\gamma)}{2\sinh(n\gamma) - 2\sinh((n+1)\gamma)} \end{aligned}\right\} \tag{4.12}$$

The input current to the ladder i_1 can now be determined by substituting these expressions into the first of the Equations (4.6)

$$\left.\begin{aligned} i_1 &= \frac{\dfrac{V}{Z_2}\exp(-(n+1)\gamma)\exp(\gamma)}{2\sinh(n\gamma) - 2\sinh((n+1)\gamma)} - \frac{\dfrac{V}{Z_2}\exp(+(n+1)\gamma)\exp(-\gamma)}{2\sinh(n\gamma) - 2\sinh((n+1)\gamma)} \\ &= \frac{V}{Z_2}\left(\frac{\sinh(n\gamma)}{\sinh((n+1)\gamma) - \sinh(n\gamma)}\right) \end{aligned}\right\} \tag{4.13}$$

This is a very useful expression from which the input impedance Z_{in} of an n-stage ladder network can be found to be

$$Z_{in} = \frac{V}{i_1} = Z_2\left(\frac{\sinh((n+1)\gamma)}{\sinh(n\gamma)} - 1\right) \tag{4.14}$$

The simplest type of LC ladder is shown in Figure 4.4. Putting $Z_1 = \mathbf{s}L$ and $Z_2 = /1\mathbf{s}C$, the input impedance of this ladder using Equation (4.14) will be given by

$$\left.\begin{aligned} \mathbf{Z}_{in}(n, \mathbf{s}) &= \frac{V}{\mathbf{s}\mathbf{I}_1(\mathbf{s})} = \frac{1}{\mathbf{s}C}\left(\frac{\sinh((n+1)\gamma)}{\sinh(n\gamma)} - 1\right) \\ \cosh(\gamma) &= 1 + \left(\frac{LC}{2}\right)\mathbf{s}^2 \end{aligned}\right\} \tag{4.15}$$

4.2.2 General Transform Equations for a Ladder Network

In order to determine the transient response of a ladder network to a particular transient input it is useful to determine the relationship between the voltage and current at the kth section of the ladder in terms of the input voltage and current to the ladder in a similar way to that determined for a continuous transmission line (Equation (2.41)). This can be done by dividing the ladder into a series of identical 'T' sections where the values of the series elements are halved, as shown in Figure 4.2.

By application of Kirchhoff's voltage law two difference equations can be written

$$\left.\begin{aligned} \mathbf{V}_k(\mathbf{s}) &= \mathbf{I}_k(\mathbf{s})\left(\frac{Z_1}{2} + Z_2\right) - \mathbf{I}_{k+1}(\mathbf{s})Z_2 \\ \mathbf{V}_{k+1}(\mathbf{s}) &= -\mathbf{I}_{k+1}(\mathbf{s})\left(\frac{Z_1}{2} + Z_2\right) + \mathbf{I}_k(\mathbf{s})Z_2 \end{aligned}\right\} \tag{4.16}$$

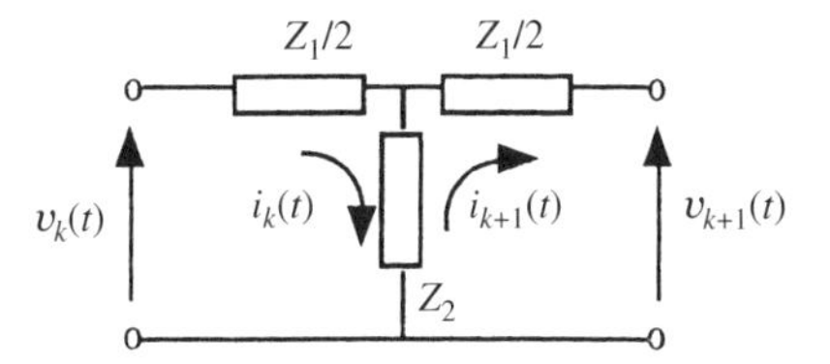

Figure 4.2 'T' section of the ladder network given in Figure 4.1

The voltages and currents have been Laplace-transformed to facilitate the solution of a given transient problem. Since this is a linear difference equation, it is now necessary to take the z-transform [3] of Equation (4.16) with

$$\left.\begin{aligned} z[\mathbf{I}_k(\mathbf{s})] &\triangleq \hat{\mathbf{I}}(z,\mathbf{s}) \\ z[\mathbf{V}_k(\mathbf{s})] &\triangleq \hat{\mathbf{V}}(z,\mathbf{s}) \end{aligned}\right\} \tag{4.17}$$

This gives

$$\begin{aligned} \hat{\mathbf{V}}(z,\mathbf{s}) &= \hat{\mathbf{I}}(z,\mathbf{s})\left(\frac{Z_1}{2}+Z_2\right) - z\hat{\mathbf{I}}(z,\mathbf{s})Z_2 + z\hat{\mathbf{I}}(0,\mathbf{s})Z_2 \\ z\hat{\mathbf{V}}(z,\mathbf{s}) - z\hat{\mathbf{V}}(0,\mathbf{s}) &= -z\hat{\mathbf{I}}(z,\mathbf{s})\left(\frac{Z_1}{2}+Z_2\right) + z\hat{\mathbf{I}}(0,\mathbf{s})\left(\frac{Z_1}{2}+Z_2\right) + \hat{\mathbf{I}}(z,\mathbf{s})Z_2 \end{aligned} \tag{4.18}$$

$\hat{\mathbf{V}}(0,\mathbf{s})$ and $\hat{\mathbf{I}}(0,\mathbf{s})$ are the input voltage and current to the ladder network. Collecting terms

$$\begin{aligned} z\hat{\mathbf{I}}(0,\mathbf{s})Z_2 &= \hat{\mathbf{V}}(z,\mathbf{s}) + \hat{\mathbf{I}}(z,\mathbf{s})\left(zZ_2 - \frac{Z_1}{2} - Z_2\right) \\ z\hat{\mathbf{I}}(0,\mathbf{s})\left(\frac{Z_1}{2}+Z_2\right) + z\hat{\mathbf{V}}(0,\mathbf{s}) &= z\hat{\mathbf{V}}(z,\mathbf{s}) + \hat{\mathbf{I}}(z,\mathbf{s})\left(z\left(\frac{Z_1}{2}+Z_2\right) - Z_2\right) \end{aligned} \tag{4.19}$$

Solving the Equations (4.19) for $\mathbf{I}(z,\mathbf{s})$ and $\mathbf{V}(z,\mathbf{s})$ gives

$$\left.\begin{aligned} \hat{\mathbf{V}}(z,\mathbf{s}) &= \frac{z\left[\hat{\mathbf{V}}(0,\mathbf{s})(z-\lambda) - \hat{\mathbf{I}}(0,\mathbf{s})Z_2(\lambda^2-1)\right]}{z^2-2\lambda z+1} \\ \hat{\mathbf{I}}(z,\mathbf{s}) &= \frac{z\left[\hat{\mathbf{I}}(0,\mathbf{s})(z-\lambda) - \hat{\mathbf{V}}(0,\mathbf{s})Z_2^{-1}\right]}{z^2-2\lambda z+1} \\ \text{where} \quad \lambda &= \left(\frac{Z_1}{2Z_2}+1\right) \end{aligned}\right\} \tag{4.20}$$

The inverse z transforms of these equations can be found by letting

$$\left.\begin{aligned} \lambda &= \cosh(\gamma) \\ \sinh\gamma &= \sqrt{\cosh^2(\gamma)-1} = \sqrt{\lambda^2-1} \end{aligned}\right\} \tag{4.21}$$

Using tables of inverse z transforms [3], this then gives

$$\left.\begin{aligned} \mathbf{V}(k,\mathbf{s}) &= \mathbf{V}(0,\mathbf{s})\cosh(k\gamma) - Z\mathbf{I}(0,\mathbf{s})\sinh(k\gamma) \\ \mathbf{I}(k,\mathbf{s}) &= \mathbf{I}(0,\mathbf{s})\cosh(k\gamma) - Y\mathbf{V}(0,\mathbf{s})\sinh(k\gamma) \\ \text{where} \quad Z &= Y^{-1} = Z_2\sqrt{\lambda^2 - 1} \end{aligned}\right\} \tag{4.22}$$

These equations, not surprisingly, are very similar to the equations relating the current and voltage at the input of a finite length of transmission line to the current and voltage at the output end of the line (Equation (2.41)). Z is the iterative impedance function and Y is the iterative admittance function. These equations are the general transform equations for a ladder network comprising a series of 'T' sections, in which the initial current and voltage on the line is zero and can be used to determine the voltage and current waveforms at the kth section of the ladder network for any given input. The equations can also be used to determine the input impedance of a ladder network of arbitrary length which is either left unconnected at its output end (i.e. it is connected to an open-circuit) or it is shorted at some arbitrary point in the ladder.

4.2.3 Input Impedance Functions of Open Circuit and Short Circuit Ladder Networks

If the ladder network discussed above is left unconnected to a load, at its output end, then no current can flow. If the ladder has k 'T' sections then from Equations (4.22)

$$\mathbf{I}(k,\mathbf{s}) = 0 = \mathbf{I}(0,\mathbf{s})\cosh(k\gamma) - Y\mathbf{V}(0,\mathbf{s})\sinh(k\gamma) \tag{4.23}$$

The input impedance is then

$$Z_{in}(\mathrm{oc}) = \frac{\mathbf{V}(0,\mathbf{s})}{\mathbf{I}(0,\mathbf{s})} = \frac{\cosh(\mathrm{k}\gamma)}{\mathrm{Y}\sinh(\mathrm{k}\gamma)} = \mathrm{Z}\coth(\mathrm{k}\gamma) \tag{4.24}$$

Similarly if the laddder is short circuited at the kth section then

$$\mathbf{V}(k,\mathbf{s}) = 0 = \mathbf{V}(0,\mathbf{s})\cosh(k\gamma) - Z\,\mathbf{I}(0,\mathbf{s})\sinh(k\gamma) \tag{4.25}$$

The input impedance now becomes

$$Z_{in}(\mathrm{sc}) = \frac{\mathbf{V}(0,\mathbf{s})}{\mathbf{I}(0,\mathbf{s})} = \frac{\mathrm{Z}\sinh(\mathrm{k}\gamma)}{\cosh(\mathrm{k}\gamma)} = \mathrm{Z}\tanh(\mathrm{k}\gamma) \tag{4.26}$$

Equations (4.24) and (4.26) are again found to be similar to the equations derived for the input impedances of finite lengths of open and short circuited transmission lines (Equation (2.46)).

It may appear initially that Equations (4.14) and (4.24) are not consistent, as they both give the input impedance of similar ladder networks which are effectively open-circuited at the final stage or output end of the ladder. However it should be recalled that in the derivation of Equations (4.24) the ladder analysed comprised a string of 'T' sections as shown in Figure 4.2. Thus the first series element in the ladder is an impedance $Z_1/2$. However, by looking at the ladder (Figure 4.1) from which Equation (4.14) is derived it can be seen that the first series element is an impedance Z_1. Thus subtraction of an impedance $Z_1/2$ from Equation (4.14) should, after rearrangement, yield Equation (4.24), i.e.

$$
\begin{aligned}
\mathbf{Z}_{in}(n, \mathbf{s}) - \frac{Z_1}{2} &= Z_2\left(\frac{\sinh((n+1)\gamma)}{\sinh(n\gamma)} - 1\right) - \frac{Z_1}{2} \\
&= Z_2\left(\frac{\sinh(n\gamma)\cosh(\gamma) + \cosh(n\gamma)\sinh(\gamma)}{\sinh(n\gamma)} - 1\right) - \frac{Z_1}{2} \\
&= Z_2(\cosh(\gamma) + \sinh(\gamma)\coth(n\gamma) - 1) - \frac{Z_1}{2} \\
&= Z_2\left(\left(\frac{Z_1}{2Z_2} + 1\right) + \sinh(\gamma)\coth(n\gamma) - 1\right) - \frac{Z_1}{2} \\
&= Z_2 \sinh(\gamma)\coth(n\gamma) \\
&= Z\coth(n\gamma)
\end{aligned}
\tag{4.27}
$$

Thus Equations (4.14) and (4.24) are consistent if $n = k$.

It is informative to use Equation (4.24) to find the input impedance of an LC ladder network of the type shown in Figure 4.4 as k approaches infinity while keeping the total inductance L_N and capacitance C_N of the network constant. As before, by putting $Z_1 = \mathbf{s}L$ and $Z_2 = 1/\mathbf{s}C$ the open circuit input impedance of the ladder using Equation (4.24) can be found to be

$$
\begin{aligned}
\mathbf{Z}_{in}(\mathrm{oc})(\mathrm{k}, \mathbf{s}) &= \mathbf{Z}\coth(k\gamma) \\
&= \frac{1}{\mathbf{s}C}\sinh(\gamma)\coth(k\gamma)
\end{aligned}
\tag{4.28}
$$

From the definition of $\cosh(\gamma)$

$$
\left.
\begin{aligned}
\cosh(\gamma) &= 1 + \frac{\mathbf{s}^2 LC}{2} = 1 + \frac{\mathbf{s}^2 L_N C_N}{2k^2} \\
\sinh(\gamma) &= \sqrt{\cosh^2(\gamma) - 1} = \frac{\mathbf{s}\sqrt{L_N C_N}}{k}\sqrt{1 + \frac{\mathbf{s}^2 L_N C_N}{4k^2}} \\
(k\gamma) &= 2k \sinh^{-1}\frac{\mathbf{s}\sqrt{L_N C_N}}{2k}
\end{aligned}
\right\}
\tag{4.29}
$$

The required limit is then given by

$$
\begin{aligned}
\lim_{k\to\infty} \mathbf{Z}_{in}(\mathrm{oc})(\mathrm{k}, \mathbf{s}) &= \lim_{k\to\infty} \frac{1}{\mathbf{s}C}\sinh(\gamma)\coth(k\gamma) \\
&= \lim_{k\to\infty} \frac{k}{\mathbf{s}C_N}\left[\frac{\mathbf{s}\sqrt{L_N C_N}}{k}\sqrt{1 + \frac{\mathbf{s}^2 L_N C_N}{4k^2}}\coth\left(2k\sinh^{-1}\frac{\mathbf{s}\sqrt{L_N C_N}}{2k}\right)\right] \\
&= \sqrt{\frac{L_N}{C_N}}\coth(\mathbf{s}\delta) = Z_N\coth(\mathbf{s}\delta) \quad \text{with} \quad \delta = \sqrt{L_N C_N}
\end{aligned}
\tag{4.30}
$$

This equation has exactly the same form as the expression derived for the input impedance of a finite length of open-circuited transmission line (Equation (2.97)), except that here δ is the transit time along the complete network and Z_N is the characteristic network impedance. Therefore, not surprisingly, the input impedance of the ladder reduces, in the limit, to the equivalent transmission line impedance function. A similar approach [4] can be used to examine the impedance function given in Equation (4.15) in the limit $n \to \infty$. Exactly the same expression as that given in the final equation of Equation (4.30) results.

4.2.4 Propagation Characteristics of an LC Ladder Network

In this section the response of an infinite LC ladder network, of the type shown in Figure 4.4, to a unit step input voltage will be investigated. This is important in the study of pulse-forming networks in that it allows the shape of pulses generated by pulse-forming networks to be calculated. In particular, it enables the shape of the tail or falling edge of the pulse to be predicted. Assuming that the network is not connected to a load, i.e. it is open circuited and it has n sections, where $n \to \infty$, then from Equation (4.22)

$$\mathbf{I}(n, \mathbf{s}) = 0 = \mathbf{I}(0, \mathbf{s}) \cosh(n\gamma) - Y\mathbf{V}(0, \mathbf{s}) \sinh(n\gamma)$$

$$\text{or} \qquad \mathbf{I}(0, \mathbf{s}) = \frac{\mathbf{V}(0, \mathbf{s})}{Z} \frac{\sinh(n\gamma)}{\cosh(n\gamma)} \tag{4.31}$$

Substituting this expression into the first of the equations in Equations (4.22) gives

$$\begin{aligned} \mathbf{V}(k, \mathbf{s}) &= \mathbf{V}(0, \mathbf{s}) \cosh(k\gamma) - \mathbf{V}(0, \mathbf{s}) \frac{\sinh(n\gamma) \sinh(k\gamma)}{\cosh(n\gamma)} \\ &= \mathbf{V}(0, \mathbf{s}) \frac{\cosh((k - n)\gamma)}{\cosh(n\gamma)} \end{aligned} \tag{4.32}$$

Writing this equation in exponential form and letting $n \to \infty$ gives

$$\begin{aligned} \mathbf{V}(k, \mathbf{s}) &= \lim_{n \to \infty} \mathbf{V}(0, \mathbf{s}) \frac{\exp((k - n)\gamma) + \exp(-(k - n)\gamma)}{\exp(n\gamma) + \exp(-n\gamma)} \\ &= \lim_{n \to \infty} \mathbf{V}(0, \mathbf{s}) \frac{\exp(-k\gamma) + \exp(-(k - 2n)\gamma)}{\exp(2n\gamma) + 1} \\ &= \mathbf{V}(0, \mathbf{s}) \exp(-k\gamma) \end{aligned} \tag{4.33}$$

This equation allows the determination of the voltage **V**(k,s) or $v(k,t)$ at the kth section of an infinite arbitrary ladder for a given input voltage **V**(0,s). From Equation (4.33) and using Equation (4.21)

$$\begin{aligned} \mathbf{V}(k, \mathbf{s}) &= \mathbf{V}(0, \mathbf{s}) \exp(-k\gamma) = \mathbf{V}(0, \mathbf{s})(\sinh(\gamma) + \cosh(\gamma))^{-k} \\ &= \frac{\mathbf{V}(0, \mathbf{s})}{\left(\sqrt{\lambda^2 - 1} + \lambda\right)^k} \end{aligned} \tag{4.34}$$

For the LC ladder and the definition of λ given in Equation (4.20)

$$\begin{aligned}\lambda &= \left(\frac{Z_1}{2Z_2}+1\right)\\ &= \left(\frac{\mathbf{s}^2LC}{2}+1\right) \qquad \text{as} \qquad Z_1=\mathbf{s}L \quad \text{and} \quad Z_2=\frac{1}{\mathbf{s}C}\\ &= \frac{LC}{2}\left(\mathbf{s}^2+\frac{2}{LC}\right)\end{aligned} \tag{4.35}$$

This can be written in a different form by introducing a new constant a given by

$$a \triangleq \frac{2}{\sqrt{LC}}$$

$$\text{i.e.} \qquad \lambda = \frac{2}{a^2}\left(\mathbf{s}^2+\frac{a^2}{2}\right) \tag{4.36}$$

Equation (4.34) then becomes

$$\begin{aligned}\mathbf{V}(k,\mathbf{s}) &= \frac{\mathbf{V}(0,\mathbf{s})}{\left(\sqrt{\lambda^2-1}+\lambda\right)^k}\\ &= \frac{\mathbf{V}(0,\mathbf{s})}{\left\{\frac{2}{a^2}\left[\left(\mathbf{s}^2+\frac{a^2}{2}\right)+\mathbf{s}\sqrt{\mathbf{s}^2+a^2}\right]\right\}^k}\\ &= \frac{\mathbf{V}(0,\mathbf{s})a^{2k}}{\left(\sqrt{\mathbf{s}^2+a^2}+\mathbf{s}\right)^{2k}}\end{aligned} \tag{4.37}$$

If a step voltage of amplitude V is injected into the ladder

$$\left.\begin{aligned}\mathbf{V}(0,\mathbf{s}) &= \frac{V}{\mathbf{s}}\\ \mathbf{V}(k,\mathbf{s}) &= \frac{Va^{2k}}{\mathbf{s}\left(\sqrt{\mathbf{s}^2+a^2}+\mathbf{s}\right)^{2k}}\end{aligned}\right\} \tag{4.38}$$

To find the way in which the step changes shape as it propagates, the inverse of this transform must be taken. This can be done by reference to tables of Laplace transforms [5] from which $v(k,t)$ is found to be

$$v(k,t) = 2kV\int_0^t \frac{J_{2k}(a\tau)}{\tau}\cdot d\tau, \qquad t\geq 0 \tag{4.39}$$

$J_{2k}(a\tau)$ is a Bessel function of order $2k$. As an example of the way in which a step changes, calculations have been performed on a ladder network with $L=1.25$ μH and $C=500$ pF. This gives a line impedance of 50 Ω and a time delay per section of 25 ns. Waveforms are drawn in Figure 4.3 at the 5th, 10th and 20th section to illustrate the way in which an input

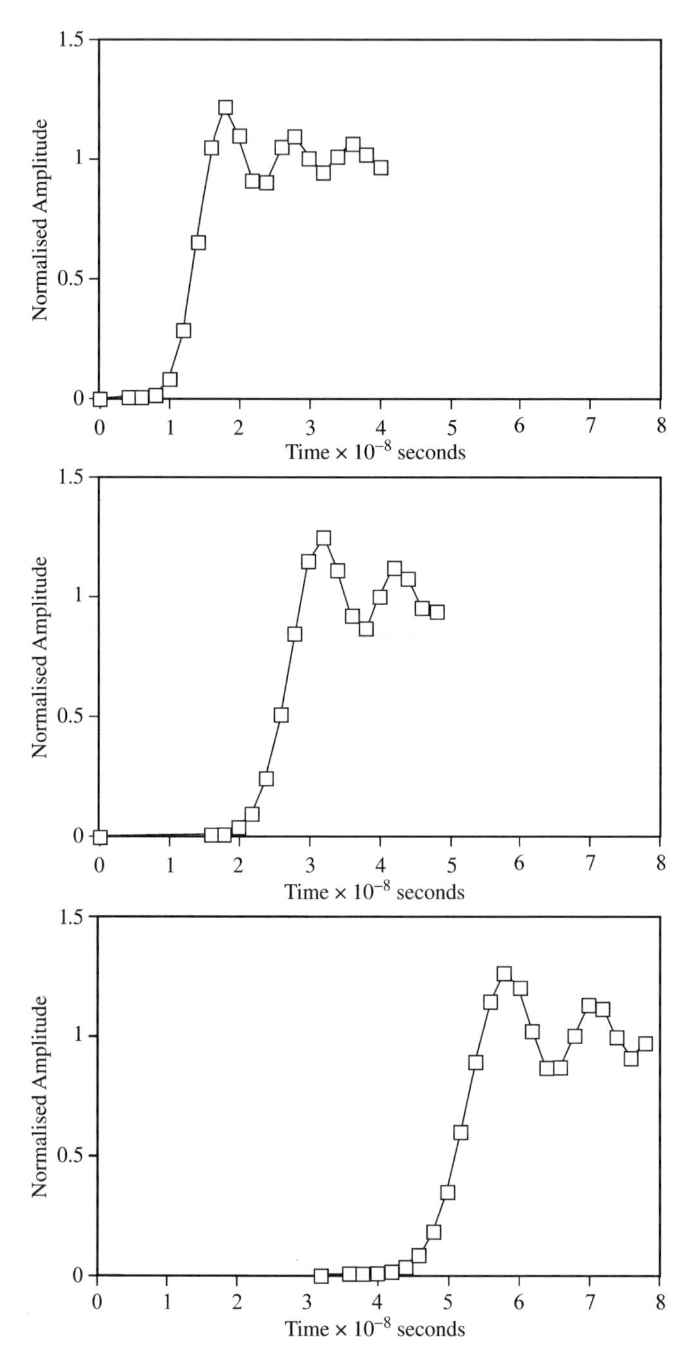

Figure 4.3 Propagation of a step of unit amplitude on a uniform LC ladder network

step changes shape on the ladder. It can be seen that as the step propagates its rise time slowly decreases while the amplitude of the oscillations produced on the leading edge remain roughly constant. Note that Equation (4.39) automatically gives a delay to the propagating pulse as can be seen from the plots given in Figure 4.3.

4.3 PULSE-FORMING ACTION OF AN LC LADDER NETWORK

Having investigated some of the basic properties of ladder networks, including *LC* ladder networks, it is now possible to describe the way in which an *LC* ladder network can be used to generate a rectangular pulse in a similar way to that using a length of open-circuit transmission line. The circuit for the basic *LC* ladder pulse-forming network is given in Figure 4.4.

The ladder has n *LC* sections and in its most common usage the capacitors are charged to a potential V and then discharged into the load via a switch. To analyse the behaviour of the circuit, when it is charged and switched on to a load, the network can be replaced by a voltage source $V/\mathbf{s}$ and an impedance $\mathbf{Z}(n,\mathbf{s})$, where $\mathbf{Z}(n,\mathbf{s})$ is defined by Equation (4.15). However to simplify the analysis, initially, it will be assumed that the ladder is infinitely long, i.e. $n \to \infty$ and the total network inductance L_N and capacitance C_N are kept constant. In this case the form of $\mathbf{Z}(n,\mathbf{s})$ given in Equation (4.30) can be used, i.e.

$$\mathbf{Z}(n,\mathbf{s}) = Z_N \coth(\mathbf{s}\delta) \tag{4.40}$$

The potential on the load is therefore given by

$$\begin{aligned} \mathbf{V}(\mathbf{s}) &= \frac{V}{\mathbf{s}} \frac{Z_l}{Z_l + Z_N \coth(\mathbf{s}\delta)} \\ &= \frac{V}{\mathbf{s}} \frac{Z_l}{Z_l + \sqrt{\dfrac{L_N}{C_N}} \coth(\mathbf{s}\sqrt{L_N C_N})} \end{aligned} \tag{4.41}$$

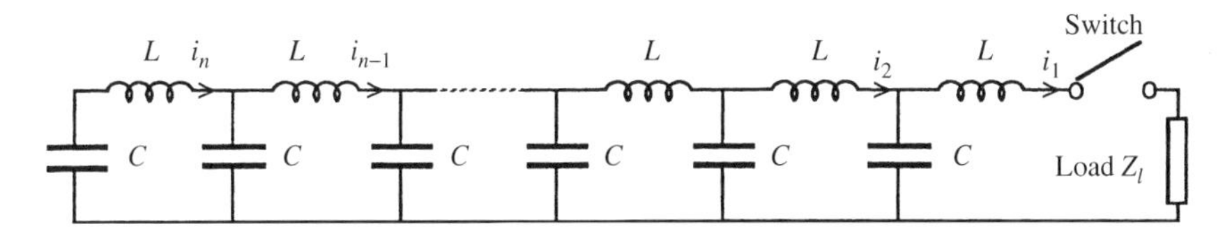

Basic LC pulse forming network

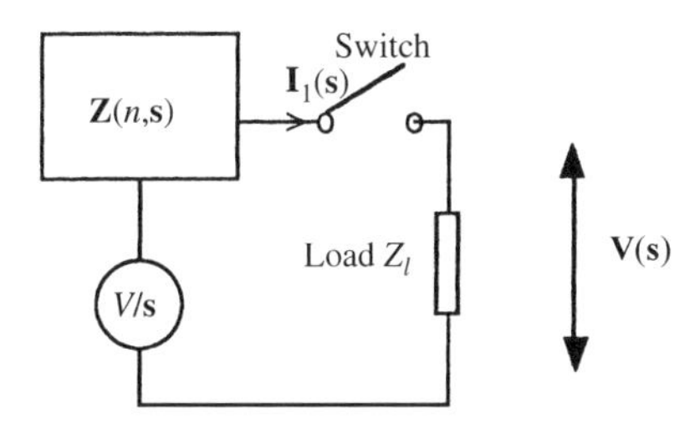

Equivalent circuit

Figure 4.4 Basic and equivalent circuit for an LC ladder pulse-forming network

If the load impedance Z_l is set equal to the network impedance Z_N then the potential on the load becomes

$$\mathbf{V}(\mathbf{s}) = \frac{V}{2}\frac{\left[1 - \exp(-2\mathbf{s}\sqrt{L_N C_N}\right]}{\mathbf{s}} \tag{4.42}$$

which is recognised as the Laplace transform of a rectangular pulse of amplitude $V/2$ and duration $2\sqrt{L_N C_N}$. This analysis is, of course, very similar to the analysis of pulse-forming action using a charged length of open-circuit transmission line. If the load is not matched to the line then exactly the same waveforms are generated as for the pulse-forming line given in Figure 3.5. This approximate analysis can be used for ladders where $n > 10$, noting that

$$\left.\begin{aligned} Z_N &= \sqrt{\frac{L_N}{C_N}} = \sqrt{\frac{L}{C}} \\ \text{and the pulse duration} \qquad t_p &= 2\delta = 2\sqrt{L_N C_N} = 2n\sqrt{LC} \end{aligned}\right\} \tag{4.43}$$

although it will not enable the detailed shape of the pulse to be predicted. This can only be done if the full form of $Z(n,\mathbf{s})$ is used. In this case the Laplace transform of the generated pulse is given by

$$\mathbf{V}(\mathbf{s}) = \frac{V}{\mathbf{s}}\frac{Z_l}{Z_l + \dfrac{1}{\mathbf{s}C}\left(\dfrac{\sinh((n+1)\gamma)}{\sinh(n\gamma)} - 1\right)} \tag{4.44}$$

In most cases this proves to be a rather difficult, if not impossible, expression to invert. However it can be used to investigate the rise-time from a given pulse forming network into a matched load, i.e. $Z_l{=}Z_N{=}\sqrt{LC}$. This is done by finding the limit of $\mathbf{Z}(n,\mathbf{s})$ as $n \to \infty$ so that the added complexity which results from the step, that propagates back into the line and is reflected at its end, is removed. However in this case it is not assumed that the total network inductance and capacitance remains constant. Thus from Equation (4.44)

$$\begin{aligned} \mathbf{V}(\mathbf{s}) &= \lim_{n\to\infty}\frac{V}{\mathbf{s}}\frac{\sqrt{\dfrac{L}{C}}}{\sqrt{\dfrac{L}{C}} + \dfrac{1}{\mathbf{s}C}\left(\dfrac{\sinh((n+1)\gamma)}{\sinh(n\gamma)} - 1\right)} \\ &= \lim_{n\to\infty}\frac{V}{\mathbf{s}}\frac{\sinh(n\gamma)}{\sinh(n\gamma) + \dfrac{1}{\mathbf{s}\sqrt{LC}}[\sinh((n+1)\gamma) - \sinh(n\gamma)]} \end{aligned} \tag{4.45}$$

Converting the sinh terms into their exponential form and rearranging gives

$$\begin{aligned} \mathbf{V}(\mathbf{s}) &= \lim_{n\to\infty}\frac{V}{\mathbf{s}}\frac{\exp(n\gamma) - \exp(-n\gamma)}{\begin{array}{c}\exp(n\gamma) - \exp(-n\gamma) + \dfrac{1}{\mathbf{s}\sqrt{LC}}\times \\ {[\exp((n+1)\gamma) - \exp(-(n+1)\gamma) - \exp(n\gamma) + \exp-(n\gamma)]}\end{array}} \\ &= \frac{V}{\mathbf{s}}\frac{1}{1 + \dfrac{1}{\mathbf{s}\sqrt{LC}}(\exp(\gamma) - 1)} \end{aligned} \tag{4.46}$$

Further rearrangement of this expression is now required in order to proceed to a transform which can be inverted, i.e.

$$\begin{aligned}\mathbf{V(s)} &= \frac{V}{\mathbf{s}}\frac{1}{1+\dfrac{1}{\mathbf{s}\sqrt{LC}}(\sinh(\gamma)+\cosh(\gamma)-1)}\\ &= \frac{V}{\mathbf{s}}\frac{1}{1+\dfrac{1}{\mathbf{s}\sqrt{LC}}\left(\sqrt{\lambda^2-1}+\lambda-1\right)}\end{aligned} \tag{4.47}$$

where $\lambda = \cosh(\gamma) = \left(1+\dfrac{Z_1}{2Z_2}\right) = \left(1+\dfrac{\mathbf{s}^2LC}{2}\right)$

As in the previous section, a constant a is introduced where a is given by

$$a \triangleq \frac{2}{\sqrt{LC}} \tag{4.48}$$

Hence Equation (4.47) is converted to

$$\begin{aligned}\mathbf{V(s)} &= \frac{V}{\mathbf{s}}\frac{1}{1+\dfrac{a}{2\mathbf{s}}\left[\dfrac{2\mathbf{s}}{a^2}\sqrt{\mathbf{s}^2+a^2}+\dfrac{2}{a^2}\left(\mathbf{s}^2+\dfrac{a^2}{2}\right)-1\right]}\\ &= \frac{V}{\mathbf{s}}\frac{a}{a+\sqrt{\mathbf{s}^2+a^2}+\mathbf{s}}\end{aligned} \tag{4.49}$$

In order to find a transform which can be inverted, the denominator of the last equation is expanded using the binomial expansion to give

$$\mathbf{V(s)} = \frac{V}{\mathbf{s}}\left[\begin{aligned}&\frac{a}{\sqrt{\mathbf{s}^2+a^2}+\mathbf{s}} - \frac{a^2}{\left(\sqrt{\mathbf{s}^2+a^2}+\mathbf{s}\right)^2}\\ &+\frac{a^3}{\left(\sqrt{\mathbf{s}^2+a^2}+\mathbf{s}\right)^3} - \frac{a^4}{\left(\sqrt{\mathbf{s}^2+a^2}+\mathbf{s}\right)^4}\\ &+\frac{a^5}{\left(\sqrt{\mathbf{s}^2+a^2}+\mathbf{s}\right)^5} - \cdots\end{aligned}\right] \tag{4.50}$$

This transform can now be inverted by reference to tables of Laplace transforms [5] to give

$$v(t) = V\left[\begin{aligned}&\int_0^t \frac{J_1(a\tau)}{\tau}\cdot d\tau - 2\int_0^t \frac{J_2(a\tau)}{\tau}\cdot d\tau\\ &+3\int_0^t \frac{J_3(a\tau)}{\tau}\cdot d\tau - 4\int_0^t \frac{J_4(a\tau)}{\tau}\cdot d\tau\\ &+5\int_0^t \frac{J_5(a\tau)}{\tau}\cdot d\tau - \cdots\end{aligned}\right] \tag{4.51}$$

In order to illustrate the use of this result, the front edge of the pulse predicted by Equation (4.51) is compared to that which is calculated by the circuit analysis programme

MICROCAP [6] for a 10-stage pulse-forming network with $L = 1.25\ \mu H$ and $C = 500$ pF. Such a network would have an impedance of 50 Ω and would generate a pulse 500 ns long. The network is discharged into a matched load. Looking at the two waveforms, it can be seen that there is an almost perfect match. However it should be noted that Equation (4.51) will never converge. This, in fact, is not a problem as each term in the series builds up the pulse sequentially from the rising edge onwards. Provided that the rising edge and early parts of the structure of the pulse are of interest, then Equation (4.51) will give accurate results. However the build up is slow and the first ten terms of the equation have to be used in order to get such a good match. It is also interesting to compare the rise-time of the front edge of the pulse (35 ns) to that which would be calculated for just one section, i.e. the rise-time of the sinusoidal waveform that would be produced if a 500 pF capacitor was discharged through a 1.25 μH inductor into a 50 Ω load. This is calculated to be 39 ns, so estimating the rise-time of the pulse produced by a given pulse-forming network, by this simple calculation, gives an overestimate of the real rise-time.

An alternative rearrangement of Equation (4.49) leads to a transform which can be inverted to give an alternative solution which must also be solved numerically. However the solution is computationally more efficient.

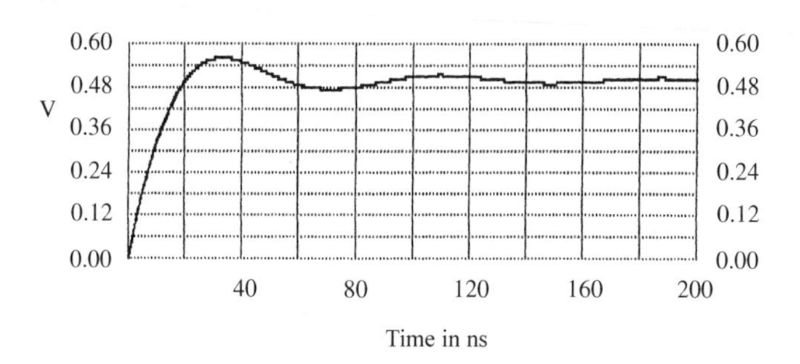

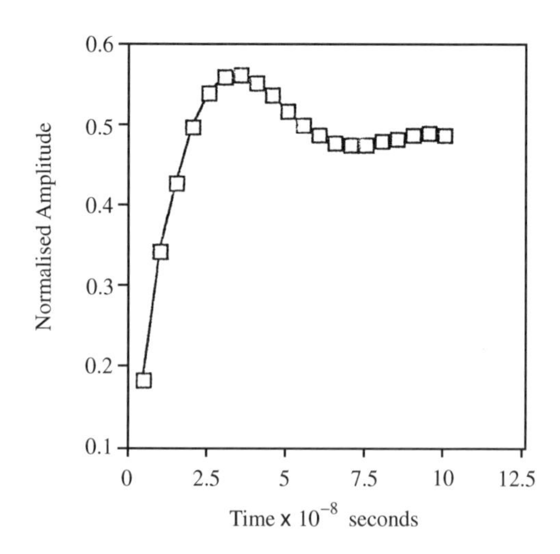

Figure 4.5 Comparison waveforms showing the front edge of a pulse predicted by Equation (4.50) and calculated using MICROCAP

From Equation (4.49)

$$\begin{aligned}\mathbf{V}(\mathbf{s}) &= \frac{Vb}{(\mathbf{s}^2 + \mathbf{s}a) + \mathbf{s}\sqrt{\mathbf{s}^2 + a^2}} \\ &= \frac{Va\left((\mathbf{s}^2 + \mathbf{s}a) - \mathbf{s}\sqrt{\mathbf{s}^2 + a^2}\right)}{2\mathbf{s}^3 a} \\ &= V\left[\frac{1}{2\mathbf{s}} + \frac{a}{2\mathbf{s}^2} - \frac{\sqrt{\mathbf{s}^2 + a^2}}{2\mathbf{s}^2}\right] \\ &= V\left[\frac{1}{2\mathbf{s}} + \frac{a}{2\mathbf{s}^2} - \frac{1}{2\sqrt{\mathbf{s}^2 + a^2}} - \frac{a^2}{2\mathbf{s}^2\sqrt{\mathbf{s}^2 + a^2}}\right]\end{aligned} \tag{4.52}$$

This can be inverted by reference to tables of Laplace Transforms [5] to give

$$v(t) = V\left[\frac{1}{2} + \frac{at}{2} - J_0(at) - \frac{a^2}{2}\int_0^t (t-u)J_0(au)\,.du\right] \tag{4.53}$$

Unfortunately the integral of the function J_0 is not known in closed form, so this Equation must be solved numerically.

4.4 THE SYNTHESIS OF ALTERNATIVE LC PULSE-FORMING NETWORKS

In the preceding sections the behaviour of a basic *LC* ladder network of the type shown in Figure 4.4 has been examined. It is possible to synthesise other types of ladder network which also produce rectangular pulses using Equation (4.40). For readers unfamiliar with the technique of network synthesis from impedance functions, a good description of the method is to be found in the book by Kuo [7]. The rational-fraction expansion of Equation (4.40) [8] is given by

$$\mathbf{Z}(n, \mathbf{s}) = Z_N \coth(\mathbf{s}\delta) = \frac{Z_N}{\mathbf{s}\delta} + \sum_{n=1}^{\infty} \frac{\dfrac{2Z_N\mathbf{s}\delta}{\pi^2 n^2}}{\dfrac{\mathbf{s}^2\delta^2}{\pi^2 n^2} + 1} \tag{4.54}$$

The first term can be rearranged as follows

$$\frac{Z_N}{\mathbf{s}\delta} = \frac{\sqrt{\dfrac{L_N}{C_N}}}{\mathbf{s}\sqrt{L_N C_N}} = \frac{1}{\mathbf{s}C_N} \tag{4.55}$$

and therefore is equivalent to a capacitor C_N, where C_N is defined above. The impedance of the parallel combination of an arbitrary inductance L_n and capacitor C_n is given by

$$Z_n = \frac{\mathbf{s}L_n \dfrac{1}{\mathbf{s}C_n}}{\mathbf{s}L_n + \dfrac{1}{\mathbf{s}C_n}} = \frac{\mathbf{s}L_n}{\mathbf{s}^2 L_n C_n + 1} \tag{4.56}$$

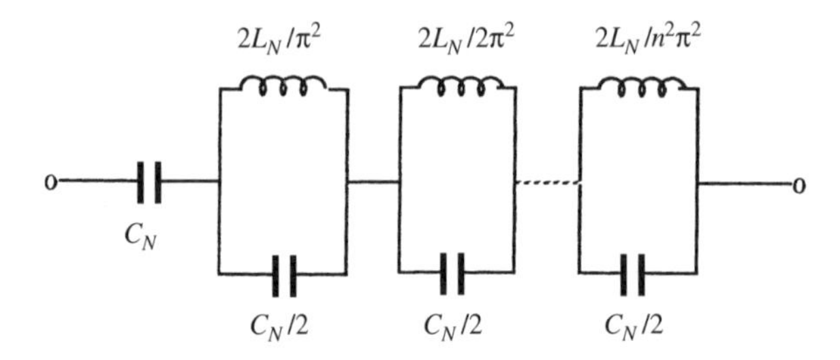

Figure 4.6 Equivalent pulse-forming network to that given in Figure 4.4

Comparison of this expression with the coefficients of the other terms in Equation (4.54) suggests that

$$\left.\begin{aligned} L_n &\equiv \frac{2Z_N\delta}{\pi^2 n^2} = \frac{2L_N}{\pi^2 n^2} \\ C_n &\equiv \frac{\delta}{2Z_N} = \frac{C_N}{2} \end{aligned}\right\} \tag{4.57}$$

Therefore another LC network can be drawn which is entirely equivalent to the LC ladder network shown in Figure 4.4. This network is given in Figure 4.6.

A second equivalent network can be found by making a rational fraction expansion [8] of the admittance equivalent to Equation (4.40), i.e.

$$\mathbf{Z}(n, \mathbf{s}) = \frac{1}{Z_N}\tanh(\mathbf{s}\delta) = \sum_{n=1}^{\infty} \frac{\dfrac{8\mathbf{s}\delta}{\pi^2 Z_N}\dfrac{\mathbf{s}}{(2n-1)^2}}{\dfrac{4\mathbf{s}^2\delta^2}{\pi^2(2n-1)^2} + 1} \tag{4.58}$$

The impedance of the series combination of an arbitrary inductance L_n and capacitor C_n is given by

$$Y_n = \frac{1}{\mathbf{s}L_n + \dfrac{1}{\mathbf{s}C_n}} = \frac{\mathbf{s}C_n}{\mathbf{s}^2 L_n C_n + 1} \tag{4.59}$$

Again, comparison with the coefficients of the terms given in Equation (4.58) suggests that

$$\left.\begin{aligned} C_n &\equiv \frac{8\delta}{(2n-1)^2\pi^2 Z_N} = \frac{8C_N}{(2n-1)^2\pi^2} \\ L_n &\equiv \frac{Z_N\delta}{2} = \frac{L_N}{2} \end{aligned}\right\} \tag{4.60}$$

Therefore a second equivalent network can be drawn as shown in Figure 4.7.

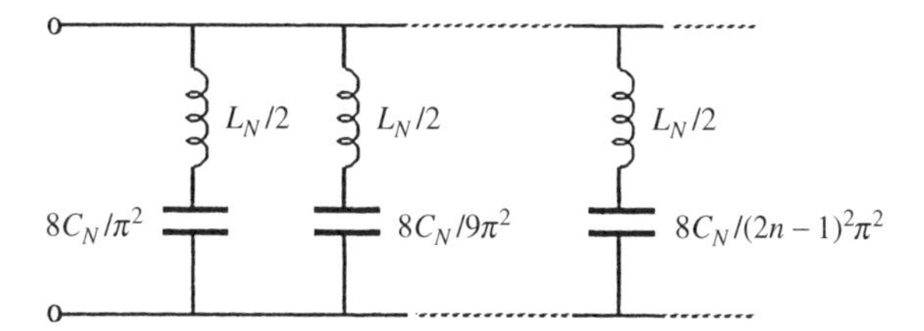

Figure 4.7 Second equivalent pulse-forming network to that given in Figure 4.4

Although the networks shown in Figures 4.6 and 4.7 are derived from the impedance function of an LC ladder network as given by Equation (4.40), closer inspection of the circuits reveals that they have limitations for practical use. For example, the circuit given in Figure 4.6 contains a number of capacitors which are connected in series. Clearly, when this network is switched on to a matched load there will be an instantaneous jump in the current to twice the matched load current. An analysis of the behaviour of the circuit in Figure 4.8, for a limited number of sections [4], reveals that there will also be a significant overshoot in potential when the charged network is discharged on to a matched load.

It has been observed that the basic LC ladder network, from which these circuits have been derived, also produces an overshoot on the leading edge of the rectangular pulse that it is designed to produce. It is clear from the integrated Bessel function

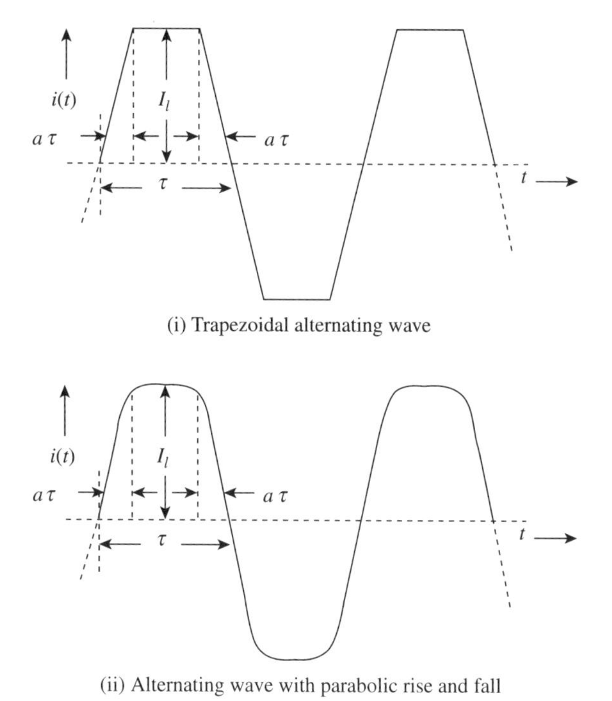

Figure 4.8 The two wave shapes analysed by Guillemin's method

series given in Equation (4.51) and the simulation results given in Figure 4.5 that an overshoot of this type will always be present for this type of pulse-forming network. As well as the overshoot, oscillations on the top of the pulse are also observed following the overshoot.

For many practical applications of this circuit such overshoots and oscillations are very undesirable. Indeed for many practical applications it is necessary to keep the top of the rectangular pulse flat so that fluctuations are kept to within 1% of the full amplitude of the pulse. An alternative explanation for this overshoot [9] is that it is caused by the attempt to simulate a discontinuity, i.e. infinitely fast rising and falling edges in a rectangular pulse, by a network which comprises a series of lumped elements. This effect is known as the Gibbs phenomenon and is caused by the inability of a Fourier series to converge in the neighbourhood of a discontinuity. On closer examination of the circuit given in Figure 4.7 it is apparent that each of the series *LC* sections will discharge into the load sinusoidally and give rise to a Fourier type component. These separate components then sum to give an approximately rectangular pulse component.

There are a number of ways to get over this problem, but perhaps the most well known is the pulse-forming network synthesis method developed by E. A. Guillemin [10].

4.4.1 Guillemin's Method

Unfortunately the original reference to Guillemin's work [10] is rather difficult to access and the only detailed account is given in the book by Glasoe and Lebacqz [4]. What follows, therefore, in this section is an abbreviated description of the method based on [4]. Guillemin argued that if the problems of overshoot and oscillation described above are due to the Gibbs phenomenon then the problem might be removed if pulses with defined rise and fall times were used to synthesise pulse-forming networks. This then removes the problem of the discontinuities at the start and end of the pulse. Such defined pulse shapes are used theoretically to generate a corresponding impedance function and from this impedance function suitable networks are synthesised. Unfortunately this proves to be a very difficult if not impossible problem to solve, and an approximate synthesis method based on Fourier analysis has to be used. It is rather surprising that the Fourier method, normally used for the solution of continuous signal analysis problems in electrical engineering, should be an effective method for solving a transient problem where Laplace transforms are normally used. It does give quite reasonable results for approximately rectangular pulses, but for more complex pulse shapes the limitation of the method becomes more apparent.

The method begins by defining the shape of the pulse for which a network is to be synthesised. Two pulse shapes originally discussed by Guillemin are the trapezoidal wave and a wave with parabolic rising and falling edges, as shown in Figure 4.8.

Waveform (i): The trapezoidal wave

The waveform is decomposed into a standard Fourier series. For readers unfamiliar with the method there are many basic mathematical texts which describe the process. A particularly good and relevant description is given in the book by Stephenson [11]. The shape of the current waveform $i(t)$ shown is odd, therefore the Fourier series contains only sine terms and being symmetrical there is no constant term. Therefore the Fourier series is given by

$$\begin{aligned} i(t) &= I_l \sum_{n=1}^{\infty} b_n \sin\frac{n\pi t}{\tau} \\ \text{where} \quad b_n &= \frac{2}{\tau}\int_0^{\tau} \frac{i(t)}{I_l} \sin\frac{n\pi t}{\tau}\,dt \end{aligned} \tag{4.61}$$

and the current waveform $i(t)$ is described by

$$\left.\begin{aligned} \frac{i(t)}{I_L} &= \frac{t}{a\tau}, && 0 \le t \le a\tau \\ \frac{i(t)}{I_L} &= 1, && a\tau \le t \le \tau - a\tau \\ \frac{i(t)}{I_L} &= \frac{\tau - t}{a\tau}, && \tau - a\tau \le t \le \tau \end{aligned}\right\} \tag{4.62}$$

Carrying out the integration for the coefficients b_n then yields

$$b_n = \frac{4}{n\pi}\frac{\sin n\pi a}{n\pi a}, \qquad \text{where} \quad n = 1, 3, 5 \ldots \tag{4.63}$$

Waveform (ii): Waveform with parabolic rise and fall

Again the waveform is odd and the Fourier series will contain sine terms only, i.e.

$$\begin{aligned} i(t) &= I_l \sum_{n=1}^{\infty} b_n \sin\frac{n\pi t}{\tau} \\ \text{where} \quad b_n &= \frac{2}{\tau}\int_0^{\tau} \frac{i(t)}{I_l} \sin\frac{n\pi t}{\tau}\,dt \end{aligned} \tag{4.64}$$

and the current waveform is now described by

$$\left.\begin{aligned} \frac{i(t)}{I_L} &= \left(2\frac{t}{a\tau} - \frac{t^2}{a^2 - \tau^2}\right), && 0 \le t \le a\tau \\ \frac{i(t)}{I_L} &= 1, && a\tau \le t \le \tau - a\tau \\ \frac{i(t)}{I_L} &= \left[1 - \left(\frac{t - \tau + a\tau}{a\tau}\right)^2\right], && \tau - a\tau \le t \le \tau \end{aligned}\right\} \tag{4.65}$$

Again, carrying out the integration for the coefficients b_n yields

$$b_n = \frac{4}{n\pi}\left(\frac{\sin\dfrac{n\pi a}{2}}{\dfrac{n\pi a}{2}}\right)^2, \qquad \text{where} \quad n = 1, 3, 5 \ldots \tag{4.66}$$

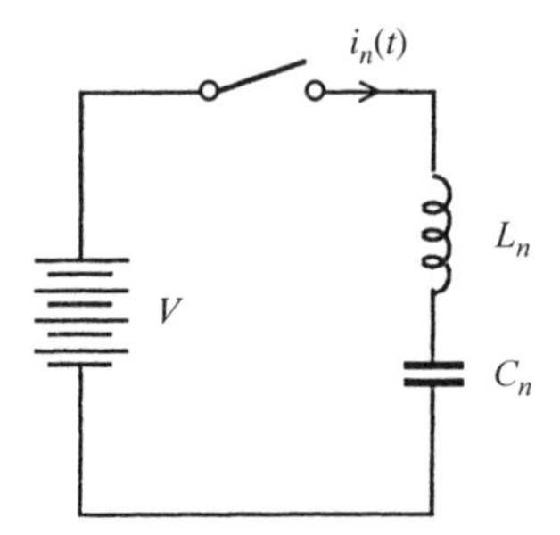

Figure 4.9 Current-generating circuit for the Fourier series

Each of the two current waveforms $i(t)$ apparently consists of a series of sine terms with amplitude b_n and frequency $n\pi/\tau$. These waveforms can be generated using the circuit given in Figure 4.9.

When the switch is closed in the circuit given in Figure 4.9 the current $i_n(t)$ that flows is given by

$$i_n(t) = V\sqrt{\frac{C_n}{L_n}} \sin \frac{t}{\sqrt{L_n C_n}} \tag{4.67}$$

The values of L_n and C_n for a given waveform can be determined by comparing the coefficients b_n for the waveform to the amplitude and the frequency of the waveform given by Equation (4.67). For example, for any given waveform

$$\left.\begin{aligned} & i(t) = I_l \sum_{n=1}^{\infty} b_n \sin \frac{n\pi t}{\tau} \\ \text{since} \quad & i(t) = \sum_{n=1}^{\infty} i_n(t) \quad \text{with} \quad n = 1, 3, 5 \ldots \\ & b_n = Z_N \sqrt{\frac{C_n}{L_n}} \quad \text{and} \quad \frac{t}{\sqrt{L_n C_n}} = \frac{n\pi t}{\tau} \quad \text{where} \quad Z_N = \frac{V}{I_l} \end{aligned}\right\} \tag{4.68}$$

and the values of L_n and C_n are found to be

$$L_n = \frac{Z_N \tau}{n\pi b_n} \quad \text{and} \quad C_n = \frac{\tau b_n}{n\pi Z_n} \tag{4.69}$$

where the appropriate expression for the coefficients b_n is substituted for the waveform chosen. Thus the network that will produce the current waveform for a given wave shape consists of a number of LC sections connected in parallel as given in Figure 4.10.

In practice, the network shown in Figure 4.10 can be difficult and expensive to build. The capacitors almost certainly will have non-standard values and the inductors will need to be built such that they have small stray capacitance. Thus network synthesis techniques are used, as in the previous section, to generate other equivalent networks which may be easier to construct. Once again the reader is referred to the excellent book by Kuo [7] for more detail on the synthesis techniques to be described. The methods are also described in the rather older book by Guillemin [12].

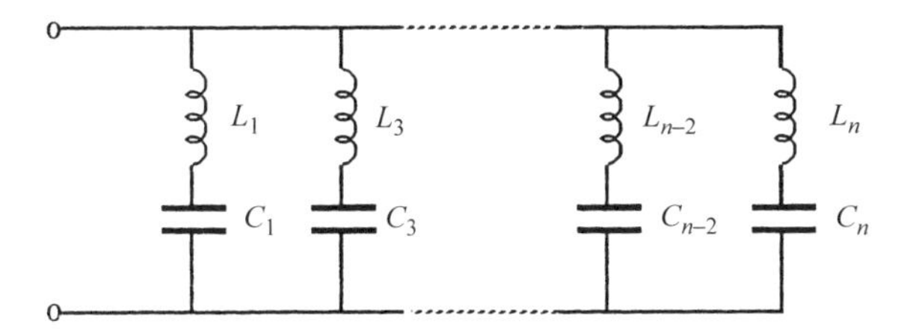

Figure 4.10 Pulse-forming network derived using Guillemin's synthesis method for any chosen waveform

The first equivalent circuit is derived by the partial fraction expansion of the impedance function **Z**(**s**) for the network shown in Figure 4.10, which is often known as Foster's method [13]. This is easily derived from the admittance function of the network, which can be written down by inspection as

$$\mathbf{Y}(\mathbf{s}) = \sum_{\nu=1,3\ldots}^{n} \frac{C_\nu \mathbf{s}}{\mathbf{s}^2 L_\nu C_\nu + 1} = \frac{1}{\mathbf{Z}(\mathbf{s})} \tag{4.70}$$

After some manipulation [4] this yields

$$\mathbf{Z}(\mathbf{s}) = \frac{1}{\mathbf{s}C_N} + \sum_{\nu=2,4\ldots}^{2n-2} \frac{\mathbf{s}L_\nu}{\mathbf{s}^2 L_\nu C_\nu + 1} + \mathbf{s}L_{2n} \tag{4.71}$$

which is recognised as the series combination of a capacitor, a number of parallel LC sections and an inductor as shown in Figure 4.12. Note that C_N and L_{2n} are found from Equation (4.70) and are given by

$$C_N = \sum_{\nu=1,3\ldots}^{n} C_\nu \quad \text{and} \quad \frac{1}{L_{2n}} = \sum_{\nu=1,3\ldots}^{n} \frac{1}{L_\nu} \tag{4.72}$$

Thus C_N is equal to the sum of all the capacitors shown in the network given in Figure 4.11 and L_{2n} is equal to all the inductors in Figure 4.10 connected in parallel.

Two further equivalent networks can be deduced using the continued fraction expansion of the impedance or admittance functions of the network given by Equation (4.70). This method is known as Cauer's method [7,14] and is achieved by dividing the numerator of the impedance or admittance function by the denominator, successively inverting the remainder

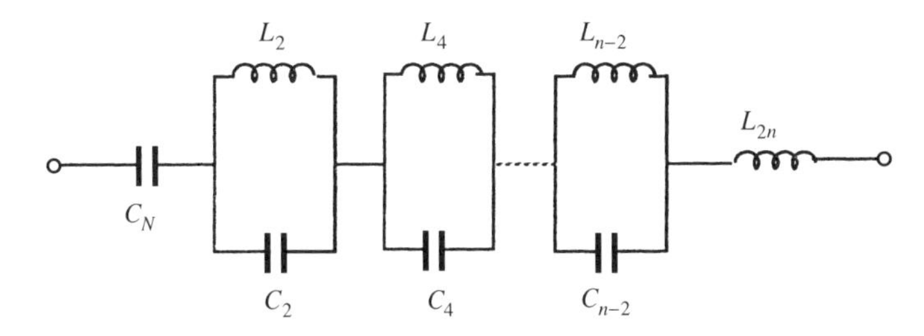

Figure 4.11 Equivalent network to that of Figure 4.10 derived using Foster's theorem

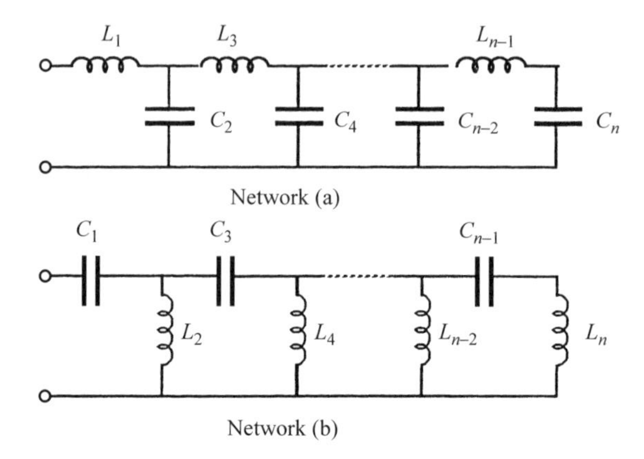

Figure 4.12 Two further equivalent circuits to that of Figure 4.10 derived using Cauer's method

until the remainder is zero. This results in a series of alternating impedances and admittances which can be recognised as components in an *LC* ladder network, i.e. for the impedance function

$$\mathbf{Z}(\mathbf{s}) = \mathbf{Z}_1(\mathbf{s}) + \cfrac{1}{\mathbf{Y}_2(\mathbf{s}) + \cfrac{1}{\mathbf{Z}_3(\mathbf{s}) + \cfrac{1}{\mathbf{Y}_4(\mathbf{s}) \ddots + \cfrac{1}{\mathbf{Z}_{n-1}(\mathbf{s}) + \cfrac{1}{\mathbf{Y}_n(\mathbf{s})}}}}} \tag{4.73}$$

This results in the two networks shown in Figure 4.12.

In the networks shown in Figure 4.12 the values of the various capacitors and inductors are found by the continued fraction expansion described. Usually this can only be accomplished using numerical coefficients.

All the above networks, derived by Guillemin's method, share the same practical disadvantage in that they all have capacitors whose sizes are different and are likely to be non-standard. This, from a manufacturing point of view, is expensive and it is highly desirable that a practical network has capacitors whose values are equal. In reference [4] a method is given by which a network can be deduced in which the size of the capacitors are equal and have a value C_N/n where C_N is the total energy storage capacitance. This is again done by manipulation of Equation (4.70) for the admittance or impedance function of a given Guillemin network. The resulting network, as shown in Figure 4.13, has a practical problem in that it contains inductors which have negative values.

This difficulty is, in practice, quite easy to remedy by simulating the negative inductors with partial mutual coupling of the inductors along the top of the circuit L_1, L_2, etc. This results in the circuit given in Figure 4.14.

In this circuit the mutual inductances $M_{12} = L_{12}$ and $M_{23} = L_{23}$, etc. It is now a relatively easy matter to construct the mutual inductances by simply partly coupling the windings of

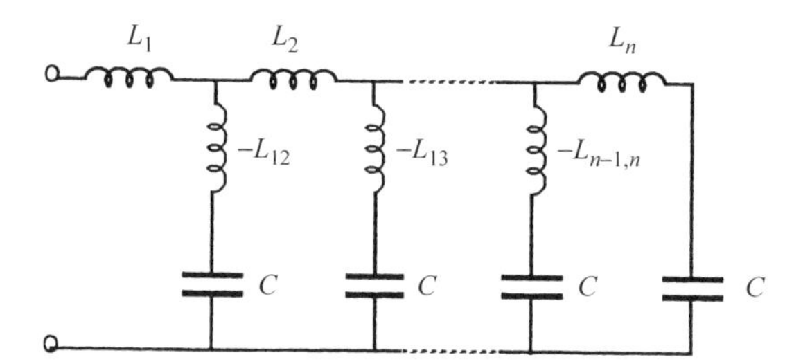

Figure 4.13 Guillemin equivalent network with equal capacitance per section

successive inductors on a common former and tapping in the capacitors at the appropriate points in the circuit. Alternatively the mutual inductance can be achieved by winding a continuous coil and connecting the capacitors at a succession of taps along the coil. The mutual inductance required, which is around 15% of the self-inductance [4], can be controlled by the ratio of the length of the coil to its diameter.

One of the advantages of the Guillemin networks is that the rise-time of the pulse can be changed simply by varying the value of the parameter a given in Figure 4.8. It is obvious that a can be varied from 0, which gives a rectangular pulse, to 1/2 which gives a triangular pulse. The pulse need not necessarily be restricted to pulses with trapezoidal or parabolic rise and falls. Indeed it is possible to generate pulses with quite elaborate structures, although the limitation of the Guillemin method becomes more noticeable in such pulses, and some fine tuning of the synthesised circuits is usually necessary to produce a good match between the pulse shape generated to that required. Indeed it is often the case that fine tuning of any Guillemin circuit is desirable particularly when using capacitors and inductors which are "non-ideal".

Figure 4.15 gives five equivalent five-section Guillemin networks which generate a pulse with trapezoidal rising and falling edges. The rise-time is 8% of the width of the pulse. To determine the values of inductance and capacitance for a given network impedance and pulse width, the inductor values should be multiplied by $Z_N\tau$ and the capacitor values by τ/Z_N.

4.4.2 Current-fed Networks

So far the pulse-forming networks that have been described are all voltage-fed in that the energy discharged in the pulse is stored in the capacitors of the network. As with pulse-forming lines, as seen in the last chapter, it is also possible to devise networks in which the

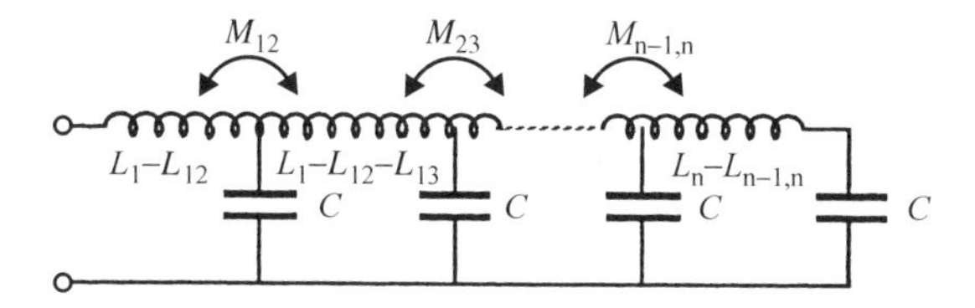

Figure 4.14 Physically realisable version of the circuit given in Figure 4.13

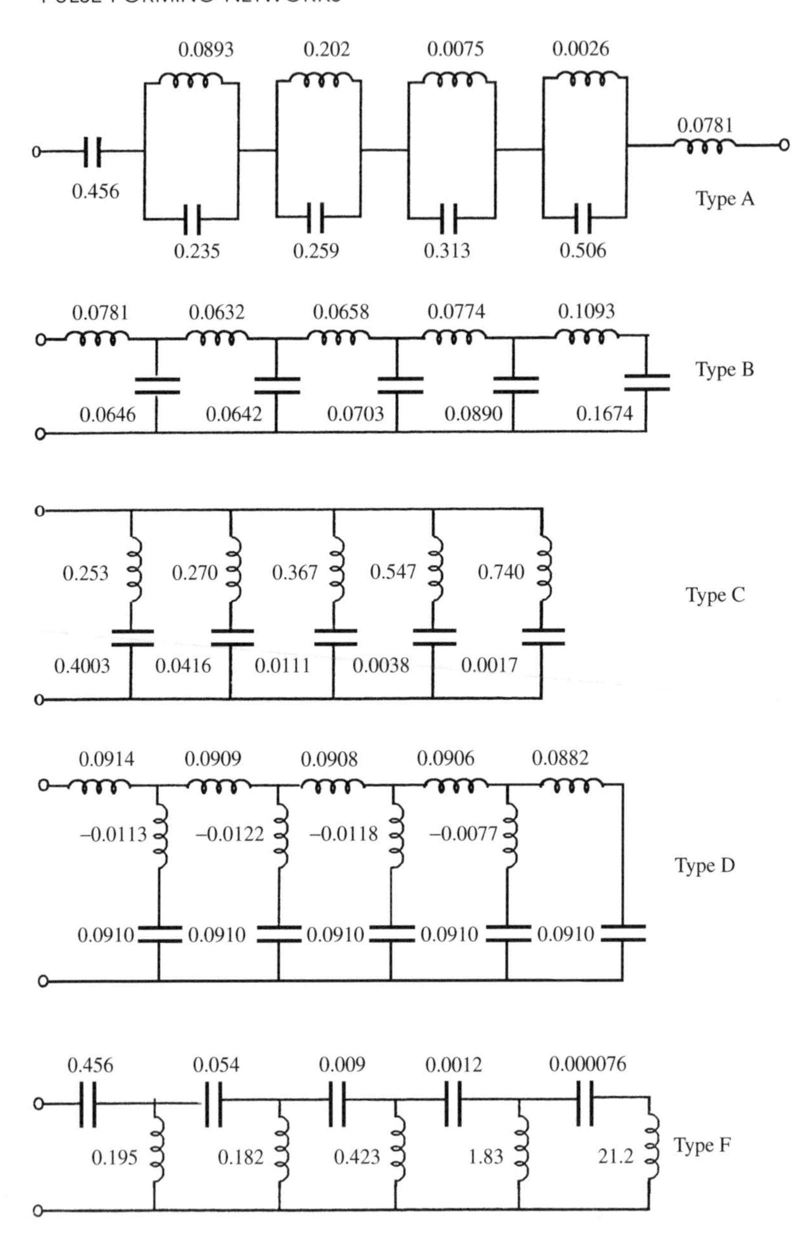

Figure 4.15 Equivalent five-section Guillemin networks

energy discharged is stored in the inductors of the network. Before investigating the impedance function that would be required for such networks, it is useful, for comparison purposes, to derive the impedance function that is required to generate a rectangular pulse in a load from a voltage-fed network. Referring to the circuit diagram given in Figure 4.4, the relationship between the current in the circuit to the potential to which the network is charged $V(0)$ is given by

$$[\mathbf{Z}(n,\mathbf{s}) + Z_l]\mathbf{I}(\mathbf{s}) = \frac{V(0)}{\mathbf{s}}$$
$$\text{or} \qquad \mathbf{Z}(n,\mathbf{s}) = \frac{V(0)}{\mathbf{I}(\mathbf{s})\mathbf{s}} - Z_l \tag{4.74}$$

From the table of Laplace transforms in the first chapter a rectangular current pulse of amplitude I and duration δ is given by

$$\mathbf{I}(\mathbf{s}) = \frac{I}{\mathbf{s}}(1 - \exp(\mathbf{s}\,\delta)) \tag{4.75}$$

Substituting this expression into Equation (4.74) gives

$$\begin{aligned}
\mathbf{Z}(n,\mathbf{s}) &= \frac{V(0)}{I(1-\exp(-\mathbf{s}\delta))} - Z_l \\
&= Z_l\left[\frac{\left(\frac{V(0)}{IZ_l}\right) - 1 + \exp(-\mathbf{s}\delta)}{1-\exp(-\mathbf{s}\delta)}\right] \\
&= Z_l\left[\frac{\left(\frac{V(0)}{IZ_l} - 1\right)\exp\left(\frac{\mathbf{s}\delta}{2}\right) + \exp\left(-\frac{\mathbf{s}\delta}{2}\right)}{\exp\left(\frac{\mathbf{s}\delta}{2}\right) + \exp\left(-\frac{\mathbf{s}\delta}{2}\right)}\right] \\
&= Z_l\left[\coth\left(\frac{\mathbf{s}\delta}{2}\right) + \frac{\left(\frac{V(0)}{IZ_l} - 2\right)\exp\left(\frac{\mathbf{s}\delta}{2}\right)}{\exp\left(\frac{\mathbf{s}\delta}{2}\right) + \exp\left(-\frac{\mathbf{s}\delta}{2}\right)}\right]
\end{aligned} \tag{4.76}$$

Putting $V(0) = 2IZ_l$ then gives

$$Z(n,\mathbf{s}) = Z_l \coth\left(\frac{\mathbf{s}\delta}{2}\right) \tag{4.77}$$

By comparison to Equation (4.30) this expression gives the input impedance of an open circuit uniform LC ladder network or transmission line with characteristic impedance Z_l and and transit time $\delta/2$ as expected.

For a current-fed network the circuit in Figure 4.16 can be used. The current-fed pulse-forming network is represented as an admittance $\mathbf{Y}(n,\mathbf{s})$, and an opening switch is used to connect the load to the pulse-forming network and current source. Once again, it is noted that a practical difficulty with the circuit is the lack of reliable repetitive opening switches which can operate at high potentials and currents. The circuit though will work at lower potentials with various types of semiconductor opening switch. Referring to the circuit it can be seen that when the switch is opened the current from the source will divide between

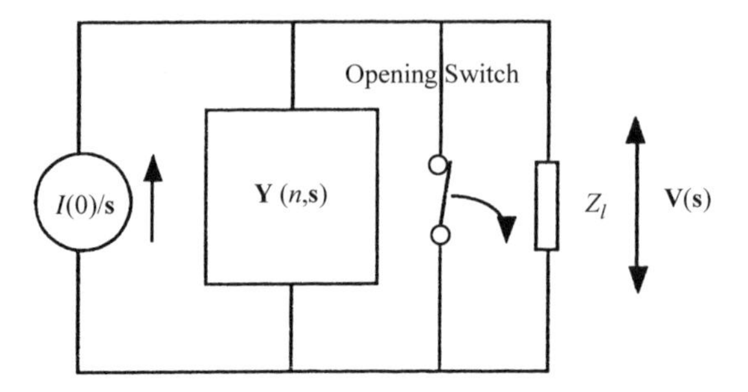

Figure 4.16 Operating circuit for a current-fed pulse-forming network

the network and the load, i.e.

$$\frac{I(0)}{\mathbf{s}} = \mathbf{V}(\mathbf{s})\left[\mathbf{Y}(n,\mathbf{s}) + \frac{1}{Z_l}\right]$$
$$\text{or} \qquad \mathbf{Y}(n,\mathbf{s}) = \frac{I(0)}{\mathbf{s}\mathbf{V}(\mathbf{s})} - \frac{1}{Z_l} \tag{4.78}$$

Again, the Laplace transform for a rectangular voltage pulse of amplitude V and duration δ is given by

$$\mathbf{V}(\mathbf{s}) = \frac{V}{\mathbf{s}}(1 - \exp(-\mathbf{s}\delta)) \tag{4.79}$$

substituting this expression into Equation (4.78) gives

$$\mathbf{Y}(n,\mathbf{s}) = \frac{1}{Z_l}\left[\frac{\frac{Z_l I(0)}{V} - 1 + \exp(-\mathbf{s}\delta)}{1 - \exp(-\mathbf{s}\delta)}\right]$$
$$= \frac{1}{Z_l}\left[\coth(\mathbf{s}\delta) + \frac{\left(\frac{Z_l I(0)}{V} - 2\right)\exp\left(\frac{\mathbf{s}\delta}{2}\right)}{\exp\left(\frac{\mathbf{s}\delta}{2}\right) - \exp\left(-\frac{\mathbf{s}\delta}{2}\right)}\right] \tag{4.80}$$

Putting $I(0) = 2V/Z_l$ gives

$$\mathbf{Z}(n,\mathbf{s}) = \frac{1}{\mathbf{Y}(n,\mathbf{s})} = Z_l \tanh\left(\frac{\mathbf{s}\delta}{2}\right) \tag{4.81}$$

which is recognised as the input impedance function of a uniform LC ladder network or transmission line with characteristic impedance Z_l and transit time $\delta/2$ which is

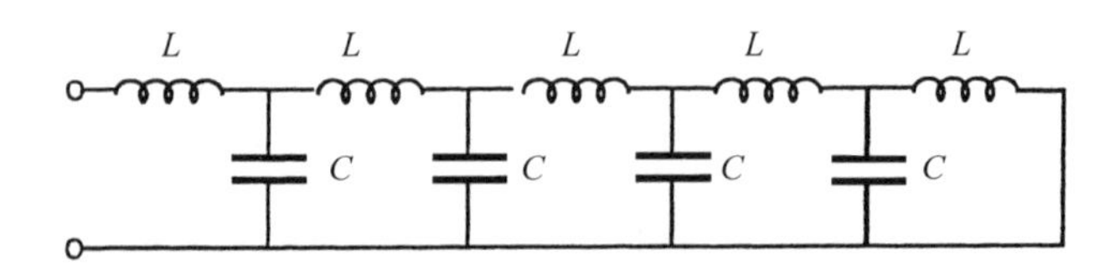

Figure 4.17 Basic uniform, current-fed LC ladder pulse-forming network

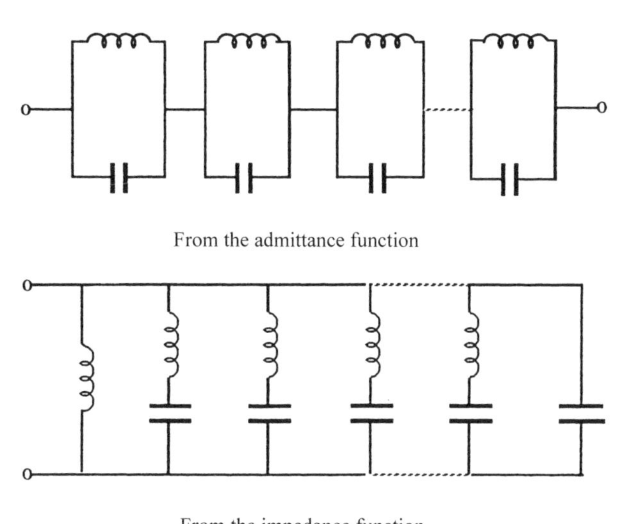

Figure 4.18 Equivalent current-fed pulse-forming networks to that shown in Figure 4.17 deduced by rational fraction expansion of the admittance and impedance function given by Equation (4.81)

short-circuited at its far end. The ladder network is shown in Figure 4.17. Thus the basic differences between current and voltage-fed pulse-forming networks are established.

4.4.3 The Synthesis of Alternative LC Current-fed Pulse-forming Networks

As with the basic, uniform, *LC* ladder pulse-forming networks that are voltage-fed, it is possible to deduce other equivalent pulse-forming networks for current-fed pulse-forming networks [4]. The starting point is the impedance function given in Equation (4.81). Again, rational fraction expansions of both the impedance and admittance functions give two equivalent networks which are shown in Figure 4.18.

4.4.4 Guillemin Type Current-fed Pulse-forming Networks

Guillemin's method can also be used to synthesise current-fed pulse-forming networks that generate pulses with specified rising and falling edges [4]. The method, in this case, relies on generating a specified alternating voltage waveform by connecting a constant current source to an unknown current-fed network. The pulse-generating circuit is the same as that shown in Figure 4.16 except that the load Z_l is removed and pulse-forming action is initiated by opening the switch.

The basic form of the current-fed pulse-forming network is a series of parallel *LC* resonant circuits, connected in series as shown in Figure 4.19. This circuit is derived, as with the voltage-fed equivalent, by making a Fourier series expansion of the alternating waveform specified and then comparing the coefficients of the series with the network

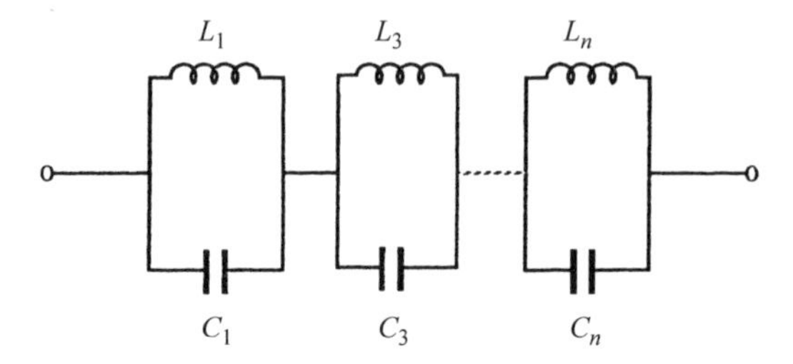

Figure 4.19 Basic form of the Guillemin current-fed network

shown in Figure 4.19. The potential across the nth parallel resonant circuit is given by

$$v_n(t) = I(0)\sqrt{\frac{L_n}{C_n}} \sin \frac{t}{\sqrt{L_n C_n}}$$

$$\text{where} \quad L_n = \frac{\tau_n b_n}{n\pi Y_N} = \frac{Z_N \tau b_n}{n\pi} \tag{4.82}$$

$$C_n = \frac{\tau Y_N}{n\pi b_n} = \frac{\tau}{n\pi b_n Z_N}$$

Further equivalent forms of this network can be deduced, in a similar way, to those deduced from the basic voltage-fed Guillemin network described earlier in the chapter. Three such networks are shown in Figure 4.20. These are all five-section networks with an 8% rise-time. To determine the values of the inductance and capacitance for a given network impedance and pulse width, the inductor values should be multiplied by $Z_N\tau$ and the capacitor values by τ/Z_n.

4.5 SOME FURTHER COMMENTS ON PULSE-FORMING NETWORKS

In this chapter various circuits have been described that will deliver a rectangular pulse into a matched load. However, in every case, there are limitations to the circuits. In the case of uniform *LC* ladder networks the circuits produce substantial overshoots on the leading edge of the pulses. In the case of the Guillemin circuits the pulses tend to ripple on what should be a flat top, and empirical modifications have to be made to the circuits in order to get the best shape possible for the pulse. No perfect synthesis method appears to exist. This is not surprising as all the methods described are attempting to simulate the behaviour of a distributed transmission line with lumped element networks. There are however a number of papers that do describe methods by which the shape of the generated pulse can be improved. For example, the paper by O'Loughlin and Sidler [15] gives a method by which the fall-time of the pulse generated by a pulse-forming network can be reduced using a synchronised switching method. Another paper by Cravey, Burks and McDuff explains a method by which the non-ideal behaviour of the components, used to build a pulse-forming network, can be compensated for to improve the shape of a generated pulse [16].

Another problem that often has to be addressed is the design of networks that will generate a flat-topped pulse into a variable impedance load. The paper by Ranon *et al.* [17] describes a method by which this may be achieved by modifying Guillemin's method. Pulse-forming networks can also be designed that will compensate for some deficiency in

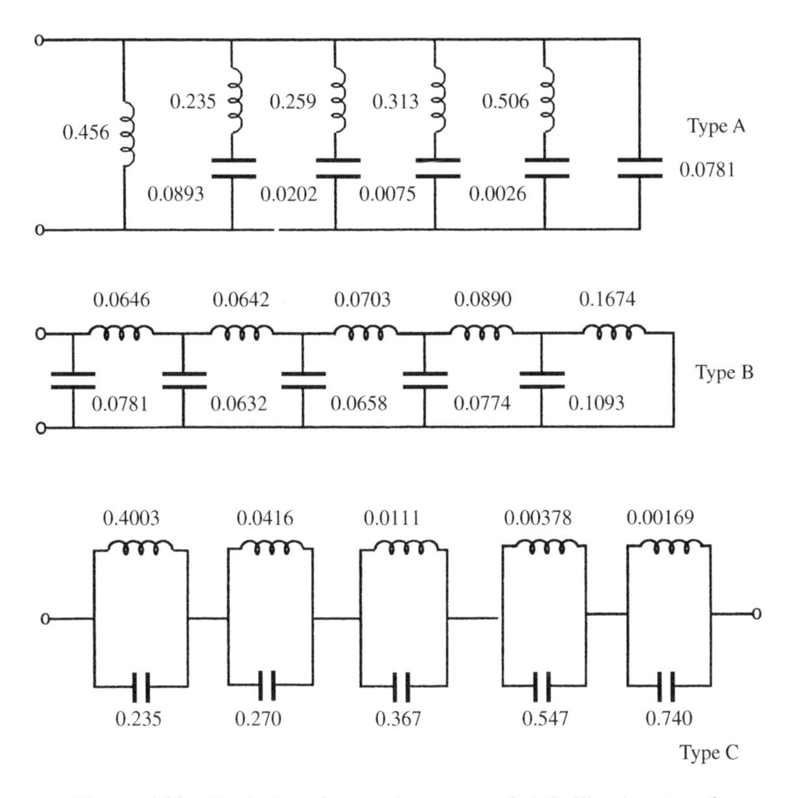

Figure 4.20 Equivalent five-section current-fed Guillemin networks

the load or circuit into which they are discharged. For example, there are a series of papers [18,19,20] which describe methods for compensating transformer pulse droop (see Chapter 5) by designing pulse-forming networks that generate pulses with rising tops.

An improvement to Guillemin's method is described by Gore and Larsen [21]. Referring to the circuit (Figure 4.10) upon which the method is based, it can be seen that it consists of a number of series *LC* sections which are connected in parallel. The component values are determined from the Fourier series for the alternating wave which is chosen to match the pulse required. The major flaw in the method is that each of these series sections discharges into the load and also interacts with the other sections in the circuit. Therefore each section does not generate a perfect sinusoidal waveform into a matched load but a damped sinusoidal waveform with some degree of back swing. Consequently the addition of these waveforms will not result in a perfect match to the pulse shape desired. In the method of Gore and Larsen a second pulse-forming network is added in parallel with the load, and it is claimed that this removes the limitation of Guillemin's method.

REFERENCES

[1] Ray Wylie C. and Barrett L. C. "Advanced Engineering Mathematics". McGraw-Hill (1985) ISBN 0-07-072188-2

[2] Bleaney B. I. and Bleaney B. "Electricity and Magnetism". Vol. 1. Oxford Science Publications (1989) ISBN 0-19-851172-8
[3] Oppenheim A. V., Willsky A. S. and Young I. T. "Signals and Systems". Prentice Hall International Edition (1983) ISBN 0-13-811175-8
[4] Glasoe G. N. and Lebacqz J. V. "Pulse Generators". Radiation Laboratory Series Vol. 5, McGraw Hill Book Company Inc. (1948)
[5] Spiegel M. R. "Laplace Transforms". Shaum's Outline Series in Mathematics, McGraw-Hill (1965) ISBN 07-060231-x
[6] "Microcap" is published by Spectrum Software, Sunnyvale, CA 94086
[7] Kuo F. F. "Network Analysis and Synthesis". John Wiley & Sons (1966) ISBN 0-471-51116-1
[8] Whittaker E. T. and Watson G. N. "Modern Analysis". Macmillan, New York (1943)
[9] Zepler E. E. and Nichols K. G. "Transients in Electrical Engineering". Chapman and Hall (1971) ISBN 412-10130-0
[10] Guillemin E. A. "A Historical Account of the Development of a Design Procedure for Pulse Forming Networks". Radiation Laboratory Report No. 43 (1944)
[11] Stephenson G. "Mathematics Methods for Science Students". Longman Scientific & Technical (1973) ISBN 0-582-44416-0
[12] Guillemin E. A. "Communication Networks". Vol. 2, Wiley, New York (1935).
[13] Foster R. M. "A Reactance Theorem". Bell System Tech. J., No. 3 (1924), 259–267
[14] Daryanani G. "Principles of Active Network Synthesis and Design". John Wiley and Sons (1976) ISBN 0-471 03523-8
[15] O'Loughlin J. P. and Sidler J. "The End of Line Tail Biter". Proceedings of the 6th IEEE Pulsed Power Conference (1987) 692–695
[16] Cravey W. R., Burkes T. R. and McDuff G. "Design of Repetitive Very High Voltage Pulse Forming Networks of Short Duration". Proceedings of the 7th IEEE Pulsed Power Conference (1989) 116–119.
[17] Ranon P. M., Pelletier P. R., O'Loughlin J. P., Weyn M. L., Baker W. L., Scott M. C. and Adler R. J. "Constant Voltage Pulse Power Driver for Variable Impedance Loads". Proceedings of the 7th IEEE Pulsed Power Conference (1989) 778–781.
[18] Ranon P. M., Hall D. J., O'Loughlin J. P., Schlicher R. L., Baker W. L., Dietz D. and Scott M. C. "Synthesis of Droop Compensated Pulse Forming Networks for Generating Flat Top, High Energy Pulses into Variable Loads from Pulsed Transformers". Proceedings of the 18th Power Modulator Symposium (1988) 54–61
[19] Ranon P. M., O'Loughlin J. P. and Chen Y. G. "Spatially Transformed Pulse Forming Networks". Proceedings of the 19th Power Modulator Symposium (1990) 69–73.
[20] Ranon P. M., Schlicher R. L., O'Loughlin J. P., Hall D. J., Weyn M. L., Pelletier P. R., Sidler J. D., Baker W. L. and Scott M. C. "Droop Compensated Pulse Forming Network Driven Pulsed Transformer Design". Proceedings of the 7th IEEE Pulsed Power Conference (1989) 113–115.
[21] Gore W. C. and Larsen T. "Linear Pulse-Forming Circuits". IRE Trans. on Circuit Theory (1956) 182–188

5 Pulse Transformers

5.1 INTRODUCTION

In this chapter the transient behaviour of conventional transformers consisting of closely coupled windings, with or without a magnetic core, will be explored. It will be assumed that the basic principles of transformer design and modelling are understood for continuous AC operation. There is a wide range of applications for pulse transformers, from very low to very high power operation, for example in digital circuits, triggering circuits, and a very wide range of pulsed power circuits. Perhaps the most important application, which initiated the development of much of the theory of pulse transformers, is pulsed radar, where a rectangular voltage pulse with an amplitude of several kV is required to drive a microwave source such as a klystron or magnetron.

There are several different uses of pulse transformers in electronic circuits. These include their use as voltage or current step-up transformers, as impedance matching devices, for DC isolation, for pulse polarity inversion and as pulse differentiators. Since rectangular pulses are often required with pulse rise-times of 50 ns or less, it is clear that pulse transformers must have a wide-band response, and the way in which the response of a given transformer, to a given input (usually a rectangular pulse) can be predicted will be discussed in detail. There are essentially two basic ways in which such a response can be determined. The first consists in decomposing the pulse into a large number of sine waves of different frequencies, finding the response of the transformer to each frequency and summing these responses to obtain the total response. This decomposition can be done by Fourier analysis, but for many transformers and circuits the analysis can prove to be very time-consuming. The method can be simplified by the arbitrary omission of frequency components with negligible amplitude and calculation of the circuit response to the remaining components. This approximation has two fairly subjective criteria: the number of frequencies to be retained and the evaluation of the frequencies for which the transformer has a poor response. The second method which is widely used [1,2,3] is to find the response of the transformer to the three parts of a rectangular pulse, namely the rising and trailing edges and the top of the pulse. Figure 5.1 gives an exaggerated illustration of the typical response of a pulse transformer to a rectangular input pulse.

As can be seen from Figure 5.1, the input pulse is distorted in several ways. First there is a loss of rise-time and there is often some overshoot and ringing on the front edge of the pulse. The top of the pulse droops and on the trailing edge of the pulse the fall time is degraded. There is also some back swing and a long negative tail appears on the end of the pulse. As will be shown later, the magnitude of the area under this tail is equal to the area under the pulse. Before the transient behaviour of pulse transformers can be investigated, it

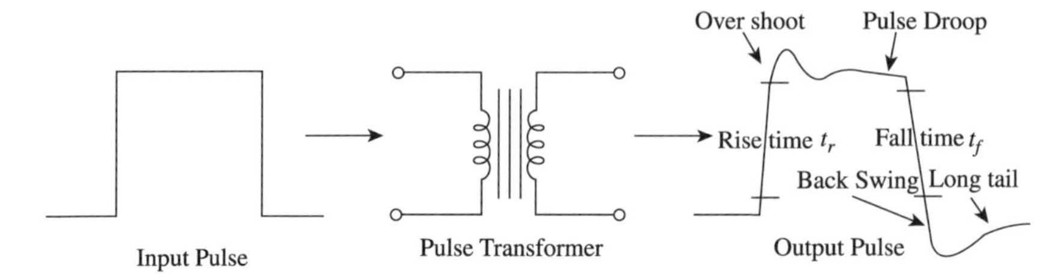

Figure 5.1 Typical response of a pulse transformer to a rectangular input pulse

is necessary to discuss some basic circuit properties that apply to all types of conventional transformers.

5.2 THE IDEAL TRANSFORMER AND THE CONCEPTS OF REFERRAL AND RELUCTANCE

Consider a basic two-winding transformer, as shown in Figure 5.2. with a primary winding consisting of n_p turns and a secondary of n_s turns. If the transformer behaves ideally then the windings have zero resistance, the flux induced in the magnetic core by the primary is totally coupled to the secondary (there is no flux leakage) and the core permeability is infinite (i.e. the net magneto-motive force or mmf needed to establish a flux in the core is zero). This latter point can be better understood if the concept of reluctance, the magnetic equivalent to resistance is introduced. Referring to the magnetic core and primary winding of the transformer depicted in Figure 5.2, application of Ampere's law to the path around the core denoted by the broken line gives

$$\oint \mathbf{H}.d\mathbf{l} = n_p i_p = Hl \tag{5.1}$$

the quantity $n_p i_p$ is known as the mmf and l is the path length. The magnetic flux density in the core B is related to the magnetic field intensity H and core flux ϕ by the relationships

$$B = \mu H = \mu_r \mu_o H = \frac{\phi}{A} \tag{5.2}$$

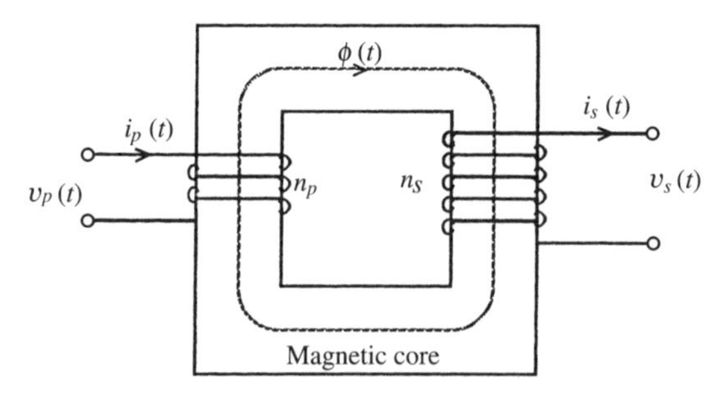

Figure 5.2 An ideal transformer

A is the cross-sectional area of the core, μ is the permeability of the core and is equal to the product of the relative permeability of the core μ_r and the permeability of free space μ_o. From Equation (5.2)

$$\phi = \frac{\mu n_p i_p}{l} A = \frac{n_p i_p}{l/\mu A} = \frac{n_p i_p}{\mathscr{R}}, \qquad \text{where} \quad \mathscr{R} = \frac{l}{\mu A} \tag{5.3}$$

$\mathscr{R}$ is known as the reluctance of the magnetic circuit, and clearly it will be equal to zero if μ is infinite. The similarity between this equation and the basic equation which relates resistivity to resistance in a resistive material explains why reluctance can be though of as being the magnetic equivalent of resistance.

The reluctance of the core can also be related to the inductance of any winding wound on it. For example, using Equation (1.3) and Faraday's law which relates a time-varying voltage $v_p(t)$ applied to the primary winding to the time-varying magnetic flux $\phi(t)$ induced in the core gives

$$v_p(t) = L_p \frac{di_p(t)}{dt} = n_p \frac{d\phi(t)}{dt} \tag{5.4}$$

Integrating this equation and using Equation (5.1), the inductances of the primary and secondary windings, L_p and L_s, can be found to be

$$L_p = \frac{n_p \phi}{i_p} = \frac{n_p BA}{i_p} = \frac{n_p \mu HA}{Hl/n_p} = \frac{n_p^2}{l/\mu A} = \frac{n_p^2}{\mathscr{R}} \quad \text{similarly} \quad L_s = \frac{n_s^2}{\mathscr{R}} \tag{5.5}$$

The flux in the core $\phi(t)$ is linked to the secondary winding, and a time-varying voltage is induced in it, i.e.

$$v_s(t) = n_s \frac{d\phi(t)}{dt} \tag{5.6}$$

Combining equations (5.4) and (5.6) gives

$$\frac{v_s(t)}{v_p(t)} = \frac{n_s}{n_p} = n \tag{5.7}$$

where n is the turns ratio.

The relationship between the primary and secondary currents can be deduced from the fact that the input power to the transformer must equal the output power, i.e.

$$\begin{aligned} v_p(t)i_p(t) &= v_s(t)i_s(t) = n v_p(t) i_s(t) \\ \therefore \quad \frac{i_s(t)}{i_p(t)} &= \frac{1}{n} \end{aligned} \tag{5.8}$$

Now consider the circuit shown on the left in Figure 5.3 where an ideal transformer is connected to an load impedance Z_l. From this circuit the relationship between the primary current and primary voltage is found to be

$$\frac{v_s(t)}{i_s(t)} = Z_l = \frac{n v_p(t)}{\dfrac{i_p(t)}{n}} = n^2 \frac{v_p(t)}{i_p(t)}$$

$$\text{or} \qquad \frac{v_p(t)}{i_p(t)} = \frac{Z_l}{n^2} \tag{5.9}$$

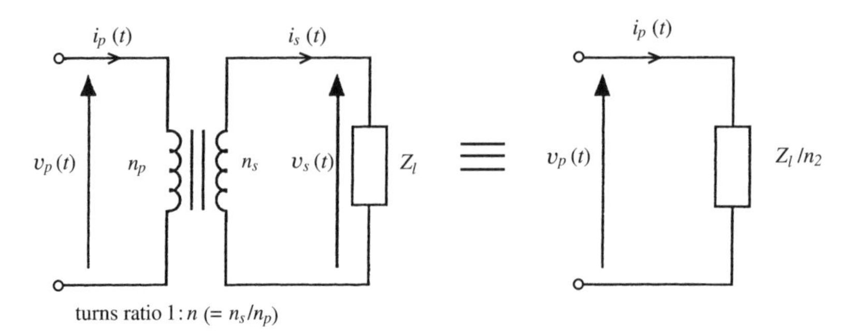

Figure 5.3 Impedance referral from secondary to primary circuit

Thus the circuit on the left can be replaced by an equivalent circuit which is shown on the right of Figure 5.3. This equivalence is important and greatly simplifies the analysis of circuits which contain transformers. The impedance Z_l is said to be referred from the secondary circuit to the primary by dividing it by the factor n^2. In using referral from the secondary to the primary circuit it is important to remember to multiply the potential on the load by n so that the voltage gain and turns ratio of the transformer are properly taken into account. This is often done by adding an "ideal transformer" to the referred circuit as shown in Figure 5.6. Components and generators can also be referred from the primary circuit to the secondary in the same way. However, in the transformer shown in Figure 5.3, an arbitrary impedance Z would be referred to the secondary by multiplying by a factor n^2.

5.2.1 Practical or Non-ideal Transformers

In the preceding section certain properties of the transformer were assumed in order that it should behave ideally and consequently simplify the basic theory described. In reality, of course, practical transformers have properties which make them non-ideal. For example, the windings have resistance, not all the flux in the core is linked to the windings, the core permeability is not infinite and the reluctance is not zero, and energy is lost in the core when it is subjected to a time-varying magnetic field. In addition, there is stray capacitance across the windings and capacitance exists between windings. All of these imperfections must be taken into account if an accurate prediction is to be made of the way in which a given transformer will behave in an electrical circuit. This can be done by deducing an equivalent circuit for the transformer in which the various imperfections are represented by circuit components that are then taken into account in carrying out any network analysis on a given circuit containing a transformer.

For example, it is physically impossible to build a transformer, such as the one shown in Figure 5.2, in which all of the flux generated by the primary winding in the core is linked to the secondary winding. To obtain a complete linkage would require the primary and secondary windings to exist in the same part of space. Thus inevitably some flux from the primary winding does not enter the core, but rather the air space between the winding and the core, and similarly not all of the core flux is linked to the secondary winding and again some is lost to the space between this winding and the core. The flux that is lost both in the primary and the secondary windings can be represented by leakage inductances L_{lp} and L_{ls}, respectively, which are connected in series with the two windings as shown in Figure 5.4(b). The inductance that results from the flux that is linked by the primary winding to the core is known as the primary magnetising inductance L_{mp}. Thus a transformer connected to a load can be represented by three equivalent circuits, as shown in Figure 5.4. An explanation and description of the first circuit Figure 5.4(a) was given in Chapter 1. The last circuit Figure 5.4(c) is deduced from the circuit shown in Figure 5.4(b) using the referral method described earlier.

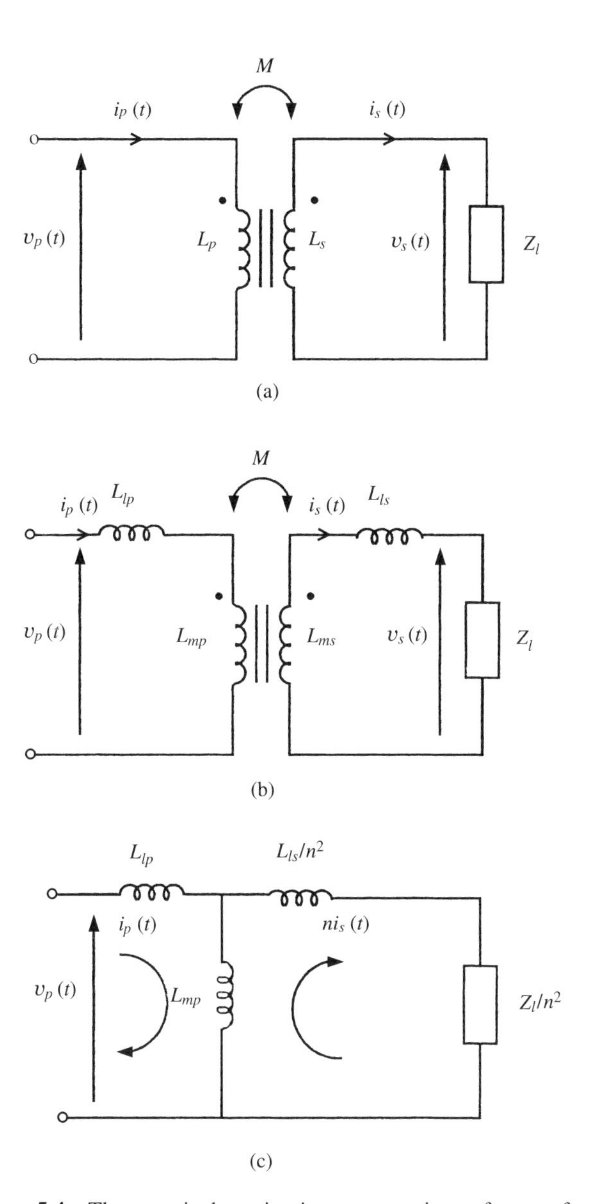

Figure 5.4 Three equivalent circuit representations of a transformer

It is instructive to check that the three circuits depicted in Figure 5.4 are indeed equivalent and what conditions must be met for this equivalence. From the Laplace equivalent of circuit (a), application of Kirchhoff's voltage law gives

$$\mathbf{V}_p(\mathbf{s}) = \mathbf{s}L_p\mathbf{I}_p(\mathbf{s}) - \mathbf{s}M\mathbf{I}_s(\mathbf{s}) \tag{5.10}$$

and

$$0 = \mathbf{s}L_s\mathbf{I}_s(\mathbf{s}) + \mathbf{I}_s(\mathbf{s})Z_l - \mathbf{s}M\mathbf{I}_p(\mathbf{s}) \tag{5.11}$$

Similarly, for circuit (c)

$$\mathbf{V}_p(\mathbf{s}) = \mathbf{s}L_{lp}\mathbf{I}_p(\mathbf{s}) + \mathbf{s}L_{mp}\mathbf{I}_p(\mathbf{s}) - \mathbf{s}L_{mp}n\mathbf{I}_s(\mathbf{s}) \tag{5.12}$$

and

$$0 = \mathbf{s}L_{mp}n\mathbf{I}_s(\mathbf{s}) + \frac{\mathbf{s}L_{ls}}{n^2}n\mathbf{I}_s(\mathbf{s}) + \frac{Z_l}{n^2}n\mathbf{I}_s(\mathbf{s}) - \mathbf{s}L_{mp}\mathbf{I}_p(\mathbf{s})$$

$$\text{or} \quad 0 = \mathbf{s}L_{mp}n^2\mathbf{I}_s(\mathbf{s}) + \mathbf{s}L_{ls}\mathbf{I}_s(\mathbf{s}) + Z_l\mathbf{I}_s(\mathbf{s}) - \mathbf{s}L_{mp}n\mathbf{I}_p(\mathbf{s}) \tag{5.13}$$

Consistency between Equations (5.10) and (5.12) requires that

$$L_p = L_{lp} + L_{mp} \quad \text{and} \quad M = \mathrm{n}L_{mp} \tag{5.14}$$

Similarly, consistency between Equations (5.11) and (5.13) requires that

$$L_s = L_{ls} + n^2L_{mp} \tag{5.15}$$

From Equations (5.14) and (5.15) it can be seen that

$$L_{mp} = \frac{M}{n}, \quad L_{lp} = L_p - \frac{M}{n} \quad \text{and} \quad L_{ls} = L_s - nM \tag{5.16}$$

Note that the coefficient of coupling k between the primary and secondary windings is defined as

$$k = \frac{M}{\sqrt{L_pL_s}} \tag{5.17}$$

n is chosen to be

$$n = \sqrt{\frac{L_s}{L_p}} \tag{5.18}$$

as could be deduced from Equation (5.5). Then from Equation (5.16)

$$L_{lp} = L_p(1-k) \quad \text{and} \quad L_{ls} = L_s(1-k) = n^2L_p(1-k)$$

$$\Rightarrow L_{lp} = \frac{L_{ls}}{n^2} \qquad \text{also} \qquad L_{mp} = kL_p \quad \text{and} \quad L_{ms} = kL_s \tag{5.19}$$

In other words, the primary leakage inductance is equal to the secondary leakage inductance, referred to the primary circuit. In the case where the coupling coefficient k has a value close to unity, as should be the case in a well-designed magnetically cored transformer, circuit (c) in Figure 5.4 is usually simplified, by approximation, still further. This is done by adding the referred secondary leakage inductance L_{ls}/n^2 to the primary leakage inductance L_{lp} to give a circuit which comprises a leakage inductance equal to twice the primary leakage inductance in series with the parallel combination of the referred load impedance and the primary inductance which, for k close to 1, will be approximately equal to the primary magnetising inductance (i.e. $L_{mp} \approx L_p$).

5.2.2 Equivalent Circuit of a Transformer

A complete equivalent circuit for a practical transformer can be constructed by including:

(i) resistors in the primary and secondary circuits, R_p and R_s, respectively, to represent the finite resistance of the windings
(ii) a resistor R_c in parallel with the primary winding to represent core loss
(iii) a capacitor C_{ps} to represent the capacitance between the primary and secondary winding
(iv) capacitors in parallel with the primary and secondary windings, C_p and C_s, respectively, to represent their self-capacitance.

in addition to the leakage and primary magnetising inductances as explained above. Such a circuit is shown in Figure 5.5. It should be noted that this circuit is still an approximation, as lumped element components are used to model inductance, capacitance, etc., which are, in reality, distributed throughout the transformer. Thus the correct positioning of components such as C_{ps}, C_p and C_s is debatable. Also the correct values of the most of the components used in the model have to be estimated. The leakage inductances can be estimated from the physical construction of the transformer although the calculation involved can be difficult. However the excellent book by Grover [4] can often provide useful formulae from which such inductances can be determined. The primary inductance and hence the primary magnetising inductance can be found using Equation (5.5). If the leakage inductances are small, as should be the case in a well-constructed transformer, then a reasonable approximation is to assume that the primary magnetising inductance is equal to the primary inductance.

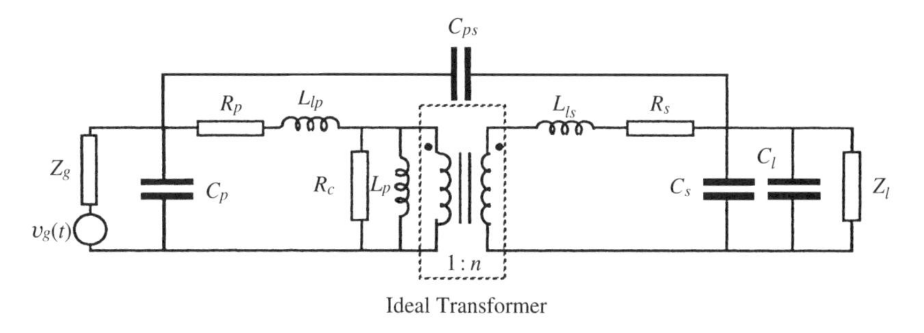

Figure 5.5 Basic equivalent circuit of a transformer

To estimate the capacitance between the primary and secondary windings C_{ps}, formulae for the capacitance between parallel plates or coaxial cylinders may be used, assuming that the windings are wound on top of each other. These formulae are given in Figure 2.1. Estimation of the self-capacitance of a winding is a little more difficult. Snelling [5] gives a good account of how this capacitance may be estimated, and also how the loss in a transformer can be determined and hence a value for R_c can be estimated. Loss in the windings, as represented by R_p and R_s, is relatively straightforward to calculate if the diameter and resistivity of the wire used to make the windings are known. At high frequencies, though, the skin effect should be taken into account as this will inevitably increase the effective resistance of the windings. Finally the circuit also includes the impedances of the generator and load and also the capacitance of the load, should this be significant.

The circuit in Figure 5.5 can be greatly simplified by adding components together and by the use of referral of components from the secondary circuit to the primary. Again it is assumed that the value of the coupling coefficient k is close to unity so that the referral process is reasonably accurate but not exact. Such an approximate simplified circuit is given in Figure 5.6. In this circuit the capacitance C is given by

$$C \approx n^2(C_s + C_l) + C_p \tag{5.20}$$

ignoring C_{ps}. The resistors R_1 and R_2 are given by

$$R_1 = R_p + Z_g \quad \text{and} \quad R_2 = \frac{Z_l + R_s}{n^2} \tag{5.21}$$

assuming that the generator and load impedances are purely resistive and that the core loss can be ignored. Finally the leakage inductance L_l is given by

$$L_l \approx L_{lp} + \frac{L_{ls}}{n^2} \tag{5.22}$$

In order to analyse the response of the transformer to a rectangular pulse the circuit given in Figure 5.6 can be used. As explained earlier, the analysis is split into three parts, namely the responses to the leading edge of the pulse, the flat top of the pulse and the trailing edge. By this means it is possible to simplify the circuit given in Figure 5.6 still further.

5.2.3 Leading Edge Response

Since the leading edge of a rectangular pulse contains the highest-frequency components of the pulse, the equivalent circuit given in Figure 5.6 can be simplified on the basis that the

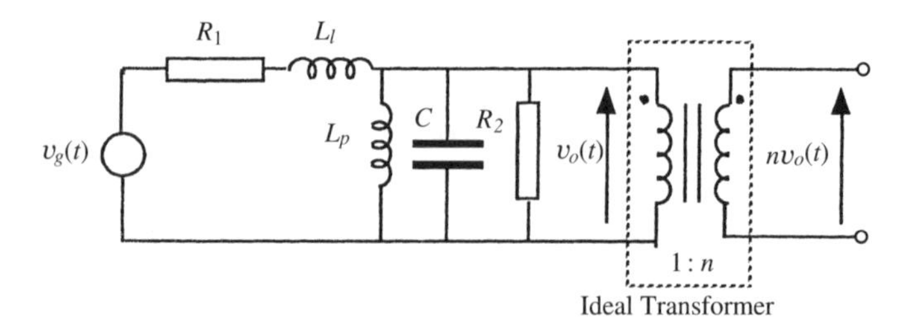

Figure 5.6 Simplified version of the transformer equivalent circuit given in Figure 5.5

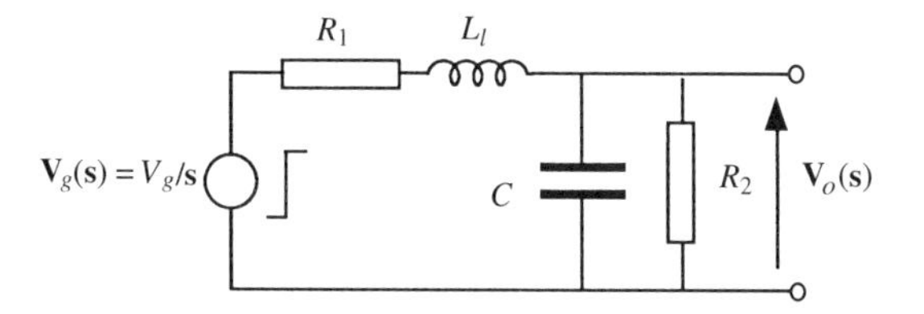

Figure 5.7 Simplified equivalent circuit to determine the rise-time response of a pulse transformer

impedance of the primary magnetising inductance L_p at these high frequencies will be very much larger than the impedance of the leakage inductance L_l and the impedance of the capacitor C. Thus L_p is neglected and the response of the simplified circuit, given in Figure 5.7, to an input step $v_g(t)$ is determined. The ideal transformer is also neglected as the output voltage $v_o(t)$ is simply multiplied by n, the turns ratio of the transformer, in order to get the correct response amplitude.

From the circuit shown the output voltage $\mathbf{V}_o(\mathbf{s})$ is found to be

$$\mathbf{V}_o(\mathbf{s}) = \frac{V_g}{\mathbf{s}} \frac{R_2 || \dfrac{1}{\mathbf{s}C}}{R_2 || \dfrac{1}{\mathbf{s}C} + R_1 + \mathbf{s}L_l}$$

$$= \frac{V_g}{\mathbf{s}} \frac{\dfrac{1}{L_l C}}{\mathbf{s}^2 + \dfrac{(R_1 R_2 C + L_l)}{R_2 C L_l}\mathbf{s} + \dfrac{R_1 + R_2}{R_2 C L_l}}$$

$$= \frac{V_g}{\mathbf{s}L_l C} \frac{1}{(\mathbf{s} + \alpha_1)(\mathbf{s} + \alpha_2)} \tag{5.23}$$

where α_1 and α_2 are given by

$$\alpha_1, \alpha_2 = \left[\left(\frac{R_1}{2L_l} + \frac{1}{2R_2 C} \right) \pm \sqrt{\left(\frac{R_1}{2L_l} + \frac{1}{2R_2 C} \right)^2 - \left(\frac{R_1 + R_2}{R_2 C L_l} \right)} \right] \tag{5.24}$$

α_1 and α_2 can be expressed in standard form [1] by putting

$$\left. \begin{aligned} a &= \frac{R_2}{R_1 + R_2} \\ T &= 2\pi\sqrt{L_l C a} \\ k &= \frac{T}{4\pi}\left(\frac{R_1}{L_l} + \frac{1}{R_2 C} \right) \end{aligned} \right\} \tag{5.25}$$

a is a resistive attenuation factor, T is a time constant and k is known as the damping factor. α_1 and α_2 are now given by

$$\alpha_1, \alpha_2 = \left[\frac{2\pi k}{T} \pm j\frac{2\pi}{T}\sqrt{(1 - k^2)} \right] \tag{5.26}$$

where α_1 and α_2 will have both a real and imaginary part if the magnitude of k^2 is less than unity. In this case the transformer response will be partly oscillatory or underdamped. If the magnitude of k^2 is greater than or equal to unity the roots will be real and the response of the transformer is said to be overdamped or critically damped, respectively, and will display no oscillatory character.

In the critically damped case where $k = 1$, the transformer response in the time domain can be determined by finding the inverse transform of the last equation in Equations (5.23). Note in this case the two roots α_1 and α_2 are now equal. The inverse transform is found by first expressing the equation in terms of partial fractions, i.e.

$$\mathbf{V}_o(\mathbf{s}) = \frac{V_g}{L_l C} \frac{1}{\alpha^2} \left[\frac{1}{\mathbf{s}} - \frac{1}{\mathbf{s} + \alpha} - \frac{\alpha}{(\mathbf{s} + \alpha)^2} \right]$$

$$\text{with} \quad \alpha_1 = \alpha_2 = \alpha = \frac{2\pi}{T} \tag{5.27}$$

The transformer response in the time domain $v_o(t)$ is thus given by

$$v_o(t) = naV_g \left[1 - \left(1 + \frac{2\pi t}{T} \right) \exp\left(-\frac{2\pi t}{T} \right) \right] \tag{5.28}$$

Note that the factor n is now included to give the correct amplitude of the transformer response as explained earlier. It is rather laborious to determine the transformer response in the cases where the circuit is underdamped ($k < 1$) or overdamped ($k > 1$). The correct responses are therefore quoted as

$k < 1$ Underdamped

$$v_o(t) = naV_g \left\{ 1 - \left[\frac{k}{\sqrt{1 - k^2}} \sin\left(\frac{2\pi t \sqrt{1 - k^2}}{T} \right) + \cos\left(\frac{2\pi t \sqrt{1 - k^2}}{T} \right) \right] \exp\left(-\frac{2\pi k t}{T} \right) \right\} \tag{5.29}$$

$k > 1$ Overdamped

$$v_o(t) = naV_g \left\{ 1 - \left[\frac{k}{\sqrt{k^2 - 1}} \sinh\left(\frac{2\pi t \sqrt{1 - k^2}}{T} \right) + \cosh\left(\frac{2\pi t \sqrt{1 - k^2}}{T} \right) \right] \exp\left(-\frac{2\pi k t}{T} \right) \right\} \tag{5.30}$$

Typical responses for Equations (5.28), (5.29) and (5.30) are given in Figure 5.8. Note that the time amplitude responses have been normalised by setting $x = t/T$ and $y = v_o(t)/naV_g$. For any of these response the rise-time t_r can be determined by finding the time taken for the response to rise from 10% to 90% of its final value. In the case of the critically damped response this response is given by

$$t_r = 0.53T = 3.35\sqrt{L_l C a} \tag{5.31}$$

This is an important equation in that it highlights one of the most important limitations of this type of pulse transformer, namely that the rise-time depends on the leakage inductance L_l of the transformer and the capacitance C. For fast rise-times it is therefore necessary to minimise the leakage inductance of the transformer and the self-capacitance of the

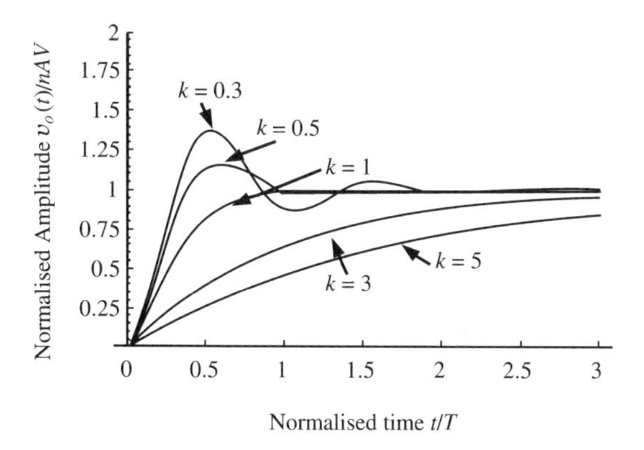

Figure 5.8 Normalised rising edge response waveforms for different values of damping factor k

transformer windings and the load. In both cases there are practical problems, particularly for transformers with large values of the turns ratio n. Clearly for large values of n the referred capacitance from the secondary circuit can be quite large, as it must by multiplied by a factor n^2 in referring it from the secondary to the primary circuit. This problem can be to some extent be reduced if some degree of overshoot in the transformer response is allowed. It is clear from the response curves in Figure 5.8 that the underdamped response profiles, where there is overshoot, rise more quickly than the overdamped response profiles. It is therefore up to the transformer designer to balance the need to reduce rise-time loss at the expense of increasing the overshoot.

This problem can be seen more clearly by rearranging the equation for the damping constant k given in Equations (5.25), i.e.

$$k = \frac{\sqrt{a}}{2}\left(\frac{R_1}{Z} + \frac{Z}{R_2}\right) \quad \text{with} \quad Z = \sqrt{\frac{L_l}{C}} \tag{5.32}$$

In other words the damping constant which determines overshoot is dependent on the impedance $Z = \sqrt{(L_l/C)}$ and the rise-time is dependent on the time constant $T \propto \sqrt{(L_l C)}$.

5.2.4 Pulse Flat Top Response

The response of the transformer to the flat top part of the rectangular input pulse can again be determined by simplifying the equivalent circuit to that given in Figure 5.9. In this case it is recognised that the lower-frequency components of the pulse must be considered such that

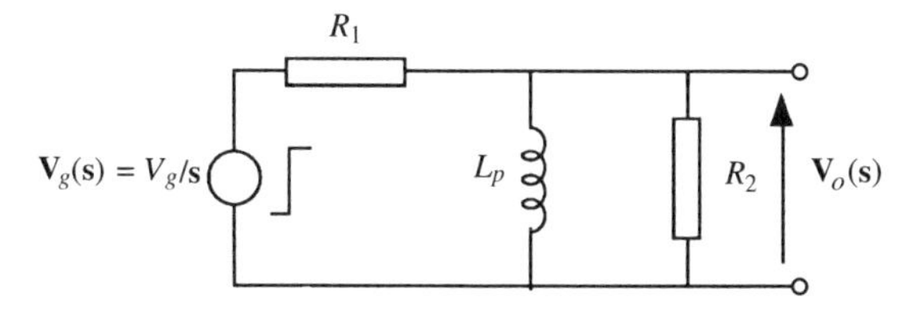

Figure 5.9 Simplified equivalent circuit to determine the flat top response of a pulse transformer

the impedance of the primary magnetising inductance falls and must be taken into account, while the impedances of the leakage inductance becomes small and the shunt capacitance large so that their effects can be neglected.

The output response of this circuit $\mathbf{V}_o(\mathbf{s})$ is given by

$$\mathbf{V}_o(\mathbf{s}) = \frac{V_g}{\mathbf{s}} \frac{\mathbf{s}L_p||R_2}{\mathbf{s}L_p||R_2 + R_1}$$

$$= \frac{aV_g}{\mathbf{s} + \dfrac{R}{L_p}}$$

$$\text{where} \quad a = \frac{R_2}{R_1 + R_2} \quad \text{and} \quad R = \frac{R_1 R_2}{R_1 + R_2} \tag{5.33}$$

The corresponding time response $v_o(t)$, allowing for the transformer turns ratio n, is again found by taking the inverse transform to get

$$v_o(t) = naV_g \exp\left(\frac{-Rt}{L_p}\right)$$

$$\text{noting that} \quad v_o(0) = naV_g \tag{5.34}$$

The output pulse therefore droops, over a pulse duration of t_p, from an amplitude of naV_g by an amount D given by

$$D = naV_g\left[1 - \exp\left(-\frac{Rt_p}{L_p}\right)\right] \tag{5.35}$$

and the fractional droop P expressed as a percentage is given by

$$P = \left[1 - \exp\left(-\frac{Rt_p}{L_p}\right)\right] \times 100\%$$

$$= \frac{Rt_p}{L_p} \times 100\% \quad \text{if} \quad \frac{Rt_p}{L_p} \ll 1 \tag{5.36}$$

It is assumed here that the pulse length is insufficiently long to cause the core of the transformer to saturate, in which case the droop would increase catastrophically. Core saturation will be discussed in a later section of this Chapter.

5.2.5 Trailing Edge Response

Once again the equivalent circuit given in Figure 5.6 can be modified to allow the trailing edge response of a pulse transformer to be determined. The response is a little more difficult to analyse as it depends on the dissipation of charge stored in the stray capacitance of the transformer C and current flowing in the primary magnetising inductance L_p at the end of the pulse. The generator can now be removed from the equivalent circuit and the energy stored in L_p and $\mathbf{C}$ set up initial conditions which must be included in the circuit as

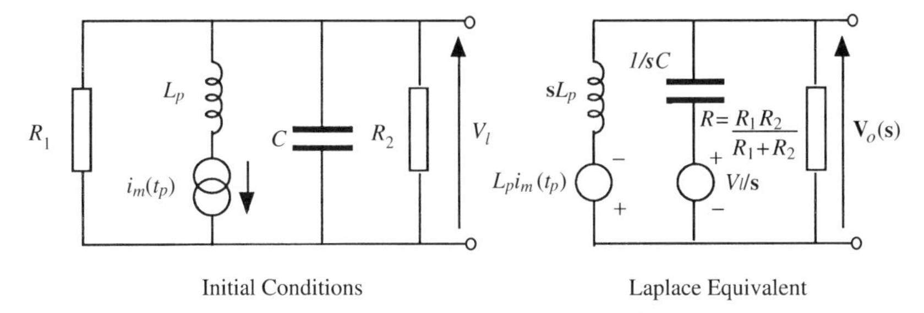

Figure 5.10 Equivalent circuits to determine the trailing edge response of a pulse transformer

current and voltage sources. The equivalent circuits for the analysis in both the time and complex frequency domains are given in Figure 5.10.

The current in the primary magnetising inductance $i_m(t_p)$ can be represented as a current source equal to the current flowing in the inductance at the end of the pulse. Similarly, the charge stored in the stray capacitance of the transformer can be represented as a step voltage source of amplitude V_l which is the voltage on the load at the end of the pulse. If it is assumed that there is little pulse droop during the pulse then the value of $i_m(t_p)$ is given by

$$i_m(t_p) = \frac{1}{L_p}\int_0^{t_p} V_l\,dt = \frac{V_l t_p}{L_p} \quad \text{if} \quad V_l \cong naV_g \tag{5.37}$$

If it is necessary to include an exponential pulse droop then, by reference to the previous section, the value of $i_m(t_p)$ is given by

$$i_m(t_p) = \frac{naV_g}{R}\left[1 - \exp\left(-\frac{Rt_p}{L_p}\right)\right] \tag{5.38}$$

For ease of analysis it will be assumed that $i_m(t_p)$ is given by Equation (5.37). Using superposition from the circuit shown the voltage across R, $\mathbf{V}_o(\mathbf{s})$ is given by

$$\begin{aligned}\mathbf{V}_o(\mathbf{s}) &= \frac{V_l}{\mathbf{s}}\frac{\mathbf{s}L_p||R}{\mathbf{s}L_p||R + \dfrac{1}{\mathbf{s}C}} - V_l t_p \frac{\dfrac{1}{\mathbf{s}C}||R}{\dfrac{1}{\mathbf{s}C}||R + \mathbf{s}L_p} \\ &= V_l\left[\frac{\mathbf{s}}{\mathbf{s}^2 + \dfrac{\mathbf{s}}{RC} + \dfrac{1}{L_pC}} - \frac{\dfrac{t_p}{L_pC}}{\mathbf{s}^2 + \dfrac{\mathbf{s}}{RC} + \dfrac{1}{L_pC}}\right]\end{aligned} \tag{5.39}$$

The common denominator in the last of the equations (5.39) can be factorised to give

$$\mathbf{V}_o(\mathbf{s}) = V_l\left[\frac{\mathbf{s}}{(\mathbf{s}+\beta_1)(\mathbf{s}+\beta_2)} - \frac{\dfrac{t_p}{L_pC}}{(\mathbf{s}+\beta_1)(\mathbf{s}+\beta_2)}\right] \tag{5.40}$$

where β_1 and β_2 are given by

$$\beta_1, \beta_2 = \left[\frac{1}{2RC} \pm \sqrt{\left(\frac{1}{2RC}\right)^2 - \frac{1}{L_pC}}\right]$$

$$= \gamma\left[1 \pm \sqrt{1 - \frac{1}{k'^2}}\right]$$

$$\text{where} \quad \gamma = \frac{1}{2RC} \quad \text{and} \quad k' = \frac{1}{2R}\sqrt{\frac{L_p}{C}} \tag{5.41}$$

k' is, as with the front edge response, a damping factor, which determines whether the trailing edge response is underdamped, critically damped or overdamped. As in the case of the front edge response, defined earlier, a time constant T' can be defined as

$$T' = 2\pi\sqrt{L_pC} \tag{5.42}$$

and therefore β_1 and β_2 become

$$\beta_1, \beta_2 = \frac{2\pi k'}{T'}\left[1 \pm \sqrt{1 - \frac{1}{k'^2}}\right] \tag{5.43}$$

The response in the time domain, once again including the turns ratio factor n, is then given by taking the inverse transform of Equation (5.40). In taking this inverse transform it is assumed that the magnitude of k'^2 is undefined in order to produce a general solution which is independent of the damping condition, i.e.

$$v_o(t) = \frac{nV_l}{\beta_2 - \beta_1}\left[\left(\beta_2 + \frac{t_p}{L_pC}\right)\exp(-\beta_2 t) - \left(\beta_1 + \frac{t_p}{L_pC}\right)\exp(-\beta_1 t)\right] \tag{5.44}$$

The response now depends on the new parameters k' and T' which define β_1 and β_2, and also the term t_p/L_pC which appears in Equation (5.44). This term can be shown to depend on the ratio of the current flowing in L_p at the end of the pulse (the magnetising current) to the current flowing in the load. This ratio is often given the symbol Δ [3] and is given by

$$\Delta = \frac{i_m(t_p)}{I_l} = \frac{V_l(t_p)}{L_p}\Bigg/\frac{V_l}{R} = \frac{t_pR}{L_p} = \frac{1}{2\gamma}\frac{t_p}{L_pC} \tag{5.45}$$

thus the trailing edge response depends not only on k' and T' but also on Δ through β_1 and β_2. Typical response profiles for different k' and Δ values are given in Figure 5.11. Once again the responses are normalised by setting $x = t/T$ and $y = v_o(t)/nV_l$.

5.2.6 Pulse Transformer Magnetic Core

The magnetic core of a transformer serves two main purposes. One is to provide flux linkage between the primary and secondary windings, particularly when the windings are separated on the core, the other is to raise the inductance of the primary winding to reduce the

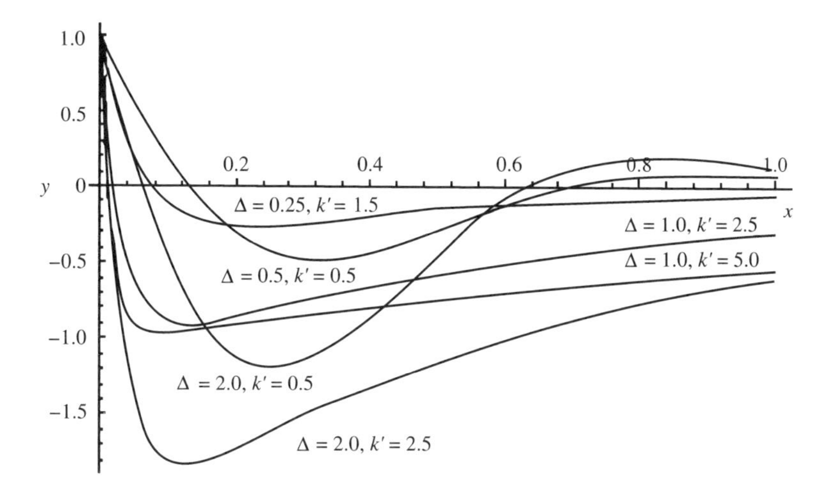

Figure 5.11 Typical falling edge response profiles for different values of k' and Δ

magnetising current. From the preceding analysis of the circuit behaviour of pulse transformers it becomes clear that, for a fast rise-time response, the leakage inductance of the transformer must be minimised and the self-capacitance of the windings kept as low as possible. For low droop and back swing the primary inductance should be kept as high as possible. Unfortunately raising the inductance of the primary tends to increase the leakage inductance. The leakage inductance can be shown to be proportional to the square of the number of turns on a winding and also the volume of the magnetic core that is used. Therefore if the number of turns of the primary is increased, in order to get a large primary inductance, the leakage inductance is also increased. Thus the requirements for fast rise-time and low droop appear to be incompatible. There is a way out of this problem, which is to increase the permeability of the core. From Equation (5.5) the inductance of a winding of n turns on a core with relative permeability μ_r is given by

$$L = \frac{\mu_o \mu_r n^2}{l_e / A_e} = \frac{\mu_o \mu_r n^2}{C_1} \tag{5.46}$$

l_e is the effective length of the magnetic core, usually taken to be the length of the core taken at the mid-point of the core (see the broken line in Figure 5.2) and A_e is the effective area of the core which must be defined if the cross-sectional area of the core varies. Manufacturers of magnetic transformer cores usually define both l_e and A_e or may quote the core factor C_1 which is also given in Equation (5.46).

From Equation (5.46) it can be seen that the inductance of a winding can be increased not only by increasing the number of turns but also by increasing the relative permeability of the core μ_r. For this reason special core materials such as Permalloy (μ_r(max) = 80,000) and Hipersil (μ_r(max) = 12,000) have been developed so that large inductances can be achieved without large numbers of turns.

Although this would seem to be a simple solution to the problem, there is another difficulty caused by the skin effect. The depth δ to which an oscillating electromagnetic field, frequency ω, will penetrate into a magnetic material with permeability μ and conductivity σ, is given by

$$\delta = \sqrt{\frac{2}{\omega\mu\sigma}} \tag{5.47}$$

Therefore, for fast pulses, the magnetic flux may not penetrate the core fully if the relative permeability of the core is high and the effective area of the core is reduced. This in turn reduces the effective permeability of the core and the inductance of a winding on the core. The effective permeability of the core and inductance of the winding will though increase with time as the penetration depth of the magnetic flux into the core increases with time. This effect can to some extent be reduced if the resistivity of the core is kept high. This can be achieved, for example, by using a ferrite core or by laminating the core.

Another difficulty is that the relative permeability of a given core material is not constant due to the nonlinear relationship between B and H in all magnetic core materials. This problem is exacerbated by the position on the BH loop that the core is operating over during a pulse. Figure 5.12 shows a typical BH loop for an initially unmagnetised magnetic core together with typical excursions that the core will make when a succession of identical unipolar pulses are applied to a winding on the core. Each pulse causes a change of flux density ΔB, and as can be seen from the figure a series of small loops are traversed which move up the BH loop until the curve of the main loop is such that a loop is traversed (from point a to point b and back) where the increase and decrease in ΔB is the same. The effective or pulse permeability μ_p is then given by the slope of the line drawn from point a to point b. Clearly, if the amplitude of the pulses applied to a winding is changed then a different loop will result and the pulse permeability μ_p will also be different. If the residual or remanent flux density B_r at the end of the pulse and the loop traversed are large enough, then there is the danger that the maximum saturation flux density B_{sat} for the core will be exceeded and the core will saturate. When this happens the core permeability will fall to a very low value and the inductance of any windings on the core will collapse

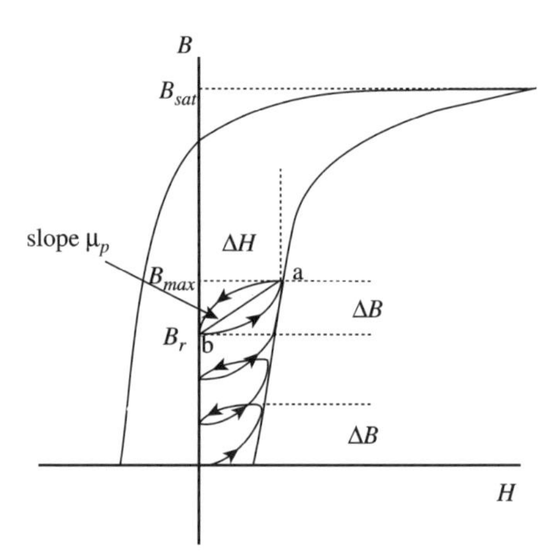

Figure 5.12 Typical excursions on a BH loop of the core of a pulse transformer resulting from a succession of unipolar pulses applied to a winding on the core

catastrophically. Typical values for B_{sat} are 250 to 350 mT for ferrites and 2 to 3 T for silicon iron cores and are temperature dependant.

The value of the flux excursion can be determined by integrating Equation (5.4). In carrying out the integration it is assumed that the amplitude of the pulse V_p, applied to a winding of n turns, is constant throughout the pulse, i.e.

$$\Delta B = \int_0^{t_p} \frac{V_p}{nA}\, dt = \frac{V_p t_p}{nA} \tag{5.48}$$

Thus to avoid core saturation

$$B_{sat} > B_r + \Delta B = B_{\max} \tag{5.49}$$

In practice there are a number of ways in which core saturation can be avoided, in addition to simply using a core material with a high value of B_{sat}. To reduce ΔB it is clear from Equation (5.48) that the number of turns or the area of the core can be increased for a fixed value of the volt-second product $V_p t_p$. To avoid the problem of the residual core flux density B_r, core reset can be used. This is done by applying a negative pulse to the winding or a separate winding, after the main pulse, whose amplitude is chosen to reduce the residual core flux density to zero.

Alternatively an air gap can be introduced into the core, which has the effect of slewing the core *BH* loop to the right such that the residual flux density B_r is reduced for a given peak flux density B_{max}. This then allows the change in flux density ΔB, during the pulse, to be increased without saturating the core. However the introduction of an air gap will also have the effect of reducing the pulse permeability and increasing the reluctance $\mathcal{R}$. It can be seen from Equation (5.5) that this will then reduce the inductance of a winding on the core. The increased reluctance $\mathcal{R}'$ for a core with an air gap of length l_a and core length l_c is given by

$$\mathcal{R}' = \frac{l_c}{\mu_o \mu_r A} + \frac{l_a}{\mu_o A} = \frac{1}{\mu_o A}\left(\frac{l_c}{\mu_r} + l_a\right) \tag{5.50}$$

and the reduced inductance of a winding L' with n turns is given by

$$L' = \frac{n^2}{\mathcal{R}'} = \frac{n^2}{\dfrac{1}{\mu_o A}\left(\dfrac{l_c}{\mu_r} + l_a\right)} \tag{5.51}$$

Since the relative permeability of, for example, a ferrite core can be at least 1000, a relatively thin air gap can have a large effect on the inductance of a given winding.

It was noted, at the start of this chapter, that one of the features of the distortion of a rectangular pulse by a pulse transformer is that a long negative tail is generated after the falling edge of the pulse. The area under this tail is, in fact, equal to the area under the pulse. Referring both to Figure 5.12 and Equation (5.48) the area A_p under the output pulse is given by

$$A_p = \int_0^{t_p} V_p\, dt = nA \int_{B_r}^{B_{\max}} dB = nA(B_{\max} - B_r) \tag{5.52}$$

Similarly the area A_t under the tail is given by

$$A_t = \int_{t_p}^{\infty} V_p \, dt = nA \int_{B_{\max}}^{B_r} dB = nA(B_r - B_{\max}) \tag{5.53}$$

Thus the sizes of the two areas are the same.

5.3 AIR-CORED PULSE TRANSFORMERS

When pulse transformers are to be operated at very high power levels, or with large volt-second products, it is clear that to avoid core saturation, large cores must be used with large numbers of turns (see Equation (5.48)). Eventually, as the power levels are raised, a point is reached where the size of the core required becomes impractical. To tackle this problem a rather different type of pulse transformer has been developed both by G. Rohwein [6] and J. C. Martin [7] in which there is no magnetic core. In order to achieve good flux linkage, between the primary and secondary windings, in the absence of a core, the windings must be wound very close to each other. This is usually achieved by using thin copper foil to wind the transformer rather than wire. The foil windings are wound tightly on top of each other with sheets of flexible plastic material between the windings to provide insulation. Nonetheless the absence of a magnetic core reduces the coupling of flux between windings, particularly in transformers with high gain. Thus the coupling coefficients in these transformers are generally lower than in magnetically cored transformers, and consequently the leakage inductance is usually a much larger fraction of the primary inductance. Another advantage of air-cored transformers is that the frequency limitations imposed on a transformer by its magnetic core are now removed. Consequently these transformers can be used at frequencies of several MHz.

Air-cored transformers are usually used in a different way to the more conventional wire-wound transformers discussed earlier in this chapter. Their predominant use is in charging capacitive loads, such as pulse-forming lines or high-voltage capacitors, to very high potentials from capacitors, charged to much lower potentials, that are then discharged into their primary windings. As such they offer a very compact and simple alternative to the Marx generator, described in Chapter 7 of this book, which is also commonly used for pulse-charging of capacitive loads. These transformers can be operated at impressively high potentials running into the MV range at energy loadings exceeding 10 kJ. The maximum energy loading depends on the width and thickness of the foil windings. An empirical rule is that such transformers can handle in excess of 1 kJ per centimetre width of winding.

There are two basic types of high voltage air-cored transformer that have been developed. The most common is the spiral-strip wound transformer developed both by Rohwein and Martin, which is shown in Figures 5.13 and 5.15. The other is the helical wire/strip type which is sometimes used with a magnetic core at lower power levels as the flux coupling between its windings can be significantly worse than that of the spiral-strip type. Its basic construction is also shown in Figure 5.13.

The main difference between the two types is the way in which insulation is achieved between the primary and secondary windings. In the helical wire/strip transformer the primary turns are wound around the secondary turns and the space between the windings is usually increased from the low to the high voltage end of the transformer to

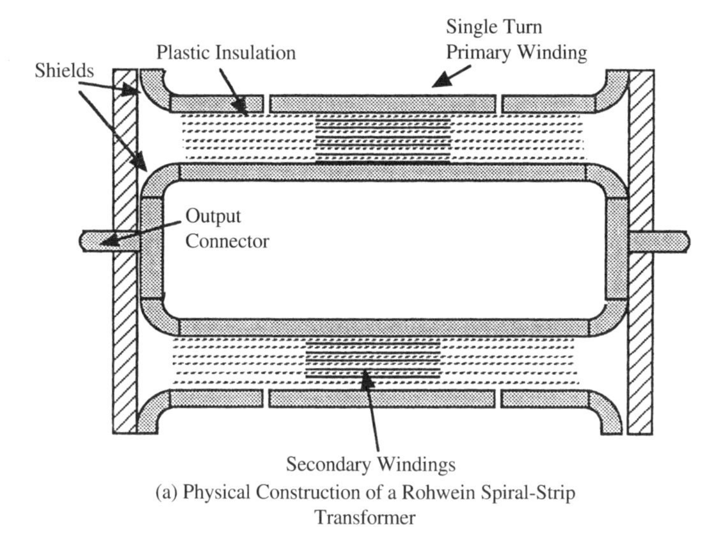

(a) Physical Construction of a Rohwein Spiral-Strip Transformer

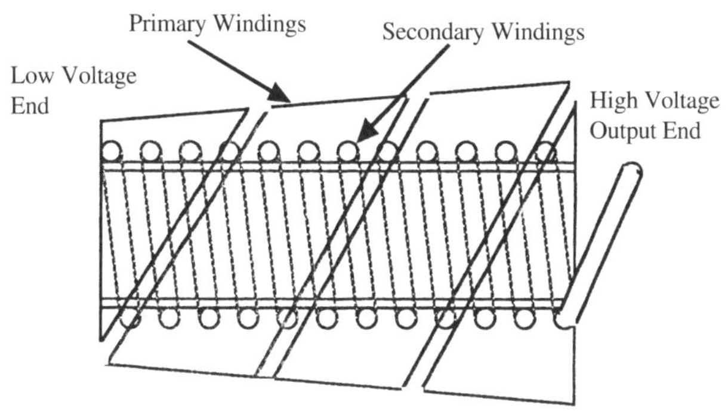

(b) Helical Wire/Strip Transformer

Figure 5.13 Physical construction of a Rohwein helical foil transformer and a spiral strip transformer

provide insulation, thereby preventing voltage breakdown. In the spiral-strip transformers the windings are made from thin copper foil, and the primary winding is either wound on the outside of the secondary winding as in the Rohwein type or in the middle of the secondary windings as in the Martin type. Thus the transformer has a radial voltage gradient across its turns.

Another difference between the Rohwein and Martin transformers is that, in the Rohwein transformer, the primary and secondary windings are isolated from one another. In the Martin transformer the winding takes the form of an autotransformer, with the primary winding usually comprising a few turns located at the centre of the single winding of the transformer, as shown in Figure 5.15. In both cases insulation is provided by plastic insulating sheet, such as polyester film, which is wound in between the foil windings. In the Rohwein and helical wire/strip transformers additional insulation is provided by impregnating and filling any air gaps with oil.

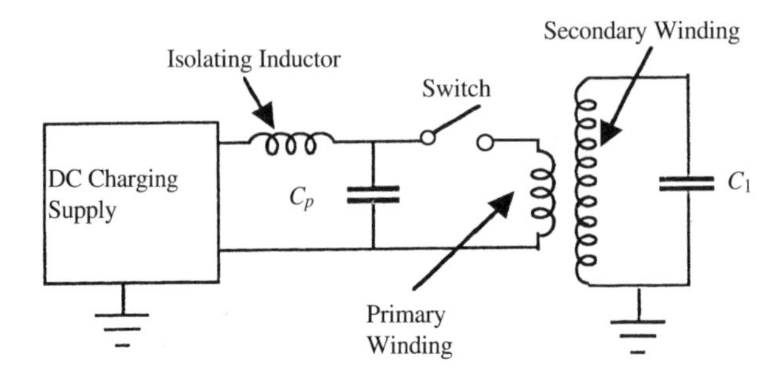

Figure 5.14 Typical operating circuit for a Rohwein helical or spiral strip transformer

A typical circuit for use with either the Rohwein spiral-strip or helical wire/strip wound transformers is given in Figure 5.14. A charged low-inductance capacitor C_p is discharged into the primary winding, usually via a low inductance rail type spark gap switch and the secondary is connected to a load capacitor C_l, which is subsequently charged to a very high potential. It is critical that the primary circuit and its components have very low stray inductance in order to minimise the total series inductance and thereby prevent loss of voltage gain in the transformer. For matched conditions C_p is usually made equal to the referred load capacitance, i.e. n^2C_l, where n is the turns ratio of the transformer.

There are two practical problems of importance in all three types of transformer. The first is electric field enhancement at the edges of the foil or strip windings which can cause breakdown between turns, and the second is the induction of eddy currents in the electric field modifying structures. These structures are used to cause the electric field to run uniformly throughout the transformer without constriction which would lead to breakdown. In the Rohwein transformer, for example, the electric field is controlled by the use of shields (see Figure 5.13) which force the electric field to maintain uniformity through the plastic insulation at the ends of the winding. Without the shields the lines of equal electric potential would tend to bend at the ends of the foil winding, causing the electric field to enhance to the point where breakdown of the insulator becomes possible. The shields must be split, otherwise they would act as loosely coupled shorted turns, which would seriously reduce the voltage gain of the transformer. However even split shields can lead to substantial eddy current loss. This can be avoided by making the shield more transparent to magnetic fields. Rohwein has successfully achieved this transparency by building the shields with sets of conducting concentric rings. Magnetic field can diffuse freely through the rings without inducing significant eddy currents in them but, again, the rings must be split to prevent the shorted turn effect.

The Martin transformer, although in many ways similar to the Rohwein transformer, has some important differences. As explained, the transformer has a single autotransformer type of winding with the primary placed at the centre of the winding. The central location results in a leakage inductance which is smaller than would be the case if the primary was placed at either end of the winding. To avoid breakdown at the ends of the foil winding, the winding is steadily increased in width from the start of the winding to the primary and then steadily reduced in width to its end. By this means the edges of the foil are never adjacent to each other. The transformer is also impregnated with a dilute solution of copper sulphate rather than oil. This causes the lines of equal potential to be resistively graded throughout the transformer windings, which again gives a much more uniform electric field thereby reducing the likelihood of break down. The only disadvantage of this impregnation technique is that it leads to resistive losses which reduce the voltage gain of the transformer, and could lead to serious heating problems were the transformer to be operated at significant pulse repetition rates.

In Figure 5.15 a typical circuit configuration is also shown for the Martin transformer. Since the primary is placed at the centre of the autotransformer winding the capacitor C_p, which is discharged into the primary, must be isolated from the charging supply as it is clear

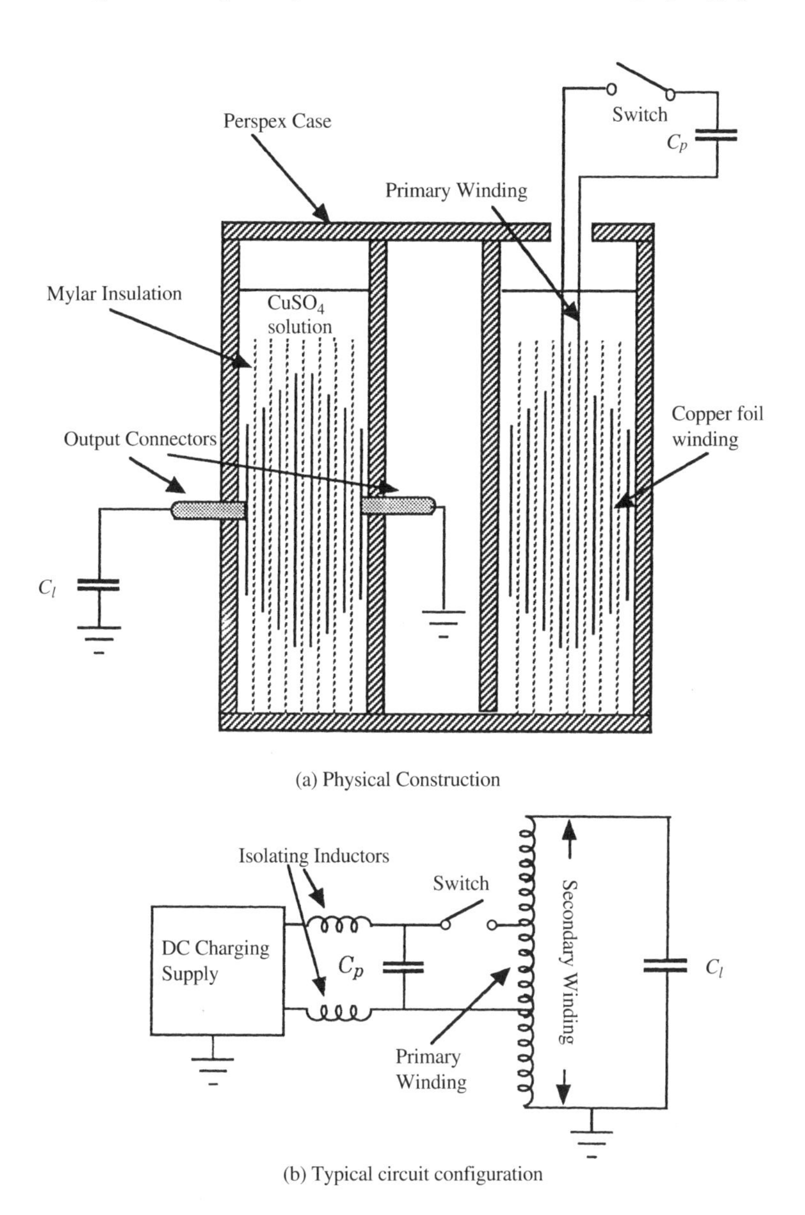

Figure 5.15 Physical construction and typical operational circuit for a Martin foil wound pulse transformer

that, when the switch is closed, the primary windings will rise to a potential equal to half the output potential at the end of the secondary winding. This is most conveniently achieved by the use of isolating charging inductors. Again, it is crucial, to maintain transformer gain, that the primary circuit and its components all have low stray inductance. Also, as before, the value of C_p is usually matched to the load inductance by making $C_p = n^2 C_l$ where n is the turns ratio of the transformer. Figure 5.16 is a photograph of a compact pulse power supply, built by the author, which comprises a primary capacitor bank, a Martin transformer and a coaxial Blumlein pulse-forming line. The tranformer is clearly visible at the bottom right hand side of the picture.

Figure 5.16 Photograph of a compact pulsed power supply that uses a Martin transformer to pulse-charge a coaxial Blumlein pulse-forming line

It should be remembered that with all three types of transformer the lack of a magnetic coil not only reduces the flux linkage between the primary and secondary windings, but also results in a primary winding inductance which can be very low. This problem can be reduced if the core of the transformer is filled with bars or rods of magnetic core material to form a "partial core" [8]. Since the core does not form a magnetic loop and effectively contains a very large air gap, the core reluctance is very large and consequently there is little chance that the core will saturate. However the inductance of the primary can by increased by a factor of at least 2 or 3, which can help to reduce the primary magnetising current.

Although the air-cored transformers described are almost exclusively used to pulse-charge capacitive loads, it is possible to use such transformers with a pulse-forming network to generate high-voltage pulses with rectangular pulse shapes [9,10]. Because the primary inductance is low it is desirable to compensate for the resulting large pulse droop by the use of a pulse-forming network, which is specifically designed to compensate for the droop. Droop-compensating pulse-forming networks for pulse transformers have been described by Ranon *et al.* [11]. Additionally filling the core of the transformer with partial magnetic cores, such as bars of ferrite, can again help to reduce the droop by raising the magnetising inductance of the primary winding. The cores will also help to improve the flux linkage and coupling coefficients between the transformer windings.

5.3.1 Analysis of Air-cored Pulse Transformer Circuit Performance

This analysis will investigate the performance of air-cored transformers when used, as is common, in circuits to charge capacitive loads. An equivalent circuit is once again drawn for the analysis but in this case, as the coupling coefficient k for air-cored transformers is not necessarily close to unity, the circuit is based around the equivalent circuit of a non-ideal transformer rather than a circuit similar to that shown in Figure 5.6. The primary capacitor C_p, initially charged to a potential V, is represented as a capacitor in series with a voltage source $V/\mathbf{s}$.

The equivalent circuit is given in Figure 5.17. In this circuit:

R_1 is the total series resistance in the primary circuit which includes the ESR of the primary capacitor C_p and the resistance of the feeds, switch and primary winding.

R_2 is the total resistance in the secondary circuit and includes the resistance of the secondary winding and any ESR in the capacitive load. It is assumed that there is no load resistance in the circuit and that any other loading resistance such as that from resistive

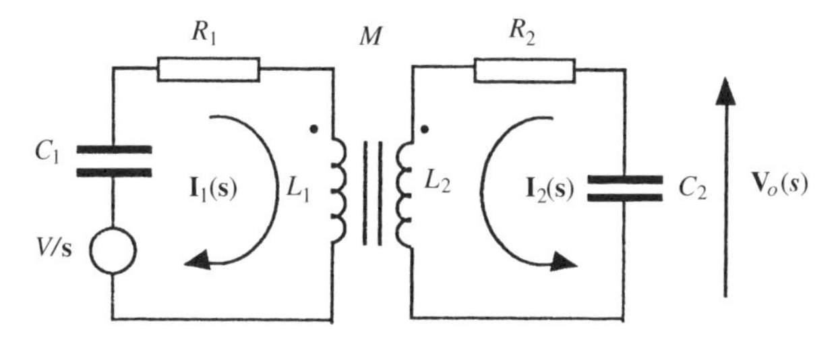

Figure 5.17 Equivalent circuit of an air-cored pulse transformer in a capacitive load pulse-charging circuit

voltage monitors or conductive transformer impregnation can be ignored. If that is not the case a parallel resistor must also be included in the equivalent circuit.

$C_1 (= C_p)$ is the primary discharge capacitor.

C_2 is the load capacitance which is normally, as explained, set equal to to $n^2 C_1$, where n is the turns ratio of the transformer.

L_1 is the inductance of the primary winding, L_2 is the inductance of the secondary winding and the non-ideal behaviour of the transformer is accounted for by including the mutual inductance of the transformer M.

From this circuit, application of Kirchhoff's voltage law results in the following equations

$$\left.\begin{aligned} \frac{V}{\mathbf{s}} &= \mathbf{I}_1(\mathbf{s})\left(\frac{1}{\mathbf{s}C_1} + \mathbf{s}L_1 + R_1\right) - \mathbf{I}_2(\mathbf{s})\mathbf{s}M \\ 0 &= \mathbf{I}_2(\mathbf{s})\left(\frac{1}{\mathbf{s}C_2} + \mathbf{s}L_2 + R_2\right) - \mathbf{I}_1(\mathbf{s})\mathbf{s}M \\ \mathbf{V}_o(\mathbf{s}) &= \frac{\mathbf{I}_2(\mathbf{s})}{\mathbf{s}C_2} \end{aligned}\right\} \tag{5.54}$$

From these equations the output voltage can be determined to be

$$\mathbf{V}_o(\mathbf{s}) = \frac{VM}{C_2} \frac{\mathbf{s}}{(L_1L_2 - M^2)\mathbf{s}^4 + (R_1L_2 + R_2L_1)\mathbf{s}^3 + \left(\frac{L_1}{C_2} + \frac{L_2}{C_1}\right)\mathbf{s}^2 + \left(\frac{R_1}{C_1} + \frac{R_2}{C_2}\right)\mathbf{s} + \frac{1}{C_1C_2}} \tag{5.55}$$

The denominator of this equation cannot be factorised so, to proceed with the analysis, it would be necessary to put in numerical values, corresponding to a particular circuit, find the roots of the polynomial in **s** in the denominator and then split the expression into partial fractions in order to find the inverse transform for the output voltage $v_o(t)$.

However it is worth simplifying the expression to get some idea of the shape of the waveform that is produced in this type of circuit. This can be done by assuming that the resistive losses in the circuit can be neglected, i.e. $R_1 = 0$ and $R_2 = 0$. In practice this is not a particularly good approximation, as the resistive losses will usually have a significant dampening effect on the output voltage waveform and cause loss of voltage gain from the transformer. The simplified expression is therefore

$$\mathbf{V}_o(\mathbf{s}) = \frac{VM}{C_2} \frac{\mathbf{s}}{(L_1L_2 - M^2)\mathbf{s}^4 + \left(\frac{L_1}{C_2} + \frac{L_2}{C_1}\right)\mathbf{s}^2 + \frac{1}{C_1C_2}} \tag{5.56}$$

From Equation (5.17) M^2 is given by

$$M^2 = k^2 L_1 L_2 \quad \text{and} \quad \therefore\ L_2L_2 - M^2 = L_1L_2(1 - k^2) \tag{5.57}$$

Therefore Equation (5.56) can be simplified to

$$\mathbf{V}_o(\mathbf{s}) = \frac{Vk}{C_2\sqrt{L_1L_2}} \frac{\mathbf{s}}{(1 - k^2)\mathbf{s}^4 + \left(\frac{1}{L_1C_1} + \frac{1}{L_2C_2}\right)\mathbf{s}^2 + \frac{1}{L_1L_2C_1C_2}} \tag{5.58}$$

This can now be factorised to give

$$\mathbf{V}_o(\mathbf{s}) = \frac{Vk}{C_2\sqrt{L_1L_2}} \frac{\mathbf{s}}{(\mathbf{s}^2 + \omega_a^2)(\mathbf{s}^2 + \omega_b^2)} \tag{5.59}$$

where the squares of the angular frequencies ω_a and ω_b of the denominator are given by

$$\omega_a^2, \omega_b^2 = \left[\frac{1}{2(1-k^2)}\left(\frac{1}{L_1C_1} + \frac{1}{L_2C_2}\right) \pm \frac{1}{2(1-k^2)}\sqrt{\left(\frac{1}{L_1C_1} + \frac{1}{L_2C_2}\right)^2 - \frac{4(1-k^2)}{L_1L_2C_1C_2}}\right] \tag{5.60}$$

By reference to tables of Laplace transforms, the inverse transform $v_o(t)$ can now be found to be

$$v_o(t) = \frac{Vk}{C_2\sqrt{L_1L_2}} \frac{\cos(\omega_b t) - \cos(\omega_a t)}{\omega_a^2 - \omega_b^2} \tag{5.61}$$

Evidently there are two oscillations of interest in the solution with different angular frequencies ω_a and ω_b. In the case where the coupling coefficient k is zero it is possible to factorise the denominator in Equation (5.58) and hence get expressions for the angular frequencies ω_a and ω_b, i.e.

$$\omega_a = \frac{1}{\sqrt{L_1C_1}} = \omega_1 \quad \text{and} \quad \omega_b = \frac{1}{\sqrt{L_2C_2}} = \omega_2 \tag{5.62}$$

Therefore, in this case, the frequencies ω_a and ω_b correspond to the resonant frequencies of the uncoupled primary and secondary circuits. In the case where the coupling coefficient is close to unity, it is clear from inspection of Equation (5.60) that one of the two frequencies will be higher than the other. If the higher of the two frequencies is arbitrarily selected to be ω_a, then the shape of the output voltage waveform $v_o(t)$ can be made by sketching the difference between the two cosine terms in Equation (5.61). Such a sketch is given in Figure 5.18 shown below.

The sketch reveals that the lower-frequency cosine term cos(ω_b) is modulated by the higher-frequency cosine term cos(ω_a). Thus the peak of the waveform occurs at an absolute

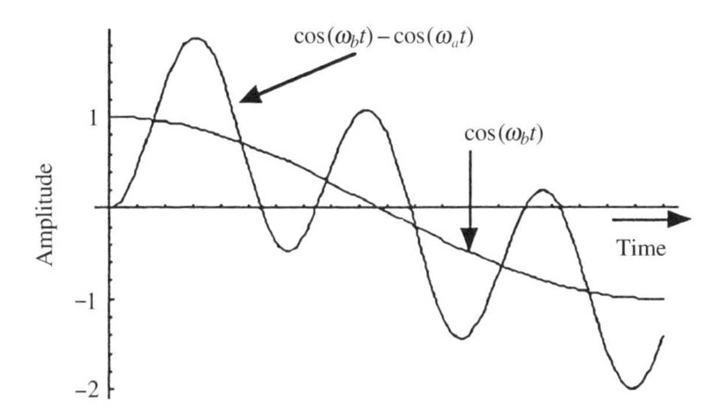

Figure 5.18 Approximate sketch of the output voltage $v_o(t)$ for the circuit given in Figure 5.15

value somewhat less than two, depending on the relative size of the two frequencies ω_a and ω_b. It is, though, important to remember that resistive damping effects have been neglected in this analysis and such damping will cause the amplitude of the output voltage waveform and the gain of the transformer to be progressively reduced as the frequency of the higher frequency component ω_a is reduced.

5.3.2 Dual Resonant Operation of Air-cored Pulse Transformers

The way in which the output voltage waveform $v_o(t)$ is determined by the relative frequencies ω_a and ω_b suggests that it may be possible to manipulate these frequencies so that there is a complete transfer of energy from the primary storage capacitor to the secondary load capacitor [12]. Examination of the waveform sketch in Figure 5.18 suggests that if the frequency of the higher-frequency component ω_a is lowered, then it would be possible to arrange that a minimum or maximum of the $\cos(\omega_a)$ waveform coincides with a minimum or maximum of the lower frequency $\cos(\omega_b)$ waveform. For example, it can be seen that the amplitude of the first positive peak of the $\cos(\omega_b)-\cos(\omega_a)$ waveform in Figure 5.18 is actually less than the amplitude of the third negative peak where there is a coincidence between the $\cos(\omega_a)$ and the $\cos(\omega_b)$ waveforms. For this to happen, then the two frequencies ω_a and ω_b would have to be related by

$$\omega_b = \frac{n\pi}{T} \quad \text{and} \quad \omega_a = \frac{(n+2m+1)\pi}{T}$$
$$m = 0, 1, 2, 3 \ldots \quad \text{and} \quad n = 1, 2, 3 \ldots \tag{5.63}$$

where T is the time at which the two minima or maxima coincide. This time is minimised if $n = 1$ and $m = 0$ which yields the condition for dual resonance, i.e. $\omega_a = 2\omega_b$. If there is to be a complete transfer of energy E from the primary storage capacitor to the secondary load capacitor then

$$E = \frac{1}{2}C_1V^2 = \frac{1}{2}n^2C_2V^2 \quad \text{or} \quad C_1 = n^2C_2 \tag{5.64}$$

Using Equation (5.18), this condition then implies that the two open circuit resonant angular frequencies of the primary and secondary circuits must be equal, i.e.

$$\omega_1 = \frac{1}{\sqrt{L_1C_1}} = \frac{1}{\sqrt{L_2C_2}} = \omega_2 \tag{5.65}$$

Substituting this relationship into Equation (5.58) gives

$$\mathbf{V}_o(\mathbf{s}) = \frac{Vk}{C_2\sqrt{L_1L_2}} \frac{\mathbf{s}}{(1-k^2)\mathbf{s}^4 + 2\omega_1^2\mathbf{s}^2 + \omega_1^4} \tag{5.66}$$

The value of the coupling coefficient k needed to achieve the dual resonance can now be found by determining the two angular frequencies ω_a and ω_b by factorisation of the denominator of Equation (5.66) and setting the condition that $\omega_a = 2\omega_b$. The two frequencies are then given by

$$\omega_a^2 = \frac{\omega_1^2}{(1-k)} \quad \text{and} \quad \omega_b^2 = \frac{\omega_1^2}{(1+k)} \tag{5.67}$$

Therefore

$$\frac{\omega_a^2}{\omega_b^2} = \frac{1+k}{1-k} = 4 \Rightarrow k = \frac{3}{5} \tag{5.68}$$

This value of the coupling coefficient k is, in practice, fairly easy to achieve, although once again it must be emphasised that this analysis has neglected the effects of resistive damping, and if damping is included a different value of k would result and perfect energy transfer from primary storage capacitor to secondary load capacitor would not be achievable. The condition on k can be extended to different n and m values in Equation (5.63) so that a general expression for k as a function of n and m can be determined to be

$$k = \frac{\left(\frac{n+2m+1}{n}\right)^2 - 1}{\left(\frac{n+2m+1}{n}\right)^2 + 1}, \quad m = 0, 1, 2, 3 \ldots \quad \text{and} \quad n = 1, 2, 3 \ldots \tag{5.69}$$

The voltage gain can be further increased by not setting L_1C_1 equal to L_2C_2. By this means a further increase in gain of 18% can be achieved [13]. This concept of tuning the primary and secondary circuits of pulse transformers into resonance can be extended to improve the efficiency of more complex circuits. For example, any stray inductance in the connections from the transformer secondary to the load can also be tuned by the addition of a capacitor [14]. The circuit will then have three characteristic frequencies which can be tuned into resonance thereby optimising the energy transfer efficiency from primary capacitor to load capacitor. Clearly the analysis for the conditions for triple resonance to occur becomes more difficult. However the use of matrix methods can simplify such an analysis, which then shows that the simplest resonant mode, which is also the mode of greatest practical interest (lowest resistive damping), is that in which the three characteristic frequencies of coupled circuits are in the ratio 1 : 2 : 3.

5.4 PULSE TRANSFORMERS WITH MULTIPLE WINDINGS

As explained in this chapter, one of the major concerns for the designers of pulse transformers is degradation or loss of rise-time of a fast rising pulse by a pulse transformer. This loss is mainly caused, in magnetically cored pulse transformers, by leakage inductance and the self-capacitance of the windings of the transformer. This problem is exacerbated in transformers where the secondary to primary turns ratio and hence voltage gain is large. Although careful physical design of pulse transformers can minimise these parameters, other considerations, such as insulation within the transformer, limit the extent to which they can be reduced. An alternative method for reducing their effect is to build transformers with multiple primary windings and there are a number of ways in which this can be done. Three possibilities are shown in Figures 5.19, 5.20 and 5.21. In all three circuits it is assumed that the coefficient of coupling between windings k is close to unity, and therefore the approximation $L_p \approx L_{mp}$ can be made.

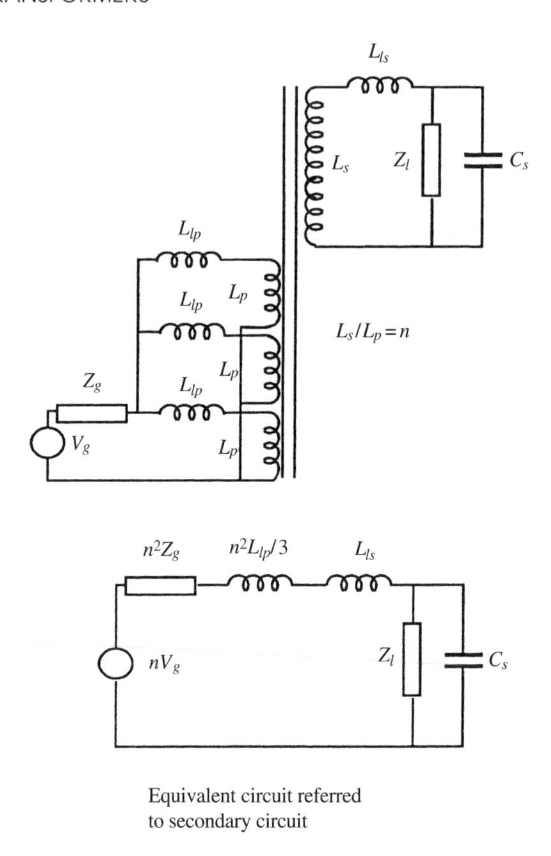

Figure 5.19 First type of pulse transformer with multiple primary windings

In the circuit shown in Figure 5.19, three primary windings are wound on a magnetic core separately from a single secondary winding. The primary circuits can be referred to the secondary circuit by multiplying the generator impedance Z_g and the primary leakage inductance L_{lp} by n^2, where n is the turns ratio between a single primary winding and the secondary. By using three primary windings rather than one, it can be seen from the equivalent circuit referred to the secondary circuit that the primary leakage inductance is reduced by a factor of 3. This will clearly help to reduce the rise-time of the transformer and, by increasing the number of primary circuits still further, will reduce the effect of the primary leakage inductance even more.

The reduction in primary leakage in the circuit shown in Figure 5.19 can be made more effective if the primary windings are wound on top of a continuous secondary winding. The close physical proximity of the primaries to the secondaries results in a higher coefficient of coupling k and consequently the primary leakage inductance is reduced to a lower value L'_{lp}. This reduced value can then be referred to the secondary circuit, as shown in the Figure. The use of three generators, while adding complexity to the circuit, reduces the effective generator impedance in the secondary circuit by a factor of three and the increased current capability of the referred generators will also help to reduce rise-time.

This circuit is particularly useful when it is desirable to build the generators from energy storage capacitors which are switched by semiconductor switches such as thyristors, GTOs

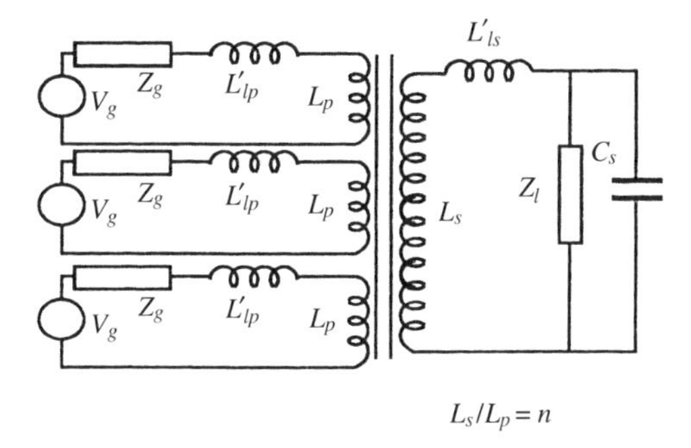

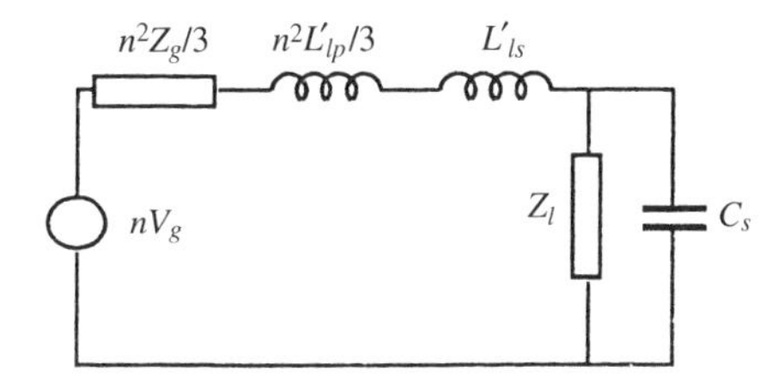

Figure 5.20 Another type of pulse transformer with multiple primary windings

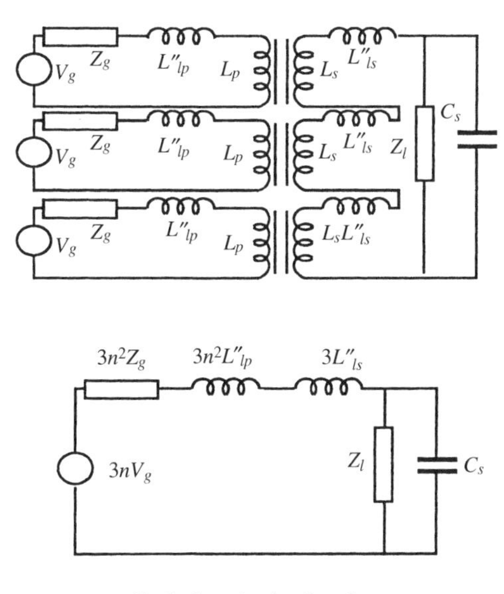

Figure 5.21 Pulse transformer built from a number of individual transformers with secondaries connected in series.

or IGBTs. Such semiconductor switches tend to have relatively low voltage ratings of a few kV, so when transformers with large voltage gains are required, many semiconductor switched generators can be connected to a series of primary windings which are wound sequentially over a long secondary winding [15]. Such a transformer need not be magnetically cored although it is desirable for air-cored devices to fill the inside of the secondary winding with a partial magnetic core to keep the inductance of the primary windings as high as possible.

In air-cored transformers of this type, strong coupling tends to be localised between an individual primary winding and the region of the secondary over which it is wound. Thus such transformers tend to operate as a series of individual transformers whose secondaries are connected in series, rather like the circuit shown in Figure 5.21. Consequently the equivalent circuit shown in this figure will be a more accurate representation of the behaviour of the transformer than the equivalent circuit given in Figure 5.20.

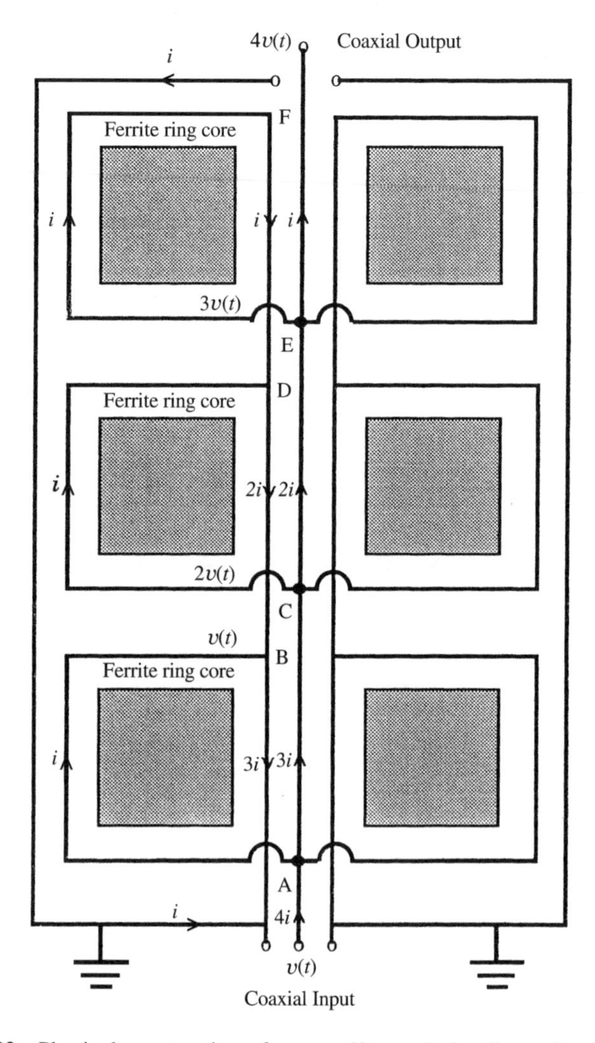

Figure 5.22 Physical construction of a wound/transmission line pulse transformer

In the circuit shown in Figure 5.21, the use of individual transformers can reduce the primary and secondary leakage inductances L''_{lp} and L''_{ls} still further, with careful design, leading to the optimum rise-time performance of the three types of transformer discussed. Note also that the output voltage of each of the secondaries adds linearly, rather than by a simple factor of n for the transformer with a single long secondary winding, as shown in Figure 5.20. Again, this type of transformer is particularly suitable for high voltage gains if the primary windings are driven by semiconductor switched generators.

5.5 HYBRID WOUND/TRANSMISSION LINE PULSE TRANSFORMER

To reduce the effects of leakage and stray inductance in pulse transformers still further, Cassel reports [16] on a rather different design which is capable of very fast rise-times. Although the transformer is built from three wound 1 : 1 transformers, which can also be thought of as auto-transformers with a turns ratio of 1 : 2, the transformers construction has very low stray inductance due to its coaxial geometry which is similar to a coaxial transmission line. A diagram of the physical construction of the transformer is given in Figure 5.22, and an equivalent circuit is given in Figure 5.23. The action of the transformer is not easy to understand so, in both diagrams, the currents that flow in the transformer and its equivalent circuit are marked and the current nodes are labelled alphabetically. The various potentials at points through the transformer are also marked.

The input to the transformer is connected to two windings around the bottom ferrite ring core. The first winding passes around and through the core returning to ground, while the other passes up through the core on to the next stage of the transformer. This winding is then

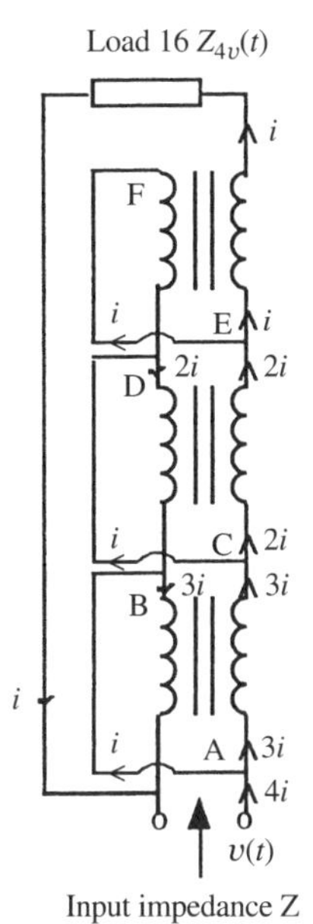

Figure 5.23 Equivalent circuit of the fast pulse transformer shown in Figure 5.20

connected to two further windings. Again, one of these windings passes around and through the second stage core back to the inside of the turn on the first core and the other passes up through the second core on to the third stage. This process repeats itself in the third stage and the final turn connects to the output of the transformer. The transformer therefore has a voltage gain of 4 and an impedance transformation from input to output of 16.

It is possible to construct other variants of the device such as an inverting transformer and an induction transformer. Also, as with most other transformers, it can be driven from the output end to the input to create a current transformer. However, because of the very low inductance that results from a single turn on the ferrite cores, the device is only really suitable for high power short pulse operation (10's of ns).

REFERENCES

[1] Millman J. and Taub H. "Pulse Digital and Switching Waveforms". McGraw-Hill (1965) ISBN 0-07-085512-9

[2] Coekin J. A. "High-Speed Pulse Techniques". Pergamon Press (1975) ISBN 0-08-018774-9

[3] Lee R. "Electronic Transformers and Circuits". John Wiley & Sons Inc.

[4] Grover F. W. "Inductance Calculations". Van Nostrand (1947) and Dover Publications (1973)

[5] Snelling E. C. "Soft Ferrites". Butterworths (1988) ISBN 0-408-02760-6

[6] Rohwein G. J. "Aircore Pulse Transformers for High Power Lasers". Laser Focus (1980) 70–74; and "A Three Megavolt Transformer for PFL Pulse Charging". *IEEE Trans. on Nuclear Science* **NS-26** (1979) 4211–4213

[7] Martin J. C., Champney P. D. and Hammer D. A. "Notes on the Construction Methods of A Martin High-Voltage Pulse Transformer". Cornell University Report (1967); and Martin J. C. and Smith I. D. "High Voltage Pulse Generating Transformer". US Patent No. 3456221 (1969)

[8] Lawson R. N. and Rohwein G. J. "A Study of Compact Lightweight High Voltage Inductors with Partial Magnetic Cores". Proceedings of the 7th IEEE Pulsed Power Conference (1989) 906–908

[9] Khan K. A. and Colclaser R. G. "Air-core Foil-wound Pulse Transformer Design Concept". Proceedings of the 9th IEEE Pulsed Power Conference (1993) 884–887

[10] Rohwein G. J., Lawson R. N. and Clark M. C. "A Compact 200 kV Pulse Transformer System". Proceedings of the 8th IEEE Pulsed Power Conference (1991) 968–970

[11] Ranon P. M., Hall D. J., O'Loughlin J. P., Weyn M. L., Baker W. L., Scott M. C. and Adler R. J. "Synthesis of Droop Compensated Pulse Foming Networks for Generating Flat Top High Energy Pulses into Variable Loads from Pulsed Transformers" Proceedings of the 18th Power Modulator Symposium (1988) 54–61

[12] Finkelstein D., Goldberg P. and Shuchatowitz J. "High Voltage Impulse System". *Rev. Sci. Instrum.* **37** (1966) 159–162

[13] Reed J. L. "Greater Voltage Gain for Tesla-Transformer Accelerators". *Rev. Sci. Instrum.* **59** (1988) 2300–2301

[14] Bieniosek F. M. "Triple Resonance Pulse Transformer Circuit". *Rev. Sci. Instrum.* **61** (1990) 1717–1719

[15] Lawson R. N. and Rohwein G. J. "An SCR Switched High Voltage High Gain Linear Transformer System". Proceedings of the 7th IEEE Pulsed Power Conference (1989) 762–765

[16] Cassel R. L. "A 200 kV 100 nsec 4/1 Pulse Transformer". Proceedings of the 20th Power Modulator Symposium (1992) 234–236

6

Transmission Line Pulse Transformers

6.1 INTRODUCTION

In this chapter three basic types of transmission line pulse transformer will be described. The first two types are constructed from a number of equal lengths of transmission line which are either laid out in a linear fashion or wound on to magnetic cores to isolate the secondary circuit from the primary. In the third type a single line is used whose cross-sectional geometry is slowly altered from the input of the line to the output. Usually this change is made by tapering the line either by increasing or decreasing the spacing between the conductors or by altering the width of the two conductors which make up the transmission line.

The use of lengths of transmission line to construct broad-band matching transformers in radio-frequency circuits is a well-established technique [1]. However, rather surprisingly, the use of transmission line transformers in power circuits, particularly pulsed power circuits, is still relatively uncommon despite the potential advantages such devices can provide in comparison to conventional magnetically cored transformers. For example, the rise-time and band-width of a conventional pulse transformer, as seen in the last chapter, can be severely limited by the effects of the leakage inductance in the transformer and the self-capacitance of its windings. Pulse droop is also a problem which can only be corrected by increasing the inductance of the primary winding. This in turn also tends to increase the leakage inductance, and consequently the rise-time of the transformer is compromised. Thus pulse droop and pulse rise-time are perhaps the two most important parameters that should be specified in characterising a particular pulse transformer design. Transmission line transformers, on the other hand, can achieve extremely fast rise-times which are, in certain circumstances, limited only by frequency-dependent loss due to skin-effect problems in the conductors of the transmission lines which are used to construct the transformer [2]. As will be shown later in this chapter, it is also possible to achieve very low pulse droop in this type of transformer. Thus if a performance figure of merit is taken as the ratio of length of pulse (within certain droop limits) to pulse rise-time, the performance of the transmission line transformer can far exceed that of conventional magnetically cored devices.

6.2 LINEAR TRANSMISSION LINE TRANSFORMERS

There are several different types of linear transmission line transformer structure that have been reported which include the stacked transmission line transformer [3,4], the stacked

Blumlein generator [5,6], and the Darlington circuit [7,8], discussed in Chapter 3. The transformers are usually constructed from either coaxial cable transmission lines or from strip-lines. The type of transformer chosen will depend on the application since each of the different types can offer special operational advantages in driving particular loads. Typical applications for linear transmission line transformers include pulsed power supplies for flash X-ray generators [9] and as step-down transformers for plasma generators [10] and to pulsed high-current electron accelerators [11]. Although transmission line transformers are usually used to produce voltage gain, transformer configurations which give both pulse inversion [12], pulse isolation [13] and current gain [14] are also possible.

It is clear that if transmission line transformers are to realise their full potential, reliable models of their operation must be available so that designs can be produced whose performance characteristics such as voltage or current gain, pulse rise-time and pulse droop can be accurately predicted. It is the purpose of this section to show how such performance parameters can be reliably predicted for the simplest forms of the transmission line transformer, in which one or more transmission lines of equal length are used to construct inverting and voltage transformers. The abbreviation TLT (for Transmission Line Transformer) will be used henceforth for this type of transformer. Also in this section the modelling of linear devices is described, in which it is assumed that the transmission line transformer is constructed using straight and parallel transmission lines of equal length that are mounted above some sort of ground plane. In the next section the use of inductive isolation is discussed where the transmission lines, used to construct the transformer, are wound inductively to reduce pulse droop.

Some modelling work has already been published on transmission line pulse transformers. Matick, for example, has made a detailed study of the way in which transformers built from parallel lines above ground planes operate [14]. He shows that the presence of parasitic secondary transmission lines within the transformer structures lead to the generation of pulses additional to the main pulse which are caused by multiple reflections within the structures. As this section will show, the effect of these secondary lines can be serious if they are not properly accounted for in the design of a given TLT. Formulae for calculating the loss of voltage gain due to secondary transmission lines between the outer conductors of coaxial lines and ground in voltage transformers have been derived by Lewis and Wells [4] with a modified version given by Wilson, Erickson and Smith [15]. This formulation will be discussed later in the chapter. A more detailed mathematical modelling method for predicting the effect of loss of gain in TLTs based on the incorporation of all the possible secondary transmission lines that can exist in a TLT has been given by Chowdorow [16]. However this model does not show how reflections propagating back and forth along the secondary lines lead to a "step-wise" pulse droop.

In this section models for a simple linear inverting transformer and also a voltage transformer are given which allow the calculation of both pulse droop rate and voltage gain, in the case of the voltage transformer. The models can be extended to deal with more complicated structures, although it will be shown that a complete model of such devices is only practical using computer techniques. In the configuration to be described (the "linear, parallel" transformer) the effect of the secondary transmission lines in the devices is the dominant consideration and a method for estimating their effect on gain and pulse droop is given. Although the modelling method specifically refers to linear TLTs, the principles and methods given should be readily adaptable to other types of transmission line transformer such as the linear stacked Blumlein TLT [5,6].

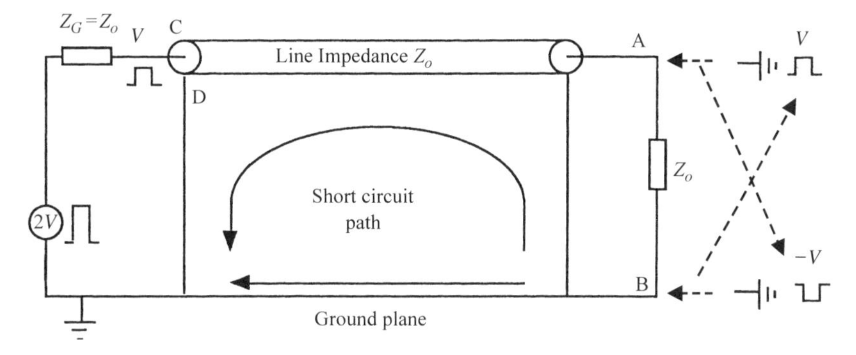

Figure 6.1 A 1 : 1 linear inverting transmission line transformer

6.2.1 The 1 : 1 Inverting Transformer

The simplest and most easily understood transformer is shown in Figure 6.1. As with the description of this device and other transformers discussed later, it is assumed that the transformer is built from a single length of coaxial line. This need not be the case and other types of transmission line can work equally well.

A rectangular pulse of duration t_p and amplitude V is injected into an arbitrary length of transmission line of characteristic impedance Z_o from a matched generator of internal impedance $Z_G = Z_o$. The outer conductor of the line is grounded. After a propagation delay due to the length of line used, a pulse of amplitude V appears at the output end of the line across a matched load Z_o. If point B is grounded then a positive pulse is produced at point A. If however, point A is grounded then a negative pulse of amplitude $-V$ will be produced at point B. This is because the inner conductor of the line is at a potential V which is positive with respect to the outer conductor. Therefore, as the inner conductor is grounded at the output end of the line, the potential of the outer conductor is forced to a negative potential $-V$. Further consideration of this circuit reveals that if point A is grounded a short circuit is set up in the circuit, i.e. point B is connected to point D through the ground plane or through the outer conductor of the transmission line as shown.

If the circuit is now redrawn, as shown in Figure 6.2, it can be seen that a second transmission line has effectively been set up between the outer conductor and the ground

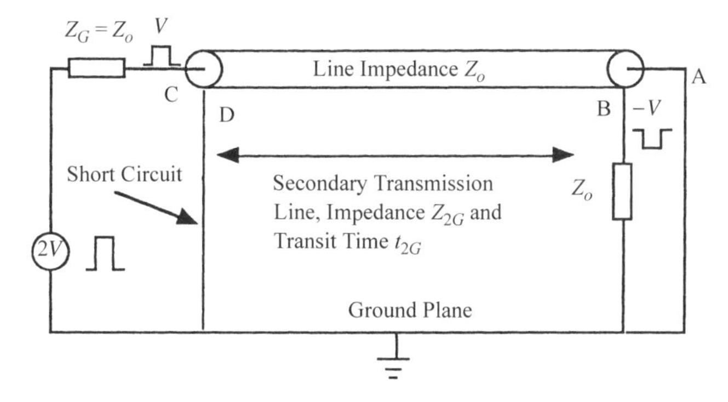

Figure 6.2 Redrawn version of the circuit shown in Figure 6.1

plane. Since this line is air-insulated the propagation velocity along the line will be quicker than the propagation in the primary coaxial line by a factor $\sqrt{\epsilon_r}$ where ϵ_r is the relative permittivity or dielectric constant of the material used to insulate the coaxial line. The relative impedance of this second line Z_{2G} is also likely to be higher than that of the primary line due to the air insulation and the distance of the outer conductor of the primary line to the ground plane.

Since point A in the circuit is grounded a negative pulse of amplitude $-V$ now propagates along the secondary line back towards the input at D. At D a short circuit exists, as shown, which has a reflection coefficient of $\rho = -1$. Thus the pulse now inverts and starts to propagate back towards the output end of the line.

When the pulse reaches point B again, it adds a positive component to the existing negative pulse at point B and causes the negative potential to reduce by an amount which depends on the relative sizes of the impedances Z_o and Z_{2G}. As Z_{2G} is normally much larger than Z_o, the reflection process continues because the secondary line is not matched at point B. The reflection coefficient at B is thus given by

$$\rho = \frac{Z_o - Z_{2G}}{Z_o + Z_{2G}} \tag{6.1}$$

As Z_{2G} is normally much larger than Z_o, the reflection coefficient at B will be negative so a new negative pulse is propagated back towards point D and the process continues. This results in a step-wise droop on the negative output pulse at point B, the width of the steps being equal to twice the transit time along the secondary line.

It is clear from this description that if the pulse being generated on the load has a width which is less than twice the propagation or transit time along the secondary line, then a negative pulse without droop will be satisfactorily produced at the output end of the line at point B. However the amplitude of the pulse will, in fact, be less than $-V$ because the load seen by the primary line is equal to Z_o in parallel with Z_{2G}. It is clear then, in this simplest form of transformer, that it is desirable to ensure that $Z_{2G} \gg Z_o$ and the duration of the pulse t_p is less than twice the transit time along the secondary line t_{2G}.

To avoid having to use excessive lengths of cable for longer pulses, the effect of the short circuit path or secondary transmission line can be suppressed by winding the primary transmission line on a magnetic core. This has no effect on the pulse propagation characteristics of the primary line but adds substantial inductance to the short circuit path from point B to point D. The impedance that results from this inductance is chosen to be large for the duration of the pulse so that little current flows from point B to point D. The inductance can be thought of as being placed in series with the line as shown in the lower diagram of Figure 6.3.

If the effect of the inductance introduced into the short-circuit path is considered rather than the behaviour of the now suppressed secondary transmission line, it is possible to assess how much inductance is required for a given amount of droop on a pulse of length t_p. This can be done by realising that the inductance is placed effectively across the load in series with the output of the line as shown in Figure 6.4.

The output voltage $v_o(t)$ on the matched load is found to be given by

$$v_o(t) = -V \exp\left(\frac{-Z_o t}{2L}\right) \tag{6.2}$$

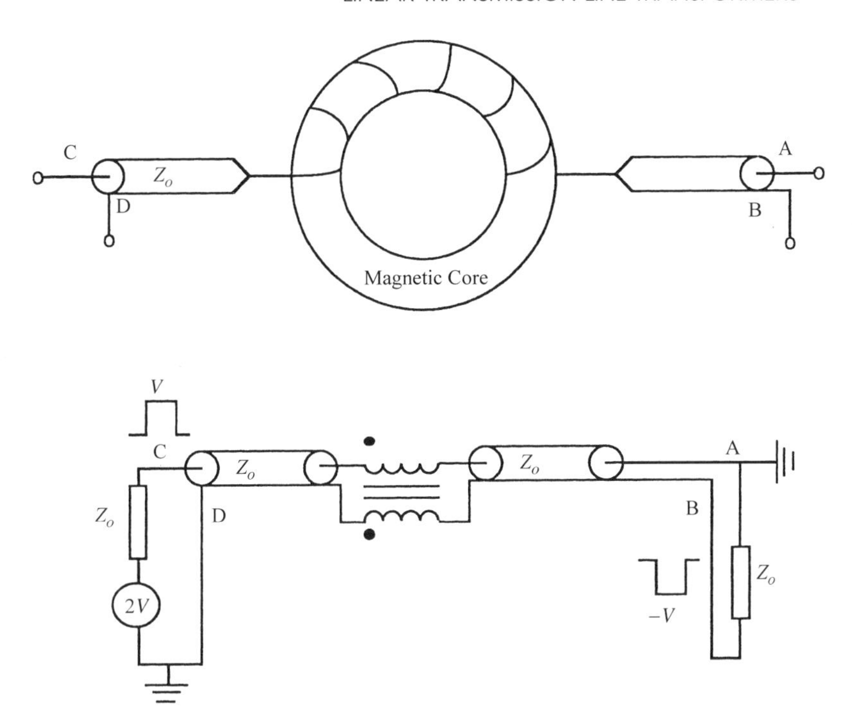

Figure 6.3 Inductive isolation of the short circuit path in the 1 : 1 inverting transmission line pulse transformer

Thus, as the current starts to increase in the inductor an exponential droop will appear on the pulse with time constant $2L/Z_o$. Clearly, if this time constant is made very large compared to the duration of the pulse t_p, the droop will be negligible.

6.2.2 The Two-stage Voltage Transformer

Figure 6.5 is a diagram of a simple two-stage voltage transmission line transformer, which consists of two equal lengths of coaxial transmission line mounted parallel to each other

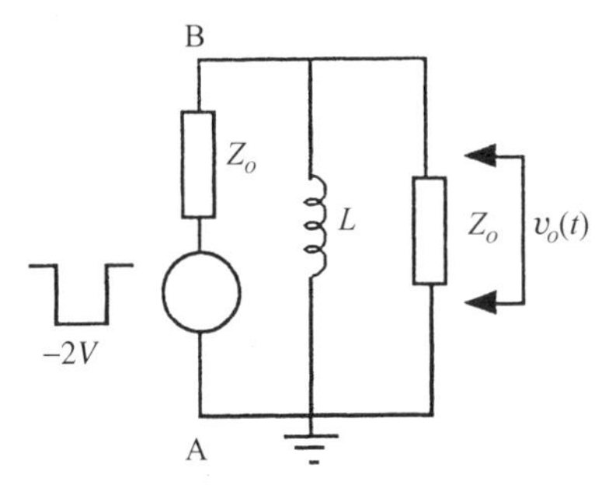

Figure 6.4 Equivalent circuit of the inverting transmission line transformer for determining pulse droop.

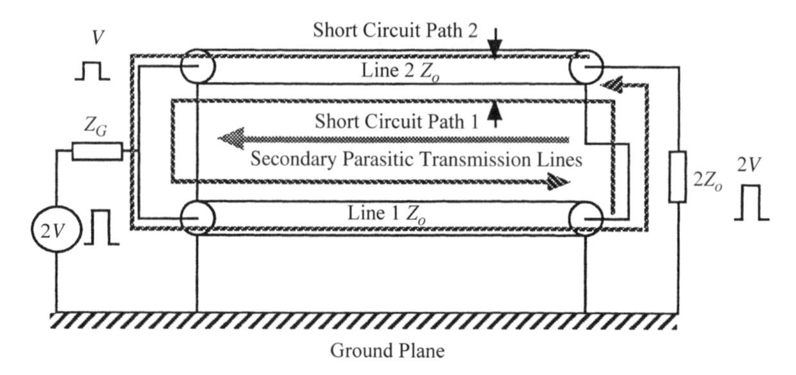

Figure 6.5 A two-stage voltage transmission line transformer

above a ground plane, with the inputs to the two lines connected in parallel and the output ends of the lines connected in series. The input impedance of the two cables connected in this way is clearly $Z_o/2$ while the output impedance is $2Z_o$. Thus the impedance transformation from input to output is a factor of 4. A rectangular voltage pulse of amplitude V and duration t_p is applied to the transformer from a pulse generator with output impedance Z_G $(= Z_o)$ and, after a propagation delay of δ, which depends both on the type of lines used and their length, the pulse emerges at the load end of the two lines. Since the lines are connected in series at the output, the potential of the outer conductor of line 2 is raised to a potential V with respect to ground, and the potential of the inner conductor of line 2 to a potential $2V$ with respect to ground; consequently a pulse of amplitude $2V$ is generated across the matched load.

However examination of the circuit shows that there is a short circuit path which runs from the inner conductor of line 1 at the output via the outer conductor of line 2 to the outer conductor of line 1, consequently shorting line 1. In the case of the linear parallel transformer this short circuit path can be considered as a secondary parasitic transmission line whose conductors are the outer conductor of line 2 and the ground plane. This is similar to the case of the 1:1 inverting transformer described in the last section. Since the outer conductors of the two cables are connected to ground at the input of the transformer, this secondary line is short-circuited at this point. This secondary line also adds an additional impedance Z_{2G} to the transformer, which is effectively connected in parallel with the output of line 1, and causes a collapse of potential with time on line 1. Similarly, further examination of the circuit shows that there is a second short-circuit path which runs from the inner conductor of line 2, at the output of the transformer, through to the inner conductor of line 1 at the input and then back again to the outer conductor of line 2 at the output. Thus line 2 is also short-circuited, and this short-circuit path can also be thought of as a parasitic transmission line which is shorted at the input end and consequently causes a collapse of potential on line 2. However it will be shown that the effect of this second parasitic transmission line can be accounted for automatically without the need to include it explicitly in the model.

If the transformer is operated in air the impedance of the first parasitic line will usually be greater than the impedance of the lines used to construct the transformer (Z_o) and the propagation delay along this line from the output of the transformer to the input will be less than δ. When the input pulse to the transformer first arrives at the output, a pulse of amplitude close to V is propagated back along the first, parasitic, secondary line to the input where the short-circuit termination gives rise to a voltage reflection coefficient of -1. Therefore the pulse is inverted and propagates back to the output of the transformer. When it reaches the output it sees a terminating impedance which is less than the secondary line impedance (giving rise to a negative reflection coefficient) and causes the potential of both the outer conductor of line 2 and the inner conductor of line 1 to fall to a value which depends on the values of the impedances of the load, the transmission lines and the secondary line. Since the reflection coefficient at the load end of the transformer will be negative, a positive pulse with an amplitude less than V will subsequently be propagated back again to the input of the transformer along the secondary line. This process repeats itself and the potential of the inner conductor of line 1 is progressively reduced in a step-wise fashion, leading to a step-wise droop on the pulse generated at the load. However if the impedance of the secondary line can be made very large in comparison to the impedance of the transmission lines making up the transformer (i.e. Z_o) or the round-trip propagation delay along the secondary line is longer than the pulse duration (i.e. long lengths of transmission line are used) then the presence of the secondary line will have little or no effect on the shape and amplitude of the pulse generated at the load.

In practice, it will usually be impossible to completely suppress the effect of the secondary transmission line in a linear transformer. This is particularly true for long pulse durations, where impractical lengths of transmission line would be needed to give a round-trip propagation time along the secondary line which was longer than the pulse length. It is also clear that, in transformers where more than two lines are used to give a larger voltage gain, there will be several parasitic secondary lines between the cables and the ground plane, and also parasitic lines between the cables themselves, as shown in Figure 6.6. Clearly loss of gain rapidly becomes much worse as the number of lines that are used is increased, due to the presence of these parasitic lines. This is mainly because the output impedance of the transformer increases as nZ_o, as the number of lines n which are used in the transformer is increased, and even with relatively small values of n this impedance will soon reach a value which exceeds the impedance of a single secondary parasitic line.

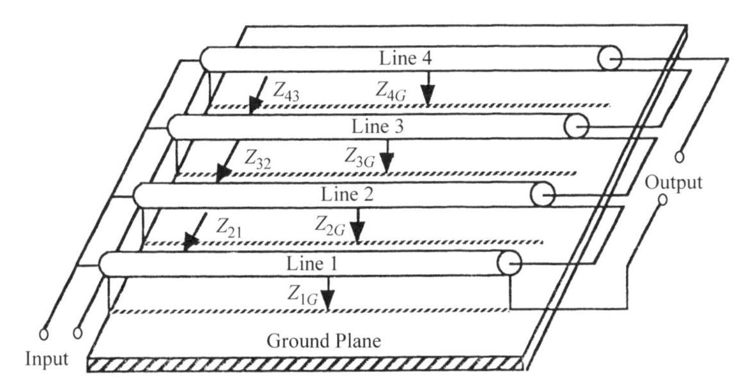

Figure 6.6 Parasitic secondary transmission lines in a four-stage voltage transformer

6.2.3 Detailed Analysis of the Two-stage Voltage Transformer

In practice the operation of the simple two-stage device is, surprisingly, much more complex than the simple description of its operation given above. In order to make an accurate analysis of the way in which the device works and to predict the output voltage waveform, it is necessary to make a detailed analysis of pulse reflections and transmission coefficients at both the input and output end of the transformer. These reflections occur because it is impossible to completely match the generator to the transformer at its input and, due to the presence of the parasitic transmission line between the outer conductor of cable 2 and the ground plane (now referred to as line 3), the device will also not be matched, at the output, if a load equal to $2Z_o$ is connected to the transformer. Similarly, consideration of the way that the device is connected at both the input and output ends show that parts of the incident pulses arriving from all the lines at both the output and input of the transformer are not only reflected but transmitted into the other lines to which they are connected. It is necessary to consider therefore the transmission coefficients between the lines at both ends of the transformer. This is relatively straightforward if the equivalent circuits for the output and input end of the transformer are used, as shown in Figure 6.7.

The voltage reflection coefficient for line 1 at the output can by inspection be seen to be given by

$$\rho_1 = \frac{(Z_{2G}||3Z_o) - Z_o}{(Z_{2G}||3Z_o) + Z_o} = \frac{2Z_{2G} - 3Z_o}{4Z_{2G} + 3Z_o} \tag{6.3}$$

Similarly the voltage reflection coefficients for lines 2 and 3 are given by

$$\rho_2 = \frac{2Z_{2G} + Z_o}{4Z_{2G} + 3Z_o} \qquad \rho_3 = \frac{3Z_o - 4Z_{2G}}{4Z_{2G} + 3Z_o} \tag{6.4}$$

The transmission coefficients which determine what fraction of the incident wave in a given line is transferred to either of the other two lines can similarly be derived from the equivalent circuits and are found to be

$$\left.\begin{aligned} \Gamma_{12} &= \frac{-2Z_{2G}}{4Z_{2G} + 3Z_o} & \Gamma_{21} &= \frac{-2Z_{2G}}{4Z_{2G} + 3Z_o} & \Gamma_{13} &= \frac{6Z_{2G}}{4Z_{2G} + 3Z_o} \\ \Gamma_{31} &= \frac{6Z_o}{4Z_{2G} + 3Z_o} & \Gamma_{23} &= \frac{-2Z_{2G}}{4Z_{2G} + 3Z_o} & \Gamma_{32} &= \frac{-2Z_o}{4Z_{2G} + 3Z_o} \end{aligned}\right\} \tag{6.5}$$

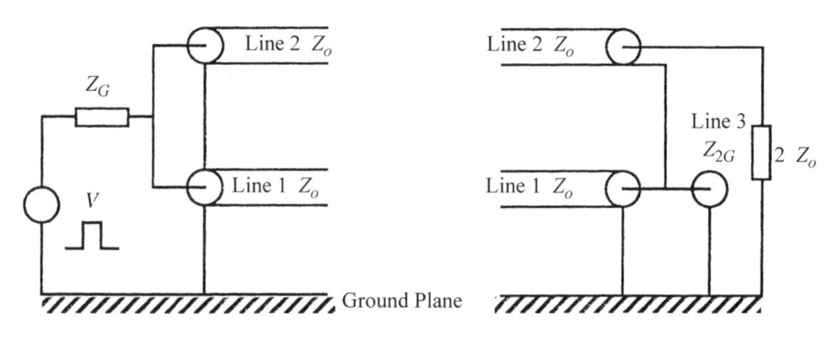

Figure 6.7 Equivalent circuits for the input and output ends of the two-stage transmission line transformer

where Γ_{12} and Γ_{13} are the fractions of the amplitude of the incident pulse in line 1 which is transmitted into lines 2 and 3, respectively, etc.

At the input to the transformer the reflection and transmission coefficients for lines 1 and 2 are the same and, again by reference to Figure 6.7, can be found to be given by

$$\rho' = \frac{-Z_o}{2Z_G + Z_o} \qquad \Gamma' = \frac{Z_G}{2Z_G + Z_o} \tag{6.6}$$

where ρ' is the reflection coefficient for incident pulses from the lines at the input of the transformer, Z_G is the generator impedance and Γ' is the transmission coefficient for line 1 into 2, and vice versa.

In order to assemble a model of the basic two line system it is helpful to simplify the problem by assuming that lines 1 and 2 are placed next to each other without any air-gap. This is done so that the propagation delay in the parasitic line 3 is very nearly equal to the propagation delays in lines 1 and 2. If lines 1 and 2 are spaced apart, the inclusion of what amounts to an air gap increases the velocity of propagation on the line which results in a more complicated structure on the output pulse due to differing propagation times on lines 1 and 2 compared to line 3. This does not affect the general applicability of the model but simplifies the description, the computer coding of the problem, and the comparison to, say, a laboratory experiment. The model is then assembled using a lattice diagram or reflection chart (discussed in Chapter 2) as shown in Figure 6.8. In the diagram it is possible to see the interaction between the propagating pulses on all three lines and their mutual interaction, both at the output and the input of the transformer using the reflection and transmission coefficients calculated from the above formulae. The diagram has been generalised by starting with a pulse amplitude of V and also by assuming that the pulse is very long in comparison to the propagation delay time of the lines. The components of the output voltage along the two central axes are added in order to calculate the pulse amplitudes after successive round-trips on the transmission lines. Thus:

After one propagation delay time δ the output voltage is

$$V_{out\,1} = (1 + \rho_2)V_{2,1}^{+} + \Gamma_{12}V_{1,1}^{+} + (1 + \rho_1)V_{1,1}^{+} + \Gamma_{21}V_{2,1}^{+} \tag{6.7}$$

after three propagation delay times 3δ the output voltage is

$$\begin{aligned} V_{out\,3} = V_{out\,1} &+ (1 + \rho_2)V_{2,3}^{+} + \Gamma_{12}V_{1,3}^{+} + \Gamma_{32}V_{3,3}^{+} \\ &+ (1 + \rho_1)V_{1,3}^{+} + \Gamma_{21}V_{2,3}^{+} + \Gamma_{31}V_{3,3}^{+} \end{aligned} \tag{6.8}$$

and after five propagation delay times 5δ the output voltage is

$$\begin{aligned} V_{out\,5} = V_{out\,3} &+ (1 + \rho_2)V_{2,5}^{+} + \Gamma_{12}V_{1,5}^{+} + \Gamma_{32}V_{3,5}^{+} \\ &+ (1 + \rho_1)V_{1,5}^{+} + \Gamma_{21}V_{2,5}^{+} + \Gamma_{31}V_{3,5}^{+} \end{aligned} \tag{6.9}$$

and so on, where expressions of the type $V_{2,1}^{+}$ refer to the incident wave (superscript $+$) on line 2 at the output after one propagation delay time δ. Using standard nomenclature reflected waves on the lattice diagram have a superscript $-$. The potentially tedious calculation of the output waveform as a function of time can be simplified if a spreadsheet computer programme is used. It is then a simple matter to change load and line impedances to explore the effect of such changes on the output voltage profile.

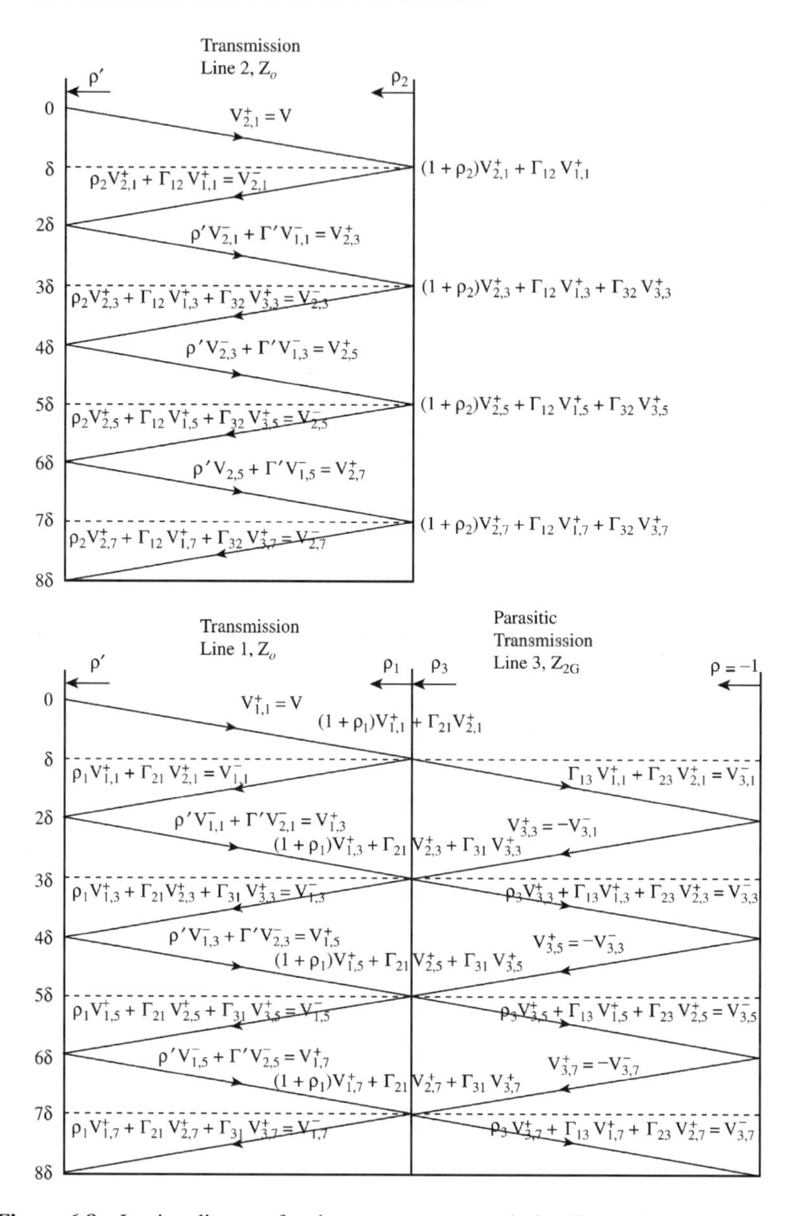

Figure 6.8 Lattice diagram for the two-stage transmission line voltage transformer

An interesting feature of the model is the analysis of the effect of the second short-circuit path or second parasitic transmission line in the structure referred to in the basic description of the two-stage voltage transformer given earlier. By reference to Figure 6.5, it can be seen that the amplitude of the pulse propagating on the second parasitic line formed by short-circuit path 2 can calculated by adding together the amplitude of the pulses propagating on line 2 and 3 and then subtracting the amplitude of the pulse propagating on line 1. This provides a convenient method of checking the validity of the model and that the effect of this line is properly accounted for by the model. This is because the amplitude of the pulse propagating on the second parasitic line towards the input must equal the amplitude of the pulse which subsequently propagates back on the line towards the load multiplied by -1, as the second parasitic line is effectively short-circuited at the input. For example, if this line is labelled line 4 then

$$\left.\begin{aligned} V_{4,1}^{-} &= V_{2,1}^{-} + V_{3,1}^{-} - V_{1,1}^{-} \\ &= \rho_2 V_{2,1}^{+} + \Gamma_{12} V_{1,1}^{+} - \rho_1 V_{1,1}^{+} - \Gamma_{21} V_{2,1}^{+} + \Gamma_{13} V_{1,1}^{+} + \Gamma_{23} V_{2,1}^{+}, \\ V_{4,3}^{+} &= V_{2,3}^{+} + V_{3,3}^{+} - V_{1,3}^{+} \\ &= (\rho' - \Gamma')(\rho_2 V_{2,1}^{+} + \Gamma_{12} V_{1,1}^{+}) + (\Gamma' - \rho')(\rho_1 V_{1,1}^{+} + \Gamma_{21} V_{2,1}^{+}) - \Gamma_{13} V_{1,1}^{+} - \Gamma_{23} V_{2,1}^{+} \end{aligned}\right\} \tag{6.10}$$

Since it is easily shown that $(\rho' - \Gamma') = -1$, then $V_{4,1}^{-} = -V_{4,3}^{+}$ as is required. As an example of the method, the waveform shown in Figure 6.9 was calculated for a two-stage voltage transformer constructed from two 10 m lengths of 50 Ω coaxial cable. The impedance of the secondary transmission line was set at 82 Ω and a pulse of amplitude 1 V was injected into the transformer.

6.2.4 Voltage Gain of Multi-stage, Linear Transmission Line Transformers

If the duration of the pulse injected into the transformer is much greater than twice the round trip time on the parasitic secondary lines then, as described, the pulse will exhibit a step-wise droop. However, a complete analysis to determine the voltage gain and output pulse shape of a multi-stage transformer, as described in the last section, would be very time-consuming. It is though possible to calculate the height of the first of step of the output pulse, in the presence of the parasitic impedances imposed on the transformer by the secondary lines, relatively easily. From such a calculation it can be shown that, as the number as the number of lines n that are stacked increases, the effect of the parasitic impedances becomes more and more serious and the voltage gain of the transformer is seriously compromised. Figure 6.10 presents an equivalent circuit from which the output voltage of the first step and hence the gain of an n-stage transformer may be estimated. In

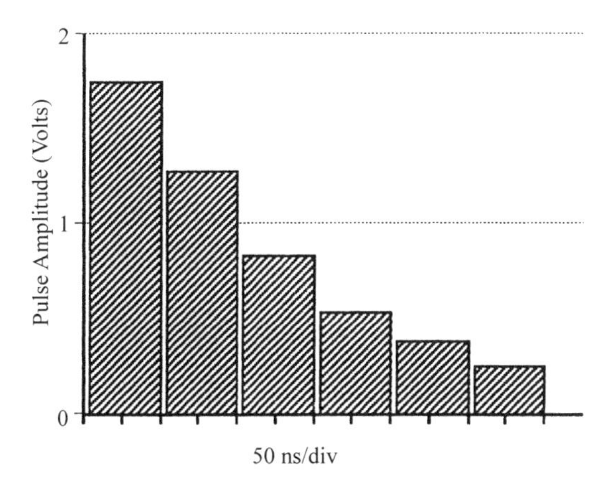

Figure 6.9 Simulated output pulse for a two-stage transmission line voltage transformer

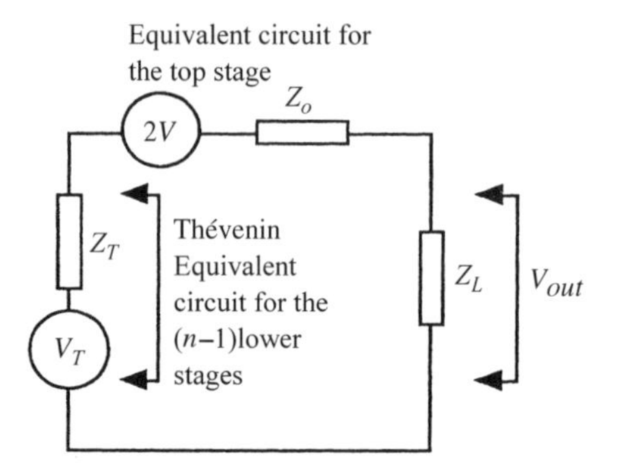

Figure 6.10 Equivalent circuit for determining the gain of an n-stage transmission line voltage transformer

this circuit the top stage or transmission line of the transformer is represented by a voltage source of amplitude $2V$ in series with the line impedance Z_o. The remaining $(n-1)$ stages or lines are represented by a Thévenin equivalent circuit made up of a voltage source of amplitude V_T and an impedance Z_T. To estimate the output voltage it is not necessary to consider the transient behaviour of the circuit as it is only the height of the first step that is of interest. Thus the problem can be reduced essentially to a DC analysis. To determine an expression for the Thévenin voltage source, the equivalent circuit in Figure 6.11 is used.

The secondary or output circuit of the transformer is represented as a ladder network of voltage sources of amplitude $2V$ connected in series with the transmission line or stage impedances Z_o. The parasitic impedance of the secondary line Z_{2G} is then connected across each stage as shown. Parasitic line impedances between lines, as shown in Figure 6.6, are ignored and it is assumed that the parasitic secondary line impedances to ground are all of equal size. From the rth stage

$$I_r = I_{r,r+1} + I_{r+1}$$

$$\therefore \quad \frac{V_r + 2V - V_{r+1}}{Z_o} = \frac{V_{r+1} + 2V - V_{r+2}}{Z_o} + \frac{V_{r+1}}{Z_{2G}} \tag{6.11}$$

which gives, after rearrangement

$$V_{r+2} - V_{r+1}\left(2 + \frac{Z_o}{Z_{2G}}\right) + V_r = 0,$$

$$\text{where} \quad r = 0, 1, 2, \ldots, (n-2) \tag{6.12}$$

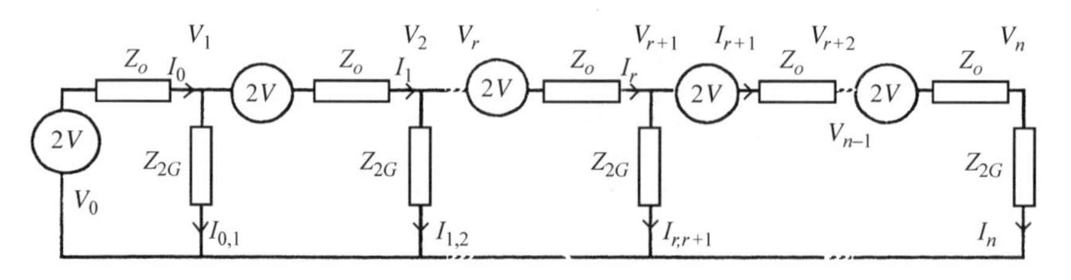

Figure 6.11 Equivalent circuit to derive an expression for the Thévenin voltage source in Figure 6.10

This is a linear constant-coefficient difference equation of the form

$$\left(\mathbf{E}^2 - \left(2 + \frac{Z_o}{Z_{2G}}\right)\mathbf{E} + 1\right)V_r = 0 \tag{6.13}$$

where $\mathbf{E}$ is the shift operator. This equation has a general solution [17] of the form

$$V_r = AM_1^r + BM_2^r \tag{6.14}$$

where A and B are constants and M_1 and M_2 are the roots of the characteristic equation of Equation (6.13) and are given by

$$M_1 = \frac{\lambda + 1}{\lambda - 1} \quad \text{and} \quad M_2 = \frac{\lambda - 1}{\lambda + 1},$$

$$\text{where} \quad \lambda = \sqrt{1 + \frac{4Z_{2G}}{Z_o}} \tag{6.15}$$

For the first and last meshes

$$\left.\begin{aligned} V_0 &= 0 = A + B \\ \text{and} \quad I_n &= \frac{V_{n-1} + 2V - V_n}{Z_o} = \frac{V_n}{Z_{2G}}, \\ \therefore \quad V_n &= (V_{n-1} + 2V)\frac{\lambda^2 - 1}{\lambda^2 + 3} \end{aligned}\right\} \tag{6.16}$$

These equations together with Equation (6.14) can be solved to derive expressions for A and B which are given by

$$A = \frac{V(\lambda - 1)^{n+1}(\lambda + 1)^{n+1}}{(\lambda + 1)^{2n+1} + (\lambda - 1)^{2n+1}} = -B \tag{6.17}$$

Thus the full solution to Equation (6.13) can be obtained by substitution of the expressions for A, B, M_1, M_2 into Equation (6.14) which gives for the $(n-1)$th stage

$$V_{n-1} = V_T = \frac{4VZ_{2G}}{(\lambda + 1)Z_o} \frac{1 - \left(\dfrac{\lambda - 1}{\lambda + 1}\right)^{2(n-1)}}{1 + \left(\dfrac{\lambda - 1}{\lambda + 1}\right)^{2n-1}} \tag{6.18}$$

The Thévenin impedance Z_T is found in a similar way. Referring to Figure 6.12, it can be seen that this impedance is the impedance looking into the ladder network of Figure 6.11, from the output end, with the voltage generators removed and replaced by short circuits.

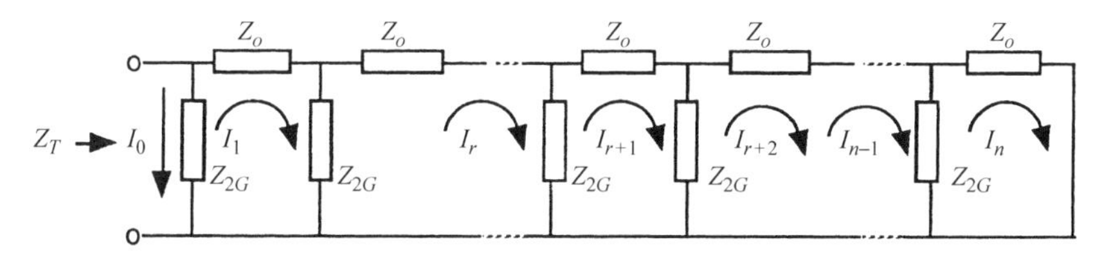

Figure 6.12 Circuit for determining the Thévenin impedance Z_T

A similar recurrence formula for current, in the rth mesh, to that derived for the voltage in the last analysis (Equation (6.12)) must apply, i.e.

$$I_{r+2} - I_{r+1}\left(2 + \frac{Z_o}{Z_{2G}}\right) + I_r = 0 \tag{6.19}$$

which gives rise to a similar difference equation with a solution

$$I_r = CM_1^r + DM_2^r \tag{6.20}$$

where M_1 and M_2 are given in Equation (6.20) above. From the first and last meshes

$$V = Z_{2G}(I_0 - I_1)$$

$$\text{and} \qquad (I_{n-1} - I_n)\,Z_{2G} = I_n Z_o \tag{6.21}$$

Substituting expressions for I_0, I_1, I_n, I_{n-1} derived from Equation (6.20) into the Equations (6.21) allows expressions for the constants C and D to be determined, i.e.

$$\left.\begin{aligned} C &= \frac{V}{2Z_{2G}}\frac{(\lambda-1)^{2n+1}}{(\lambda+1)^{2n} - (\lambda-1)^{2n}} \\ D &= \frac{V}{2Z_{2G}}\frac{(\lambda+1)^{2n+1}}{(\lambda+1)^{2n} - (\lambda-1)^{2n}} \end{aligned}\right\} \tag{6.22}$$

Applying Equation (6.20) to the first mesh gives

$$I_0 = C + D \tag{6.23}$$

Therefore, for the bottom $(n-1)$ stages of the transformer

$$\left.\begin{aligned} I_0 &= \frac{V(\lambda+1)}{2Z_{2G}}\left[\frac{1 + \left(\frac{\lambda-1}{\lambda+1}\right)^{2n-1}}{1 - \left(\frac{\lambda-1}{\lambda+1}\right)^{2(n-1)}}\right] \\ \text{and} \qquad Z_T &= \frac{V}{I_0} = \frac{2Z_{2G}}{(\lambda+1)}\left[\frac{1 - \left(\frac{\lambda-1}{\lambda+1}\right)^{2(n-1)}}{1 + \left(\frac{\lambda-1}{\lambda+1}\right)^{2n-1}}\right] \end{aligned}\right\} \tag{6.24}$$

Thus, from the circuit shown in Figure 6.10, the output voltage V_{out} for an n-stage transmission line voltage transformer is given by

$$V_{out} = (V_T + 2V)\frac{Z_L}{Z_L + Z_T + Z_o} \tag{6.25}$$

and the voltage gain can now be calculated. The equations for the output voltage of the transformer are similar to that published by Lewis and Wells [4], who point out that for an infinite number of stages the expression for the Thévenin voltage V_T reduces to

$$V_{T\infty} = \frac{4VZ_{2G}}{(\lambda+1)Z_o} \tag{6.26}$$

From this expression and Equation (6.18) they then show that, for a transformer with nine stages and a realistic estimate of the ratio of the secondary mode impedance to line impedance, the gain of the transformer would be 95% of the gain of a transformer with an infinite number of stages. They also show that the output voltage of the transformer is just over half what it would be if the secondary mode impedance were infinite. Clearly, in this type of transformer the parasitic secondary mode impedances must be kept as high as possible.

6.3 WOUND TRANSMISSION LINE TRANSFORMERS

In the earlier sections of this chapter, on pulsed transmission line transformers with linear parallel constructions, it was explained that such devices are only really suitable for applications where the transformers are to be used with pulses whose durations are of the same order or less than the transit time along the transmission lines from which the transformer is constructed. Furthermore such devices are also only practical as voltage transformers when the desired gain of the transformer is relatively small, i.e. the number of lines built into a stacked configuration is low, because of the presence of parasitic secondary transmission lines within the structure of the transformer, which can severely reduce its voltage gain and hence efficiency. To avoid this problem, the standard technique is to wind the lines, used to construct the transformer, inductively with or without the aid of a magnetic core, as shown in Figure 6.13. This has the effect of suppressing propagation in the parasitic secondary lines by raising their effective impedance. By this means it is quite possible to build very efficient broad-band transformers in which very little electrical energy is wasted within the structure of the transformer, even for pulse durations which are at least one order of magnitude greater than the transit time along the lines from which the transformer is built. As with linear transmission line transformers, work has been published on inductively isolated, pulse inversion and pulse isolation transformers [18,19], current transformers [20] and voltage transformers [21,22]. The inductive winding technique is also applicable to other types of transmission line transformer such as the stacked Blumlein pulse generator [23]. Mesyats gives a detailed analysis of the performance of transmission line transformers with inductive isolation [24] which relies on modelling the transformer as a ladder network consisting of a string of voltage generators and line impedances. This approach leads to very complicated approximate expressions for the pulse response of such transformers in terms of their rise-time and pulse droop capability. The model presented here is much simpler to use and appears to be totally accurate, giving for the first time a method by which the behaviour and performance of this potentially important class of transformers can be analysed.

In order to model this type of transformer, an entirely different approach must be taken to that used for linear transformers, and it is shown that a transformer model can be established which is, in many ways, analogous to the lumped element model used to analyse the performance of conventional pulse transformers wound on a magnetic core (described in the preceding chapter) [25,26]. The model presented allows the designer of such devices to work out where inductive isolation should be put, how much isolation is required at each stage of a transformer to achieve acceptable pulse droop, and the benefits of using mutually coupled isolation windings rather than separated windings on individual cores. The model

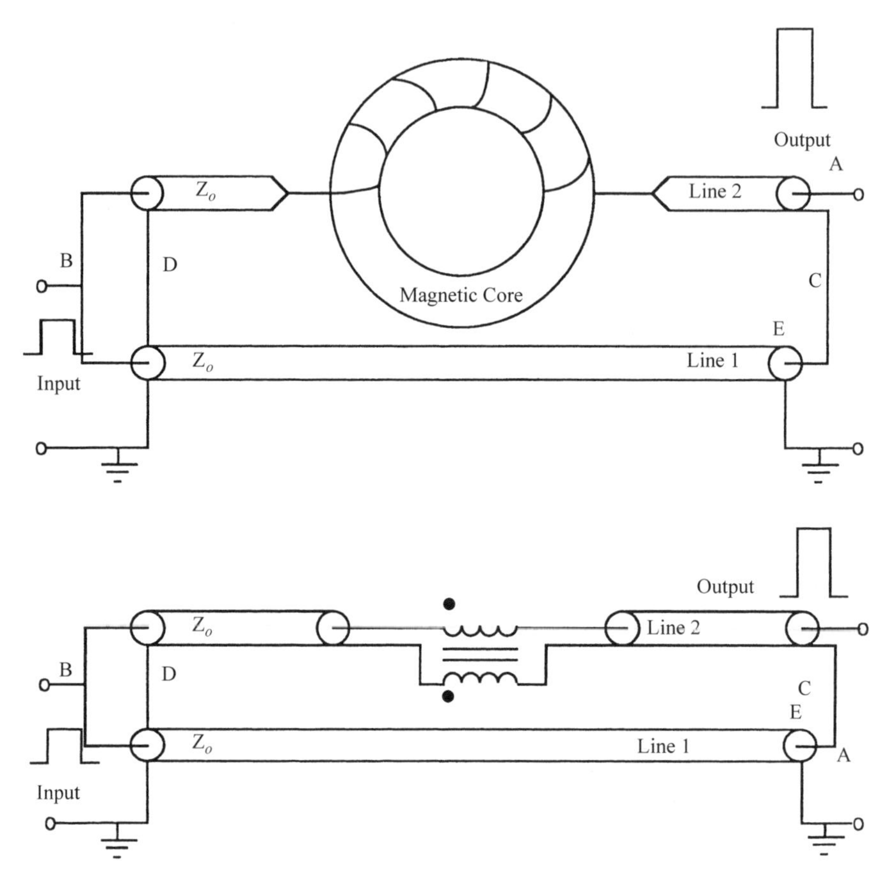

Figure 6.13 Circuit diagram and equivalent circuit of a simple two-stage wound transmission line transformer

also shows what needs to be done to minimise pulse rise-time and allows the designer to assess, reliably, the size and type of magnetic core and the number of turns for the transmission line windings required for optimum inductive isolation. From the model it is clear that the performance of a properly engineered transformer can easily exceed that of conventionally wound pulse transformers, particularly in minimising pulse-shape distortion and maximising gain-band-width response.

6.3.1 Basic Operation

It is assumed throughout this section that the voltage transmission line transformers to be described are constructed from lengths of standard coaxial cable transmission line and that inductive isolation is provided by winding the lines on toroidal ferrite cores. There is no reason why other types of transmission line or magnetic core should not be used; the choice is subject only to practical or other physical constraints. It is also important, prior to discussing these transformers, to define the term n-stage transformer, as distinct from the number of lines used to construct a given device. An n-stage transformer here is a device in which each stage contributes a single multiple of the input voltage to the output voltage of the transformer, i.e. an n-stage transformer has an ideal voltage gain of n. In most cases the number of stages is equal to the number of lines used to build a given transformer. However it is quite possible, for example, to build transformers with quite low output impedances by connecting a number of transmission lines in parallel at each stage of the transformer [22]. Clearly, in such transformers the number of transmission lines used to build the transformer will not be equal to the number of stages as defined above.

Figure 6.13 is a circuit diagram of a two-stage voltage transformer in which inductive isolation is provided by winding the top stage of the transformer on a toroidal magnetic core. The connection of the two lines used to make the transformer is standard, i.e. the lines are connected in parallel at the input to the transformer and in series at the output, and the lines are of equal physical length. Ideally such a transformer would have a voltage gain of 2 as, in this case, the number of stages and lines are equal. An electrical equivalent circuit of the arrangement is also given in Figure 6.13, in which the effect of the inductive winding is represented as a 1:1 transformer connected in series with the inner and outer conductors of line 2 at the centre of the line. The purpose of the winding can be seen by reference to both circuits in Figure 6.13. Without the winding it is apparent that there would be two short-circuit paths within the device, which would severely degrade its operation. Line 1 is shorted via a path which starts at point C on its inner conductor to point D on the outer conductor of line 2 and back to point E on the outer conductor of line 1. Similarly, line 2 is shorted via a path which starts from point A on the inner conductor of line 1 to point B on the inner conductors of both lines and then back to point C on the outer conductor of line 2. The purpose of winding line 2, on a magnetic core, is to make these short-circuit paths inductive. It is clear from the bottom circuit in Figure 6.13 that both short-circuit paths pass just once through the inductance provided by the winding while the normal TEM propagation mode of line 2 remains unaffected. Provided that the impedance of the inductance in the short-circuit paths remains large, during the pulse, in comparison to the impedance of the lines used to construct the transformer, then very little current will pass through the short-circuit paths and the output pulse from the transformer will show very little pulse droop or distortion. Good isolation of the output of the transformer from the input can be achieved in this way, provided the size of the inductance is large enough. In the case of pulse inversion, pulse isolation and current pulse transformers, this type of inductive isolation is equally valid and the modelling technique to be described can easily be adapted to deal with these variants.

It is readily deduced, by consideration of the two-stage transformer shown in Figure 6.13, that in the case of multi-stage voltage transformers, every line in the structure, apart from the bottom one, must be wound inductively to ensure that all the short-circuit paths in the transformers are suppressed. It is not immediately clear, however, whether the same amount of isolation should be used on each line, i.e. should the inductance of the windings in the transformer be kept constant at each stage or should it vary from stage to stage or is there some benefit in mutually coupling the windings between stages?

The modelling method to be described allows these questions to be resolved so that the design of a given transformer can be optimised to minimise pulse droop and distortion. The basis of the model is to refer components connected to the output of the transformer and parasitic stray impedances in the secondary circuit to the input primary circuit in a similar way to a conventional magnetically cored pulse transformer, in which components or impedances are referred from the secondary circuit to the primary or vice versa by dividing or multiplying their values by a factor of n^2, where n is the turns ratio of the transformer (see Chapter 5). In the case of transmission line transformers, of the type described, the same technique can be used but, in this case, components are referred by a factor of n^2, where in this case n is the number of stages used in the transformer. At first glance, this seems to be a reasonable technique given that the input impedance of the simple two-stage device shown in Figure 6.13 is $Z_o/2$ and the output impedance is $2Z_o$, so that the ratio of output impedance to input impedance is 4, i.e. 2^2. However it is not immediately clear how the mutually coupled inductance due to each winding should be analysed, or what the effect will be of any stray capacitance in the windings of these transformers.

6.3.2 Model Development

To explain the theory of the model, the method is applied to the detailed analysis of the operation of a simple three-stage voltage transformer, as shown in Figure 6.14. Assuming that a pulse of amplitude V is input to the transformer, examination of the distribution of the

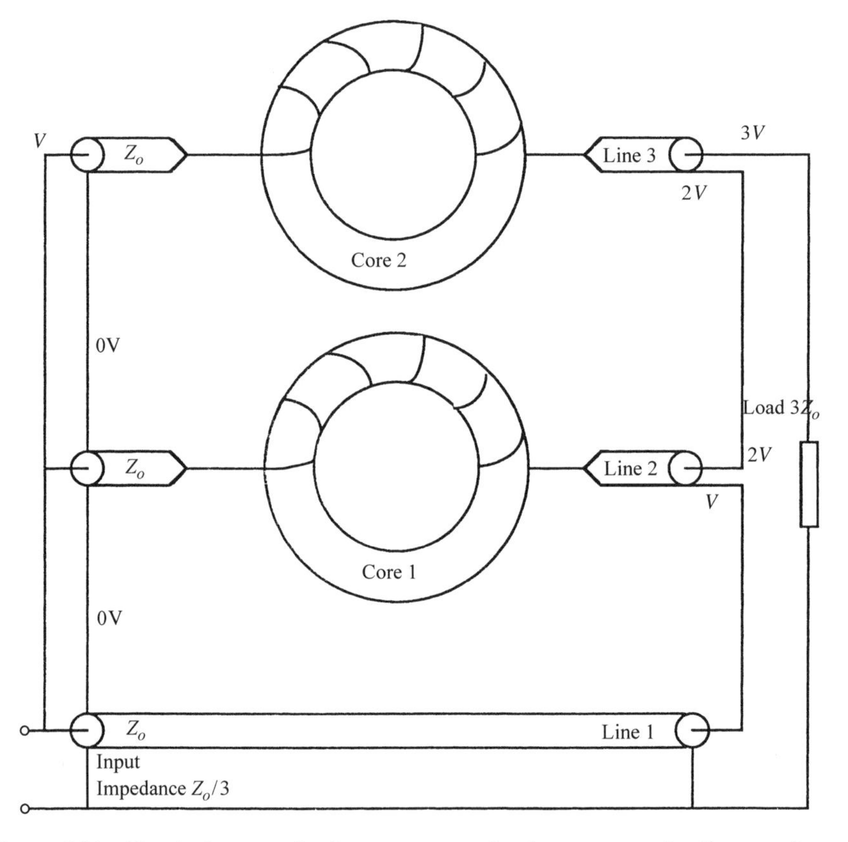

Figure 6.14 Circuit diagram of a three-stage wound voltage transmssion line transformer

potentials, at different points within the structure, reveals that, initially, there is a drop in potential of $2V$ across both the inner and outer conductors of line 3 from output to input. Similarly, there is a drop in potential of V across both the inner and outer conductors of line 2. Based both on this observation and an examination of the way in which the mutual inductances are connected within the structure of the transformer, an equivalent Laplace circuit may be drawn of the output side of the transformer as given in the top circuit in Figure 6.15. In this diagram each of the transmission lines is represented as a voltage step source of amplitude $2V$ in series with an impedance Z_o, being the impedance of each of the lines. The mutually coupled inductors represent the two windings in the transformer and are connected such that the potential drop across the windings is correctly represented by

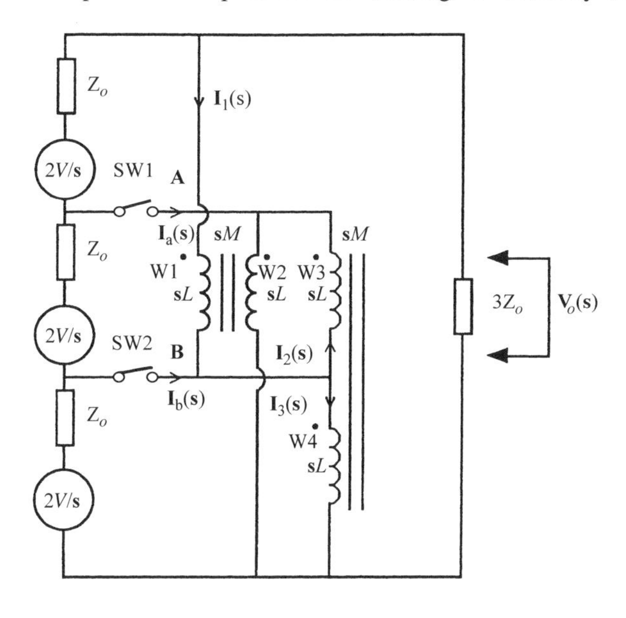

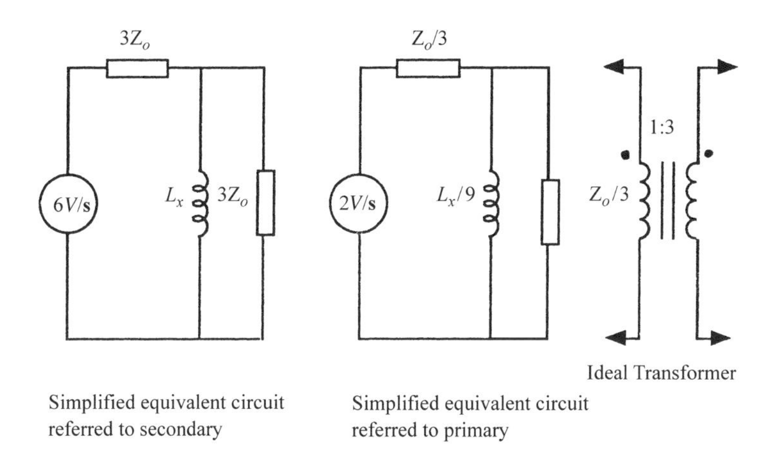

Figure 6.15 Equivalent circuits of the three-stage voltage transformer given in Figure 6.11

reference to Figure 6.14. Switches SW1 and SW2 are included solely to help explain the development of the model. It is assumed that the number of turns on both windings is the same so that the inductances of all four inductors are equal. If it is also assumed that the coupling between both conductors in each of the lines is perfect, i.e. $M = L$, then by mesh analysis or otherwise it is relatively straightforward to derive an expression for the output voltage of the transformer, as a function of time, which appears across the matched load impedance $3Z_o$.

This voltage is given by

$$v_o(t) = 3V \exp\left[-\frac{5}{6}\frac{Z_o t}{L}\right] \tag{6.27}$$

Thus the output voltage will droop exponentially at a rate which depends on both the impedance of the transmission lines and the inductance of the windings. Analysis of the circuit also reveals that the currents $\mathbf{I}_a(\mathbf{s})$ and $\mathbf{I}_b(\mathbf{s})$ are zero, therefore the network of two coupled mutual inductors can effectively be uncoupled from the intermediate connections to the lines by opening switches SW1 and SW2. If the effective inductance L_x, represented by the two coupled mutual inductors, is then calculated, the circuit can be simplified to that shown at bottom left in Figure 6.15 by putting L_x in parallel with the load.

Using the referral method discussed earlier, an equivalent circuit referred to the primary circuit can also be drawn, where the impedances and the inductor are referred by a factor of 3^2. An ideal transformer with a turns ratio of 3 is also included as a reminder that the output voltage of the transformer will be 3 times the input. Continuing the analysis further reveals that the inductance L_x has a value of $9L/5$ and if this value is used to derive an expression for the output voltage as a function of time for either the simplified secondary or primary circuits, an expression results which is identical to Equation (6.27).

The modelling method can be developed further by consideration of the top circuit in Figure 6.15. From analysis of the circuit, with switches SW1 and SW2 closed, the potentials at points A and B can be found to be $\frac{2}{3}\mathbf{V}_o(\mathbf{s})$ and $\frac{1}{3}\mathbf{V}_o(\mathbf{s})$, respectively, where $\mathbf{V}_o(\mathbf{s})$ is the output potential across the matched transformer load $3Zo$. From the circuit

$$\left.\begin{aligned}
\frac{2}{3}\mathbf{V}_o(\mathbf{s}) &= \mathbf{I}_2(\mathbf{s})\mathbf{s}L + \mathbf{I}_1(\mathbf{s})\mathbf{s}M = \mathbf{s}L(\mathbf{I}_1(\mathbf{s}) + \mathbf{I}_2(\mathbf{s}))\\
\frac{1}{3}\mathbf{V}_o(\mathbf{s}) &= \mathbf{I}_3(\mathbf{s})\mathbf{s}L - \mathbf{I}_2(\mathbf{s})\mathbf{s}M = \mathbf{s}L(\mathbf{I}_3(\mathbf{s}) - \mathbf{I}_2(\mathbf{s}))\\
\mathbf{I}_1(\mathbf{s}) &= \mathbf{I}_2(\mathbf{s}) + \mathbf{I}_3(\mathbf{s})
\end{aligned}\right\} \tag{6.28}$$

These equations can be solved to derive expressions for the various currents through the mutual inductor network $\mathbf{I}_1(\mathbf{s})$ to $\mathbf{I}_3(\mathbf{s})$, i.e.

$$\mathbf{I}_1(\mathbf{s}) = \frac{5\mathbf{V}_o(\mathbf{s})}{9\mathbf{s}L}, \qquad \mathbf{I}_2(\mathbf{s}) = \frac{\mathbf{V}_o(\mathbf{s})}{9\mathbf{s}L}, \qquad \mathbf{I}_3(\mathbf{s}) = \frac{4\mathbf{V}_o(\mathbf{s})}{9\mathbf{s}L} \tag{6.29}$$

From these currents, the impedances of equivalent inductors Z_{1-4} can be determined for each of the four windings W1–4 in the circuit. These are

$$\left.\begin{aligned}
\mathbf{Z}_1(\mathbf{s}) &= \frac{\frac{2}{3}\mathbf{V}_o(\mathbf{s})}{\mathbf{I}_1(\mathbf{s})} = \frac{\frac{2}{3}\mathbf{V}_o(\mathbf{s})}{\frac{5}{9\mathbf{s}L}\mathbf{V}_o(\mathbf{s})} = \frac{6}{5}\mathbf{s}L \\
\mathbf{Z}_2(\mathbf{s}) &= \frac{\frac{2}{3}\mathbf{V}_o(\mathbf{s})}{\mathbf{I}_2(\mathbf{s})} = \frac{\frac{2}{3}\mathbf{V}_o(\mathbf{s})}{\frac{1}{9\mathbf{s}L}\mathbf{V}_o(\mathbf{s})} = 6\mathbf{s}L \\
\mathbf{Z}_3(\mathbf{s}) &= \frac{\frac{1}{3}\mathbf{V}_o(\mathbf{s})}{-\mathbf{I}_2(\mathbf{s})} = \frac{\frac{1}{3}\mathbf{V}_o(\mathbf{s})}{-\frac{1}{9\mathbf{s}L}\mathbf{V}_o(\mathbf{s})} = -3\mathbf{s}L \\
\mathbf{Z}_4(\mathbf{s}) &= \frac{\frac{1}{3}\mathbf{V}_o(\mathbf{s})}{\mathbf{I}_3(\mathbf{s})} = \frac{\frac{1}{3}\mathbf{V}_o(\mathbf{s})}{\frac{4}{9\mathbf{s}L}\mathbf{V}_o(\mathbf{s})} = \frac{3}{4}\mathbf{s}L
\end{aligned}\right\} \tag{6.30}$$

Hence the network with mutual inductors shown in Figure 6.15 can then be replaced by an equivalent network containing four separate inductors, as shown in Figure 6.16. The appearance of an inductor with a negative value is not uncommon in the circuit analysis of circuits containing mutually coupled inductors [27]. These inductors can now be referred to the primary circuit using an n^2 factor where n, as before, represents the number of stages across which the inductors are mounted. This results in the middle equivalent circuit shown in Figure 6.16. It is a simple matter to show that the parallel combination of the four inductors has a value of $9L/5$ as expected. However if the inductors that are connected across one and two stages are combined, an even simpler equivalent circuit results (lower circuit, Figure 6.16).

In Figure 6.17 the results of an identical analysis of a four-stage transformer are presented in the form of an equivalent secondary circuit containing inductors, comparable to the top circuit in Figure 6.16, an equivalent circuit referred to the primary, and a simplified equivalent primary circuit where inductors connected across one, two and three stages are combined. The combined equivalent circuits (bottom circuits in Figures 6.16 and 6.17) now provide the basis for another equivalent circuit from which the performance of a wide range of wound TLTs can be very simply determined. The observation that a four-stage transformer can be reduced to an equivalent primary circuit with three parallel inductors (L, $L/4$ and $L/9$), suggests that the analysis of this type of transmission line voltage transformer can now be greatly simplified by considering only the inductive paths which exist in the transformer from the outer conductors of each of the lines to ground, at the output side of the transformer.

The method can be illustrated by reference to Figure 6.18 which shows a four-stage transformer. Ignoring resistive losses and losses in the cores, but including the stray capacitance across each of the windings, leads to an equivalent circuit shown at the top of Figure 6.18. The inductive paths to ground are placed in the circuit on the basis of the analysis presented above. If the components on the output side of the transformer are now

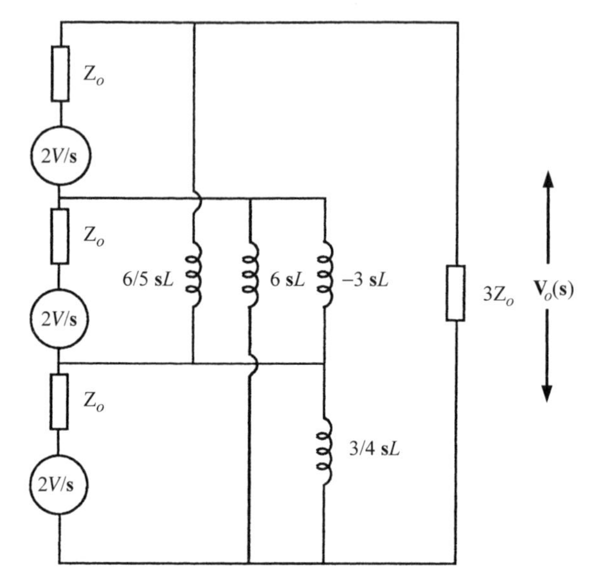

Reduced equivalent circuit of the 3-stage voltage transformer

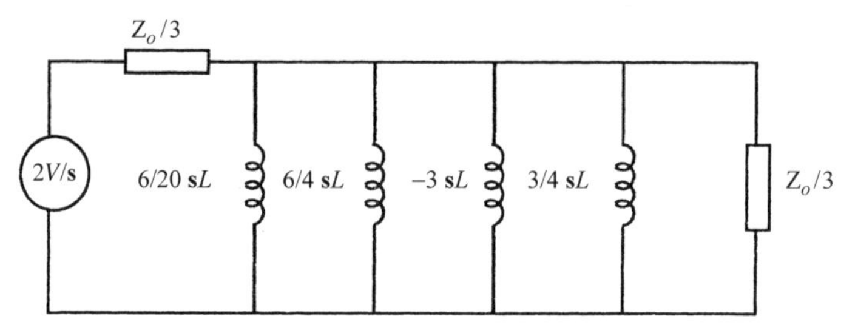

Reduced equivalent circuit referred to primary

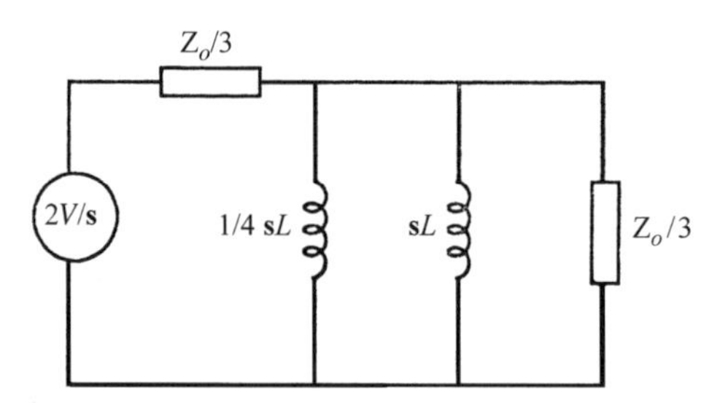

Reduced equivalent circuit referred to primary simplified

Figure 6.16 Simplified equivalent circuits of the three-stage voltage transformer of Figure 6.15

referred to the input or primary circuit with n^2 factors, which are determined by the stage to which each component is connected, an equivalent circuit can be drawn, as given in the middle circuit of Figure 6.18. This circuit can be further simplified by calculating the equivalent inductance and capacitance of the parallel combination of the inductors and capacitors, as shown in the lower circuit in Figure 6.18. This very simple circuit gives the

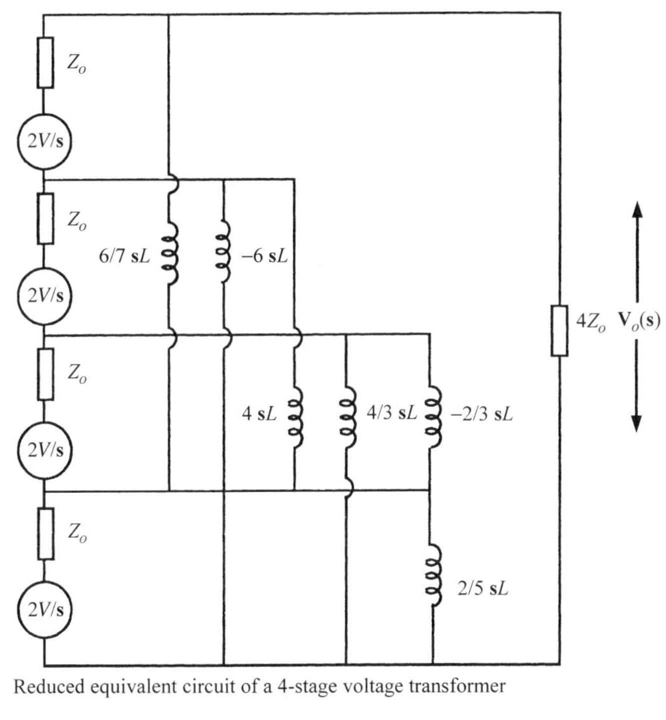

Reduced equivalent circuit of a 4-stage voltage transformer

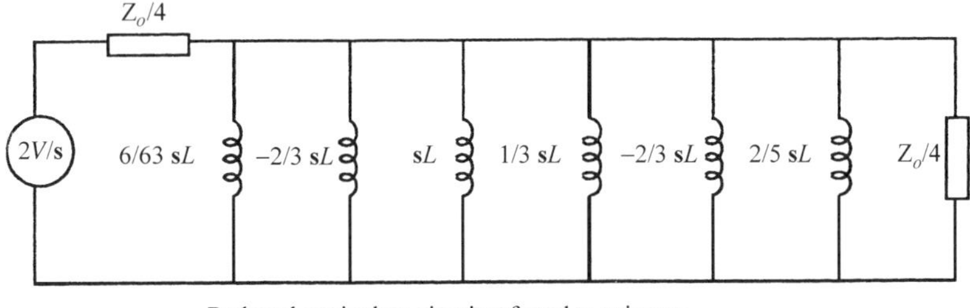

Reduced equivalent circuit referred to primary

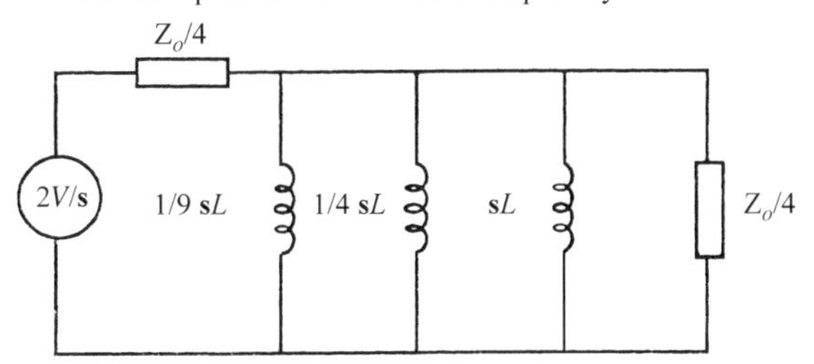

Reduced equivalent circuit referred to primary simplified

Figure 6.17 Equivalent and simplified circuits of a four-stage voltage transformer

same expression for the output voltage of the transformer as a function of time, for any given input, as is also derived from the analysis of the four-stage Laplace equivalent circuit analogous to that shown in Figure 6.15, for a three-stage device. From this simple model a generalised expression for the output voltage of an n-stage transformer of this type, when

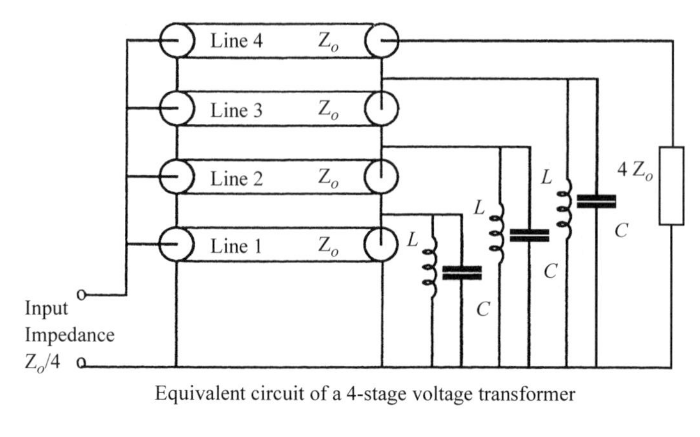

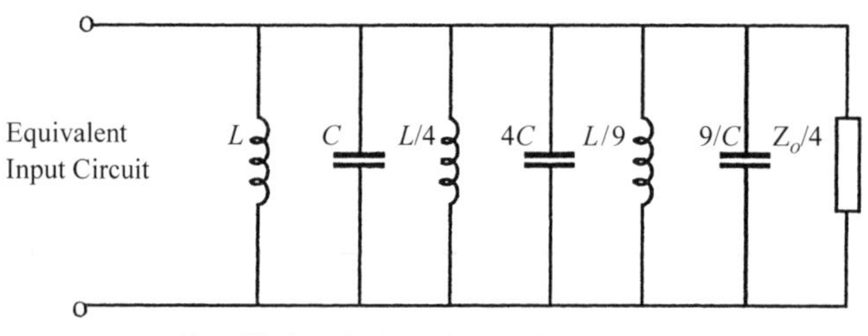

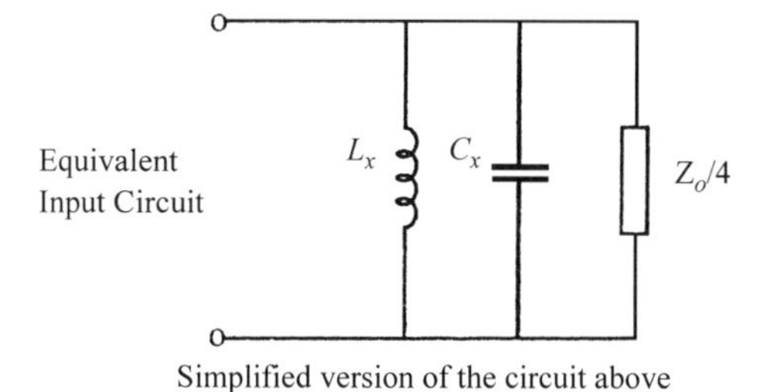

Figure 6.18 Revised equivalent circuit for a four-stage voltage transfomer with two futher simplified circuits

driven from a voltage step source of amplitude V and ignoring stray capacitance, is found to be

$$v_o(t) = nV \exp\left(-\frac{Z_o t}{2nL}\sum_{0}^{n}(n-1)^2\right) \tag{6.31}$$

Such an expression allows pulse droop to be determined and if the stray capacitance is included, as in Figure 6.18, the rise-time of the transformer can also be calculated.

This simple method of modelling and analysing the behaviour of transmission line voltage transformers highlights an important feature of these devices concerning the droop rate of flat-topped input pulses. It can be seen from the middle circuit in Figure 6.18 that the most serious cause of droop is the inductive path to ground of the inductor connecting the outer conductor of the highest stage in the transformer to ground. It is therefore clear that the inductive isolation required, at each stage of the transformer, to minimise droop should be increased relative to its nearest lower neighbour, at each succesive stage of the transformer. Similarly it can also be seen that in order to minimise rise-time it is imperative that the stray capacitance associated with the inductive windings of the top stages of these transformers should be kept as low as possible.

6.3.3 Mutually Coupled Windings

There are many other winding techniques which can be used to provide inductive isolation in these transformers, so is it possible to achieve more effective isolation using windings other than the basic technique described so far? In this section the potential benefits of mutually coupled windings are demonstrated and it is shown that slower droop rates can be achieved using this method. For comparison purposes the analysis of the performance of the three-stage voltage transformer in Figure 6.19 is carried out. In this device the top stage now has two windings, one of which is mutually coupled to the winding of the second stage. An equivalent circuit may be drawn in a similar way to that used to produce the top circuit in Figure 6.15 although, in this case, there are effectively four coupled windings on one core and two on the other. Again, detailed analysis shows that the effect of the inductance of the coupled windings can be partly decoupled from the circuit to produce an equivalent inductance (L_x) which can be placed in parallel with the load. However in this case the inductance is found to be $9L/2$, and consequently the output voltage of this transformer into a matched load, when fed from a voltage step source of amplitude V, is found to be

$$v_o(t) = 3V \exp\left(-\frac{Z_o t}{3L}\right) \tag{6.32}$$

which clearly has a slower droop rate than the three-stage transformer analysed previously.

Furthermore, consideration of the distribution of potential throughout the transformer structure, as indicated in Figure 6.19 together with detailed mesh analysis of the equivalent circuit, suggests again that it is necessary to consider only the inductive paths to ground through the outer conductors of the transmission lines. Thus the inductive path to ground, in this transformer, consists solely of two inductors connected in series, as shown in middle circuit in Figure 6.19. Reflection of these inductors to the primary circuit, as before, then results in a very simple equivalent circuit which makes analysis of the pulse rise-time and pulse droop rate response of the transformer a very simple task. This winding technique can be extended in an n-stage device by winding all but the bottom line on one core, winding all but the bottom two lines on the next core and so on until the top line is wound alone on the last core. The equivalent circuit which results from such a winding technique gives rise to the general form for an n-stage transformer shown in the lower circuit of Figure 6.19; the expression for the output voltage of an n-stage transformer of this type is

$$v_o(t) = nV \exp\left(-\frac{(n-1)Z_o t}{2nL}\right) \tag{6.33}$$

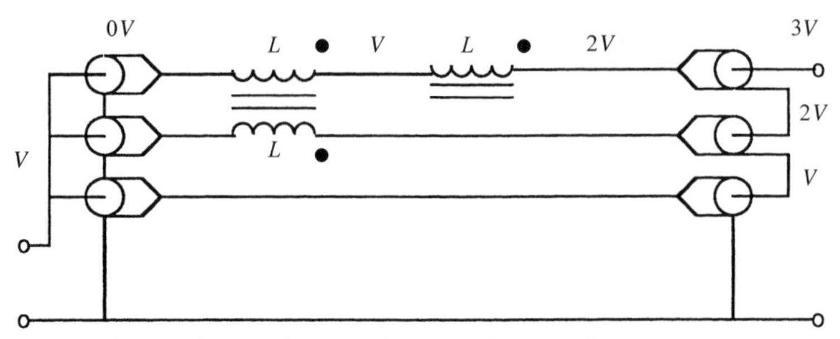

Diagram of a mutually coupled 3-stage voltage transformer

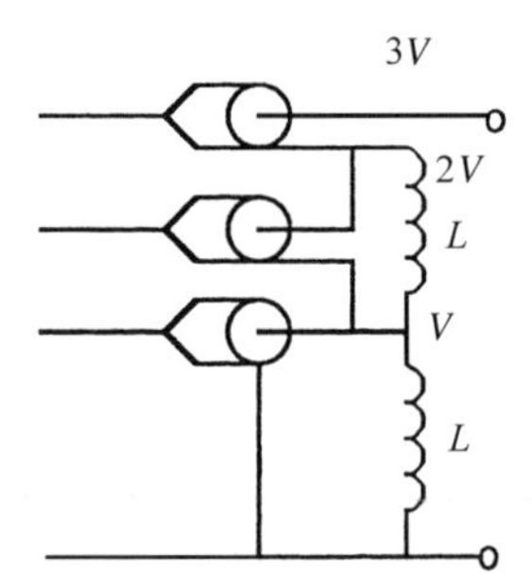

Equivalent circuit of the transformer output

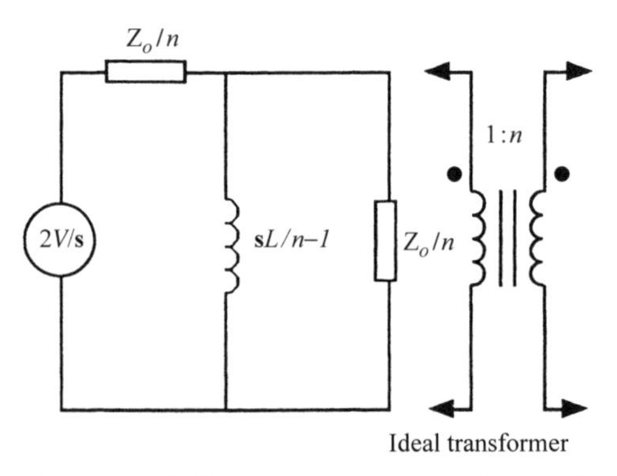

Ideal transformer

Equivalent circuit of an n-stage transformer referred to the primary

Figure 6.19 Circuit diagram of a mutually coupled three-stage voltage transformer with simplified equivalent circuits

6.3.4 Frequency Response Analysis

The analysis and modelling method presented has so far only been applied to the examination of the transient response of wound TLTs. However the method can be used, in modified form, to determine the AC frequency response of these devices. This is done by combining all the n transmission lines used to build the transformer into one transmission

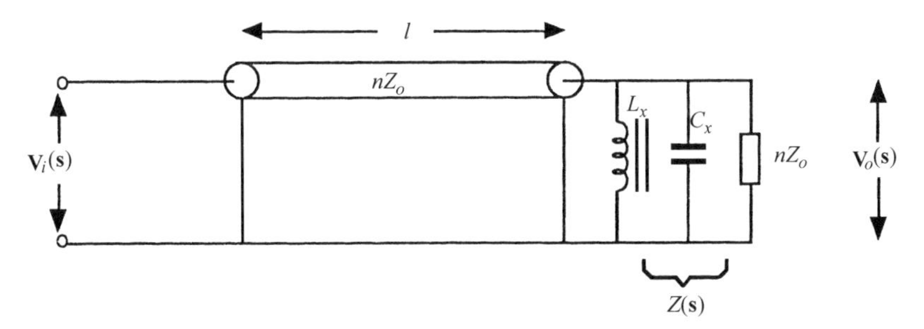

Figure 6.20 Equivalent circuit of a transmission line voltage transformer used to determine its frequency response

line with characteristic impedance *nZo*. The output of this line is connected to the combined equivalent components and load referred to the secondary circuit as shown in Figure 6.20. From the standard matrix relationship connecting the input and output currents and voltages of a length of line l and propagation constant $\gamma(\mathbf{s})$, i.e.

$$\begin{pmatrix} \mathbf{V}_i(\mathbf{s}) \\ \mathbf{I}_i(\mathbf{s}) \end{pmatrix} = \begin{pmatrix} \cosh \gamma(\mathbf{s})l & Z_o \sinh \gamma(\mathbf{s})l \\ Z_o^{-1} \sinh \gamma(\mathbf{s})l & \cosh \gamma(\mathbf{s})l \end{pmatrix} \begin{pmatrix} \mathbf{V}_o(\mathbf{s}) \\ \mathbf{I}_o(\mathbf{s}) \end{pmatrix} \tag{6.34}$$

the transfer function from the input to the output of the line can be derived as

$$\mathbf{T}(\mathbf{s}) = \frac{\mathbf{V}_o(\mathbf{s})}{\mathbf{V}_i(\mathbf{s})} = \frac{\mathbf{Z}(\mathbf{s})}{Z_o \sinh \gamma(\mathbf{s})l + \mathbf{Z}(\mathbf{s}) \cosh \gamma(\mathbf{s})l} \tag{6.35}$$

$\mathbf{Z}(\mathbf{s})$ is the impedance of the parallel combination of the combined equivalent inductance to ground L_x, the stray capacitance C_x and load in the secondary circuit. From this expression it is then possible to determine the frequency response of the transformer, provided an estimate of the propagation constant $\gamma(\mathbf{s})$ for the cables can be made.

A good example of the use of this technique is to be found in [28] which describes the performance of a 10-stage, 200 kV wound TLT. A circuit diagram of this transformer is given in Figure 6.21 and a photograph of the device in Figure 6.22. From the circuit diagram, it can be seen that full advantage of the mutual winding technique has been made to produce a transformer that has very low droop and a good low-frequency response. The transformer was wound with 10 lengths of coaxial cable, each of which was 110 m long, on to a large PVC tube. Large numbers of ferrite cores were used on each stage to increase the inductive isolation required between the secondary and primary circuits. Because the winding lengths for each stage of the transformer are different, as can be seen in the diagram, sections of cable which are not wound in the transformer are "stored" on a second winding which can be seen on the right of the photograph. The transformer was driven by a 5 Ω pulse-forming line which was switched by a low inductance, rail, spark gap and produced a fast rising 20 kV and 1 μs rectangular pulse.

Typical input and output waveforms are shown in Figure 6.23, from which it can be seen that the rise-time of the transformer is 65 ns. This figure is very close to that which is

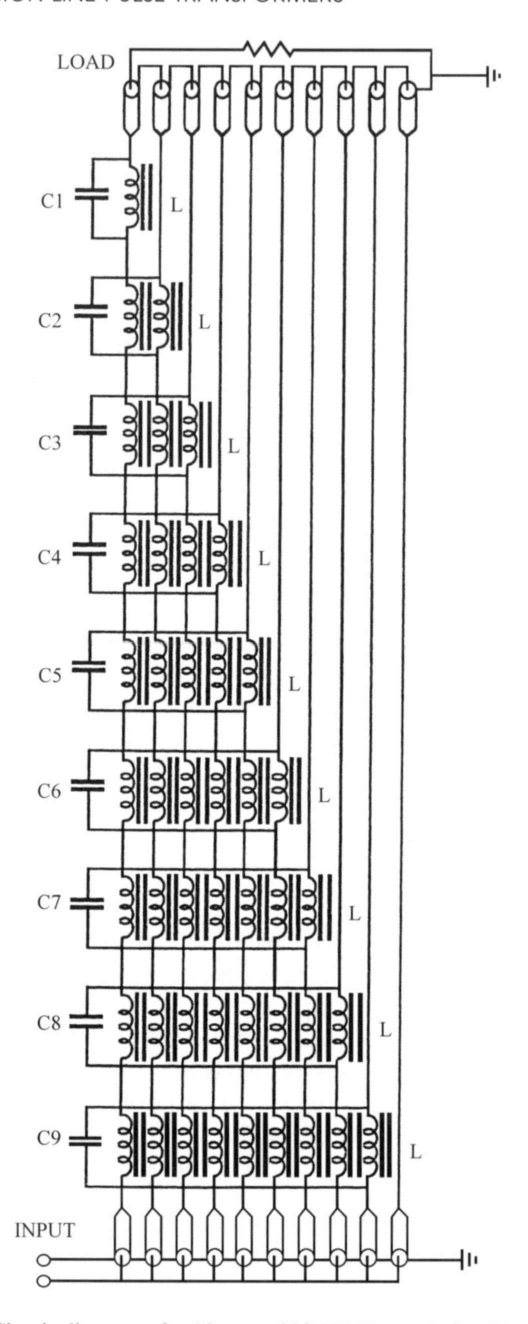

Figure 6.21 Circuit diagram of a 10-stage 200 kV Transmission Line Transformer

predicted by referral of the estimated stray capacitance in the transformer into the primary circuit by the method described earlier in this chapter. Similarly the frequency response of the transformer, as shown in Figure 6.24, is very close to that which is estimated using the method described above.

Figure 6.22 Photograph of the 200 kV Transmission Line Transformer

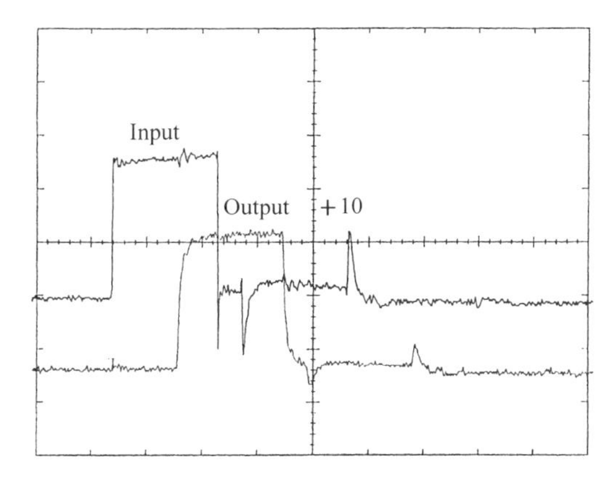

Figure 6.23 Typical input and output waveforms for the transmission line transformer. The pulse on the left is the input pulse and the one on the right is the output. The time scale per division is 500 ns and the amplitude of the output waveform has been reduced by a factor of 10

6.4 TAPERED TRANSMISSION LINE TRANSFORMERS

The last type of transmission line transformer to be described in this chapter is the tapered line transformer. A comprehensive review of such devices as pulse transformers is given in the book by Lewis and Wells [4]. Although they are not often used as pulsed transformers, they do find application as a matching devices particularly in large pulsed power systems and in microwave circuits [29]. It is interesting to note that, although this type of transformer is quite different in both construction and operation to all of the pulse transformers described so far, it still suffers from rise-time loss (due to the skin effect [2]) and pulse droop limitations. It also has the disadvantage that it can only be used at very high

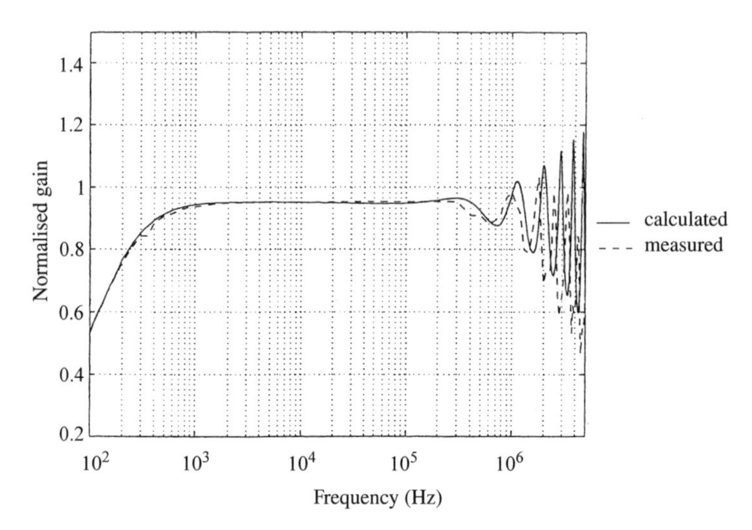

Figure 6.24 Calculated and measured frequency response curves for the transmission line transformer

frequencies or for very short pulses of the order of a few 10s of nanoseconds or less. Fast rise-times are achievable and such transformers can be used at very high power levels. They also do not need magnetic cores and are relatively easy to construct. However pulse inversion and isolation are not possible with this type of transformer.

The transformers are constructed from transmission lines in which the capacitance and inductance per unit length are varied with position along the line. The most obvious way to do this, in the case of a strip line, is to vary the spacing between the strip conductors or their width along the line, as shown in Figure 6.25. It is also possible to change the capacitance by varying the dielectric constant of the insulating material between the lines, although in practice this may be difficult to achieve.

Since the inductance L and capacitance C per unit length now depend upon the distance z along the line the basic line, equations for voltage and current (Equations (2.11) and (2.13)) must be modified to give

$$\left.\begin{aligned} \frac{d^2\mathbf{V}(z,\mathbf{s})}{dz^2} - \frac{1}{L(z)}\frac{dL(z)}{dz}\frac{d\mathbf{V}(z,\mathbf{s})}{dz} - \mathbf{s}^2 L(z)C(z)\mathbf{V}(z,\mathbf{s}) = 0 \\ \frac{d^2\mathbf{I}(z,\mathbf{s})}{dz^2} - \frac{1}{C(z)}\frac{dC(z)}{dz}\frac{d\mathbf{I}(z,\mathbf{s})}{dz} - \mathbf{s}^2 L(z)C(z)\mathbf{I}(z,\mathbf{s}) = 0 \end{aligned}\right\} \tag{6.36}$$

These equations assume that the line is loss-free. A characteristic impedance of the line can be defined as before (Equation (2.17)) but in this case the impedance is also a function of position z, i.e.

$$Z_0(z) = \sqrt{\frac{L(z)}{C(z)}} \tag{6.37}$$

It is also useful to define a time delay per unit length along the line $T(z)$ which is given by

$$T(z) = \sqrt{L(z)C(z)} \tag{6.38}$$

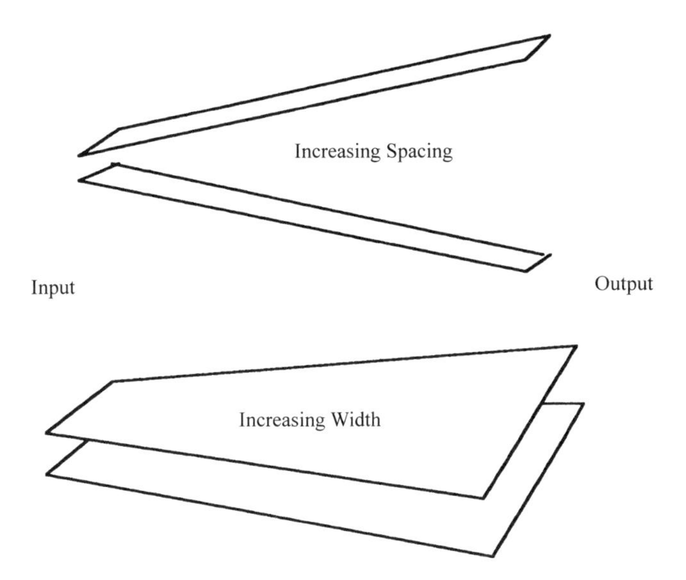

Figure 6.25 Two types of linearly tapered transmission line transformer

In the case of tapered lines, in which the insulating medium between the lines is not changed along the line, $T(z)$ will remain constant with distance z since the velocity of propagation of an electromagnetic wave along the line will be constant and depends solely on the dielectric constant of the medium used to insulate the line.

An approximate solution to the Equations (6.41) has been given by Slater [30] who suggested that a trial solution for the voltage equation of the form

$$\mathbf{V}(z,\mathbf{s}) = A(z)\exp\left(-\int_0^z \gamma(z',\mathbf{s})\,dz'\right) \equiv A(z)\exp\left(-\mathbf{s}\int_0^z T(z')\,dz'\right) \tag{6.39}$$

would be reasonable after consideration of the equivalent solutions for a uniform line given by Equation (2.77) with the substitution of the propagation constant $\gamma\,(z,\mathbf{s})$ by a time delay factor given, for example, by Equation (2.87). Here the time delay to a given position on the line, distance z from the input, can be determined by integrating Equation (6.38). After substitution of this expression into Equations (6.36) two differential equations result, in which certain terms can be neglected so that an expression for $A(z)$ can be determined. Thus Equation (6.39) becomes

$$\mathbf{V}(z,\mathbf{s}) = \mathbf{V}(0,\mathbf{s})\sqrt{\frac{Z_0(z)}{Z_0(0)}}\exp\left(-s\int_0^z T(z')\,dz'\right) \tag{6.40}$$

and the equivalent solution for the current equation is given by

$$\begin{aligned}\mathbf{I}(z,\mathbf{s}) &= \mathbf{I}(0,\mathbf{s})\sqrt{\frac{Z_0(0)}{Z_0(z)}}\exp\left(-s\int_0^z T(z')\,dz'\right)\\ &= \frac{\mathbf{V}(0,\mathbf{s})}{Z_0(0)}\sqrt{\frac{Z_0(0)}{Z_0(z)}}\exp\left(-s\int_0^z T(z')\,dz'\right)\end{aligned} \tag{6.41}$$

In deriving these expressions it is assumed that

$$\left|\frac{\lambda\, dL}{L\, dx}\right| \ll 2\pi, \quad \left|\frac{\lambda\, dC}{C\, dx}\right| \ll 2\pi \tag{6.42}$$

where λ is the wavelength of a continuous sinusoidal input or the longest wavelength component of a pulsed input to the tapered line. These approximations suggest, for the solution given in Equation (6.40) to be valid, that the rate of tapering of the line must be such that fractional change in L or C over a distance which is a fractional part of a wavelength must be very much less than 2π. For example, over a distance equivalent to 10% of a wavelength the percentage change in inductance must be very much less than 60%, i.e. the rate of tapering must be slow. The solution given by Equation (6.40) also assumes that there is no backward wave, i.e. the tapered line is infinite (see Equation (2.64)). The impedance at a given position on the line can be determined by dividing the voltage at that point by the current. Therefore, for the forward travelling components of voltage and current given by Equations (6.40) and (6.41)

$$\frac{\mathbf{V}(z, \mathbf{s})}{\mathbf{I}(z, \mathbf{s})} = Z_0(z) \tag{6.43}$$

From Equation (6.40) it is interesting to note that the ratio of the voltage at a distance z from the input to the input voltage is equal to the square root of the ratio of the impedance at a distance z to the impedance at the input to the taper, if the time delay factor is ignored. This is entirely consistent with transformer action, as has been seen in the equivalent behaviour of wound pulse transformers and transmission line transformers.

The need for a slow rate of tapering can also be deduced by consideration of the reflection coefficient at the input of a tapered line transformer. Figure 6.26 shows a continuous taper of length L which is approximated by a series of steps of differential length dz with a differential change in impedance of dZ_0. The impedance at the input of the taper is $Z_0(0)$ and the output is connected to a matched resistive load Z_l.

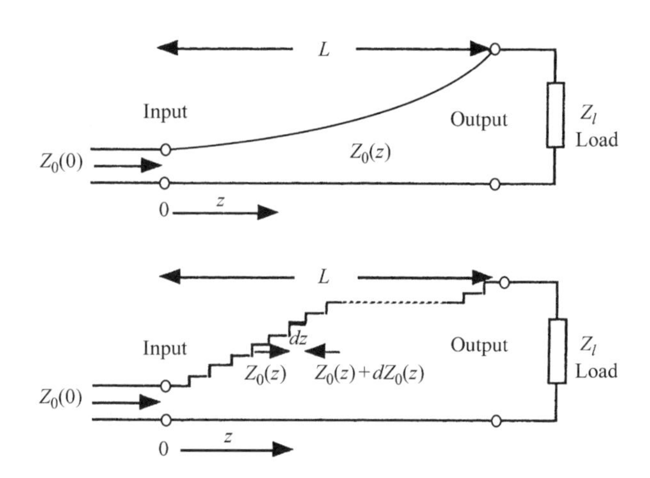

Figure 6.26 Tapered line approximation

Considering a step distance z from the input to the taper the differential reflection coefficient due to this step is given by

$$d\rho(z) = \frac{Z_o(z) + dZ_0(z) - Z_o(z)}{Z_o(z) + dZ_0(z) + Z_o(z)} \approx \frac{dZ_0(z)}{2Z_0(z)} = \frac{1}{2}\frac{d}{dz}(\ln Z_0(z))\, dz \tag{6.44}$$

At the input to the taper the contribution to the reflection coefficient by this step is given by

$$d\rho(0) = \exp(-2j\beta z)\frac{1}{2}\frac{d}{dz}(\ln Z_0(z))\, dz \tag{6.45}$$

where it is assumed that the tapered transmission line is loss-free and has a phase constant $\beta = 2\pi/\lambda$. The total reflection coefficient at the input to the line $\rho(0)$ can then be determined by summing all the individual contributions from each of the steps over the length of the taper L, i.e.

$$\rho(0) = \frac{1}{2}\int_0^L \exp(-2j\beta z)\frac{d}{dz}(\ln Z_0(z))\, dz \tag{6.46}$$

If the variation of $Z_0(z)$ with z is known, then an expression for the reflection coefficient $\rho(0)$ is readily determined. For example, if the line has an exponential taper then ln $Z_0(z)$ varies linearly and $Z_0(z)$ varies exponentially with distance z from an impedance $Z_0(0)$ to Z_l, i.e.

$$Z_0(z) = Z_0(0)\exp\left[\frac{z}{L}\ln\left(\frac{Z_l}{Z_0(0)}\right)\right]$$

$$\text{or} \qquad \ln\left(\frac{Z_0(z)}{Z_o(0)}\right) = \frac{z}{L}\ln\left(\frac{Z_l}{Z_0(0)}\right) \tag{6.47}$$

Substitution of this equation into Equation (6.51) gives

$$\begin{aligned}\rho(0) &= \frac{1}{2}\int_0^L \frac{\ln\left(\frac{Z_l}{Z_0(0)}\right)}{L}\exp(-2j\beta z)\, dz \\ &= \frac{1}{2}\exp(-j\beta L)\ln\left(\frac{Z_l}{Z_0(0)}\right)\frac{\sin\beta L}{\beta L}\end{aligned} \tag{6.48}$$

A plot of the normalised magnitude of the reflection coefficient $\rho(0)$ as a function of βL is given in Figure 6.27, from which it can be seen that when L is greater than $\lambda/2$, the reflection coefficient at the input to the taper is quite small. Hence for a pulsed input to a tapered line transformer it is important to ensure that the wavelength λ of the lowest-frequency component of the pulse satisfies this requirement for efficient operation of the transformer. This again is equivalent to specifying a maximum rate at which the tapering of impedance with distance should occur for a given pulse length.

6.4.1 The Exponentially Tapered Transmission Line Transformer

Clearly there are many ways in which the tapering of impedance in a tapered line transformer can be specified. However, in practice, the analysis of the behaviour of

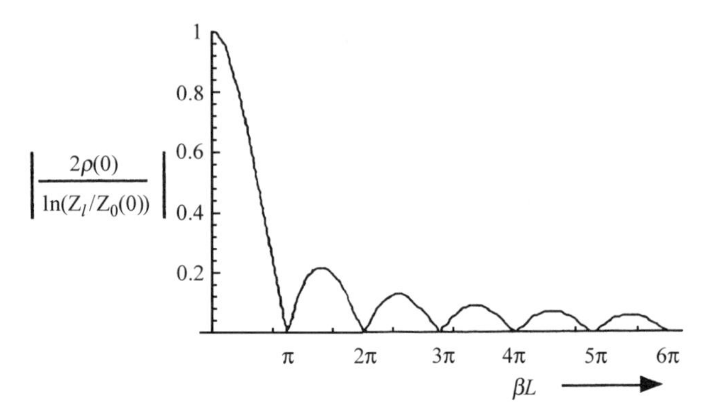

Figure 6.27 Input reflection coefficient for an exponential tapered line transformer

exponentially tapered lines is one of the easiest to deal with mathematically. Indeed, exact solutions to Equations (6.36) are possible for exponentially tapered lines [31,32]. The detail of such an exact analysis lies beyond the scope of this book, however it is interesting to carry out an approximate analysis of the way in which a pulse is distorted as it propagates along an exponentially tapered transmission line transformer. For the exponential line it is usually assumed that the inductance and capacitance per unit length vary according to

$$L(z) = L(0)\exp(kz) \qquad \text{and} \qquad C(z) = C(0)\exp(-kz) \tag{6.49}$$

where $L(0)$ and $C(0)$ are the inductance and capacitance at the start of the taper and k is a constant which can be either positive or negative. The impedance will therefore also vary exponentially along the line as

$$Z_0(z) = \sqrt{\frac{L(z)}{C(z)}} = \sqrt{\frac{L(0)}{C(0)}}\exp(kz) = Z_0(0)\exp(kz) \tag{6.50}$$

Assuming that the time delay per unit length $T(z)$ remains constant (see Equation (6.38)) then

$$T(z) = \sqrt{L(z)C(z)} = \sqrt{L(0)C(0)} = T(0) \tag{6.51}$$

At this point a new parameter $\delta(z)$ is introduced, which is the total time delay from the start of the tapered line to a point position co-ordinate z on the line. $\delta(z)$ is then given by

$$\delta(z) = \int_0^z T(z')\,dz' \qquad \text{and} \qquad \frac{d\delta}{dz} = T(z) \tag{6.52}$$

An alternative definition of a tapered line can now be written as

$$Z_0(\delta) = Z_0(0)\exp(k\delta/T(0)) \tag{6.53}$$

where points on the line are measured from the input by the delay parameter δ rather than z. Equation (6.53) reduces to Equation (6.50), provided $T(z)$ is constant and $\delta = T(0)z$.

Equations (6.36) can now be re-written in terms of δ to give

$$\left.\begin{aligned}\frac{1}{\mathbf{s}^2}\frac{d^2\mathbf{V}(\delta,\mathbf{s})}{d\delta^2}-\frac{1}{\mathbf{s}^2}\frac{d\ln Z_0(\delta)}{d\delta}\frac{d\mathbf{V}(\delta,\mathbf{s})}{d\delta}-\mathbf{V}(\delta,\mathbf{s})=0\\ \frac{1}{\mathbf{s}^2}\frac{d^2\mathbf{I}(\delta,\mathbf{s})}{d\delta^2}+\frac{1}{\mathbf{s}^2}\frac{d\ln Z_0(\delta)}{d\delta}\frac{d\mathbf{I}(\delta,\mathbf{s})}{d\delta}-\mathbf{I}(\delta,\mathbf{s})=0\end{aligned}\right\} \tag{6.54}$$

Using Equation (6.53) these equations then become

$$\left.\begin{aligned}\frac{1}{\mathbf{s}^2}\frac{d^2\mathbf{V}(\delta,\mathbf{s})}{d\delta^2}-\frac{k}{\mathbf{s}^2T(0)}\frac{d\mathbf{V}(\delta,\mathbf{s})}{d\delta}-\mathbf{V}(\delta,\mathbf{s})=0\\ \frac{1}{\mathbf{s}^2}\frac{d^2\mathbf{I}(\delta,\mathbf{s})}{d\delta^2}+\frac{k}{\mathbf{s}^2T(0)}\frac{d\mathbf{I}(\delta,\mathbf{s})}{d\delta}-\mathbf{I}(\delta,\mathbf{s})=0\end{aligned}\right\} \tag{6.55}$$

The solution to the voltage equation in Equations (6.55) is

$$\left.\begin{aligned}&\mathbf{V}(\delta,\mathbf{s})=\mathbf{V}_+(\delta,\mathbf{s})+\mathbf{V}_-(\delta,\mathbf{s})\\ \text{where}\quad&\mathbf{V}_+(\delta,\mathbf{s})=\mathbf{V}_+(0,\mathbf{s})\exp\left[\frac{\delta\left(\frac{k}{2}-\mathbf{q}(\mathbf{s})\right)}{T(0)}\right]\\ &\mathbf{V}_-(\delta,\mathbf{s})=\mathbf{V}_-(0,\mathbf{s})\exp\left[\frac{\delta\left(\frac{k}{2}+\mathbf{q}(\mathbf{s})\right)}{T(0)}\right]\\ &\mathbf{q}(\mathbf{s})\equiv\sqrt{\mathbf{s}^2T(0)^2+k^2/4}\end{aligned}\right\} \tag{6.56}$$

Similar solutions are found for the current on the taper. The subscripts $+$ and $-$, as before, represent forward and backward travelling waves, respectively, and $\mathbf{V}_+(0,\mathbf{s})$ and $\mathbf{V}_-(0,\mathbf{s})$ are the forward and backward voltage signals at the input to the line. To investigate these solutions it is helpful to look at three approximations. The first is to consider only high frequencies where

$$\begin{aligned}\mathbf{s}T(0)\gg k/2\\ \Rightarrow\ \mathbf{q}(\mathbf{s})\approx\mathbf{s}T(0)\end{aligned} \tag{6.57}$$

$\mathbf{s}$ then appears only in the solutions for the voltage on the taper as a factor $\exp(\pm\mathbf{s}\delta)$, which is simply a time delay from the input to the taper to points with position coordinate δ. The voltage transformation n for forward travelling waves can be determined from Equations (6.53) and (6.56) to be

$$n=\frac{\mathbf{V}_+(\delta,\mathbf{s})}{\mathbf{V}_+(0,\mathbf{s})}=\exp\left(\frac{\delta k}{2T(0)}\right)=\sqrt{\frac{Z_0(\delta)}{Z_0(0)}} \tag{6.58}$$

as expected. For the second approximation the expression for $\mathbf{q(s)}$ in Equations (6.61) are expanded by the binomial expansion to give

$$\mathbf{q(s)} \approx \mathbf{s}T(0) + \frac{k^2}{8\mathbf{s}T(0)} \tag{6.59}$$

The solution for the forward travelling voltage wave given in Equations (6.56) therefore becomes

$$\mathbf{V}_+(\delta,\mathbf{s}) = \mathbf{V}_+(0,\mathbf{s})\exp\left(\frac{\delta k}{2T(0)}\right)\exp(-\mathbf{s}\delta)\exp\left(\frac{\delta k^2}{8\mathbf{s}T(0)^2}\right) \tag{6.60}$$

Expansion of the last exponential term in Equation (6.65), neglecting higher-order terms, then gives

$$\mathbf{V}_+(\delta,\mathbf{s}) = \mathbf{V}_+(0,\mathbf{s})\exp\left(\frac{\delta k}{2T(0)}\right)\exp(-\mathbf{s}\delta)\left(1 - \frac{\delta k^2}{8\mathbf{s}T(0)^2}\right) \tag{6.61}$$

In Equation (6.61) the first exponential term is recognised from Equation (6.58) as the voltage transformation ratio, the second is a delay term and the third represents pulse droop. If a rectangular pulse of amplitude V_0 and duration τ is injected into the line then

$$\left.\begin{aligned} \mathbf{V}_+(0,\mathbf{s}) &= V_0\frac{[1-\exp(-\mathbf{s}\tau)]}{\mathbf{s}} \\ \mathbf{V}_+(\delta,\mathbf{s}) &= V_0\frac{[1-\exp(-\mathbf{s}\tau)]}{\mathbf{s}}\exp\left(\frac{\delta k}{2T(0)}\right)\exp(-\mathbf{s}\delta)\left(1-\frac{\delta k^2}{8\mathbf{s}T(0)^2}\right) \end{aligned}\right\} \tag{6.62}$$

Without inverting this expression to find $v_+(\delta,t)$, it is possible to deduce the shape of the pulse, after propagation delay δ, by inspection. A sketch of the input and propagated pulse is given in Figure 6.28, where it has been assumed that the tapering leads to an increase in line

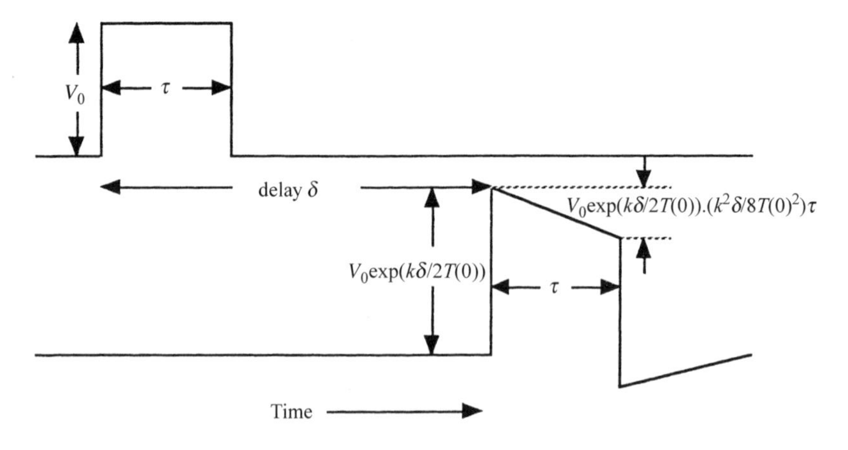

Figure 6.28 Input and propagated pulses on a tapered transmission line transformer

impedance and hence pulse amplitude. Note that the very last term in the expression for $\mathbf{V}_+(\delta,\mathbf{s})$ when multiplied by $1/\mathbf{s}$ gives a negative going ramp or linear droop.

The last approximation involves looking at the propagation of low frequencies on the tapered line transformer. In this case

$$\mathbf{s}T(0) \ll k/2$$

$$\text{and putting } \mathbf{s} = j\omega, \qquad j\omega T(0) \ll k/2 \tag{6.63}$$

and $\mathbf{q}(\mathbf{s})$ now becomes a real quantity. This means that low frequencies are not propagated and are strongly reflected. Thus $\omega T(0) = k/2$ becomes a cut-off condition, although it must be remembered that there still exists a DC connection from the input of the taper to the output. Thus the cut-off condition is not complete as it would be, say, in the case of a high-pass filter or waveguide.

From Equations (6.58) and (6.62) it is possible to derive an expression for the percentage fall in amplitude P at the end of the pulse.

$$\frac{\delta}{\tau} = \frac{50\ln^2(n)}{P}$$

$$\text{where } P \text{ is given by} \qquad P = \frac{\delta k^2 \tau}{8T(0)^2} \times 100\% \tag{6.64}$$

Thus the delay on a propagating pulse divided by its duration can give an estimate of pulse droop if the voltage transformation ratio n is known. This equation provides a very useful design guide for assessing pulse droop on tapered transmission lines.

Finally, to correctly terminate the tapered line and hence achieve optimum electrical efficiency it is necessary to know the reflection coefficient at an arbitrary load termination. This can be done if the voltage to current ratios are known for both the forward and backward voltage and current waves. This is achieved using Equations (6.56) and the equivalent solutions for the current distribution on the tapered line, which are given by simply changing the sign of k in the equations and dividing $\mathbf{V}_+(0,\mathbf{s})$ and $\mathbf{V}_-(0,\mathbf{s})$ by $Z_0(\delta)$. Thus the forward and backward voltage to current ratios are given by

$$\left.\begin{aligned}\frac{\mathbf{V}_+(\delta,\mathbf{s})}{\mathbf{I}_+(\delta,\mathbf{s})} &= Z_0(\delta)\left(1+\frac{k}{2\mathbf{s}T(0)}\right)\\ \frac{\mathbf{V}_-(\delta,\mathbf{s})}{\mathbf{I}_-(\delta,\mathbf{s})} &= -Z_0(\delta)\left(1-\frac{k}{2\mathbf{s}T(0)}\right)\end{aligned}\right\} \tag{6.65}$$

The reflection coefficient $\rho_l(\mathbf{s})$ at an arbitrary load Z_l located after a delay δ_l is given by

$$\left.\begin{aligned}\rho_l(\mathbf{s}) &= \frac{\mathbf{V}_-(\delta_l,\mathbf{s})}{\mathbf{V}_+(\delta_l,\mathbf{s})}\\ Z_l(\mathbf{s}) &= \frac{\mathbf{V}_+(\delta_l,\mathbf{s})+\mathbf{V}_-(\delta_l,\mathbf{s})}{\mathbf{I}_+(\delta_l,\mathbf{s})+\mathbf{I}_-(\delta_l,\mathbf{s})}\end{aligned}\right\} \tag{6.66}$$

Thus by substitution

$$\rho_l(\mathbf{s}) = \frac{\left(1 - \dfrac{k}{2\mathbf{s}T(0)}\right)Z_l - Z_0(\delta_l)}{\left(1 + \dfrac{k}{2\mathbf{s}T(0)}\right)Z_l + Z_0(\delta_l)} \tag{6.67}$$

For a nominal match at the output to the tapered line

$$\left.\begin{aligned} Z_l &= Z_0(\delta_l) \\ \rho_l(\mathbf{s}) &\approx -\frac{k}{4\mathbf{s}T(0)} \end{aligned}\right\} \tag{6.68}$$

which gives a problem for pulses as $\rho_l(\mathbf{s})$ is frequency-dependent through $\mathbf{s}$.

REFERENCES

[1] Ruthroff C. L. "Some Broad-Band Transformers". *Proceedings of the IRE* **47** (1959) 1337–1342

[2] Dreher T. "Cabling fast pulses? Don't trip on the steps". The Electronic Engineer (1969) 71–75

[3] Lewis I. A. D. "Some Transmission Line Devices for Use with Millimicrosecond Pulses". *Electronic Engng* **27** (1955) 448–450

[4] Lewis I. A. D. and Wells F. H. in "Millimicrosecond Pulse Techniques". Pergamon Press (1953) 109–111

[5] Fitch R. A. and Howell V. T. S. "Novel Principle of Transient High-Voltage Generation". *Proc. IEE* **111**(4) (1964) 849–855

[6] Bhawalkar J. D., Davanloo F., Collins C. B., Agee F. J. and Kingsley L. E. "High Power Repetitive Stacked Blumlein Pulse Generators Producing Waveforms with Pulse Duration Exceeding 500 nsec". Proceedings of the 9th IEEE Pulsed Power Conference (1993) 857–860

[7] Glasoe G. N. and Lebacqz J. V. "Pulse Generators". MIT Radiation Laboratory Series Vol. 5, McGraw Hill Book Company Inc. (1948) 464–465

[8] Smith I. D. "Principles of the Design of Lossless Tapered Transmission Line Transformers". Proceedings of the 7th IEEE Pulsed Power Conference (1989) 103–107

[9] Coogan J. J., Davenloo F. and Collins C. B. "Production of High-Energy Photons from Flash X-ray Sources Powered by Stacked Blumlein Generators". *Rev. Sci. Instrum.* **61**(5) (1990) 1448–1456

[10] Bulan V. V., Grabovskii E. V., Kalenskii V. A., Korolev V. V., Koba Yu. V., Liksonov V. I., Luki A. A., Nedoseev S. L., and Yampol'skii I. R. "Transmission-Line Pulse Transformer". *Instum. Exp. Tech (USA)* **31** (1988) 1197–1200

[11] Eccleshall D. and Temperley J. K. "Transfer of Energy from Charged Transmission Lines with Applications to Pulsed High–Current Accelerators". *J. Appl. Phys.* **49**(7) 3649–3655

[12] Rochelle R. W. "A Transmission-Line Pulse Inverter". *Rev. Sci. Instrum.* **23**(6) (1952) 298–300

[13] Homma A. "High-Voltage Subnanosecond Pulse Transformer Composed of Parallel-Strip Transmission Lines". *Rev. Sci. Instrum.* **70** (1999) 232–236

[14] Matick R. E. "Transmission Line Pulse Transformers—Theory and Applications". *Proc. IEEE* **56** (1968) 47–62

[15] Wilson C. R., Erickson G. A. and Smith P. W. "Compact, Repetitive, Pulsed Power Generators based on Transmission Line Transformers". Proceedings of the 7th IEEE Pulsed Power Conference (1989) 108–112

[16] Chodorow A. M. "The Time Isolation High-Voltage Impulse Generator". Proc. IEEE Letters (1975) 1082–1084
[17] Wylie C. R. and Barrett L. C. "Advanced Engineering Mathematics". McGraw Hill (1985) ISBN 0-07-072188-2
[18] Winningstrad C. N. "Nanosecond Pulse Transformers". IRE Trans. on Nuclear Science (1959) 26–31
[19] Millman J. and Taub H. "Pulse, Digital and Switching Wavefoms". McGraw Hill (1965) ISBN 0-07-085512-9
[20] Andersson L., Radermacher E. and Rubbia C. "A Pulse Transformer for Large Wire Spark Chambers". *Nuc. Instrum. and Methods* **75** (1969) 341–342
[21] Smith P. W. and Wilson C. R. "Transmission Line Transformers for High Voltage Pulsed Power Generation". Proceedings of the 17th IEEE Power Modulator Symposium (1986) 281–285
[22] Pirrie C. A., Maggs P. N. D. and Smith P. W. "A Repetitive, Thyratron Switched, 200 kV, Fast Rise-Time Pulse Generator based on a Stacked Transmission Line Transformer". Proceedings of the 8th IEEE Pulsed Power Conference (1991) 310–314
[23] Sommerville I. C, MacGregor S. J. and Farish O. "An Efficient Stacked-Blumlein HV Pulse Generator". *Meas. Sci. Technol.* **1** (1990) 865–868
[24] Mesyats G. A., Nasibor A. S. and Kremev V. V. "Formation of nanosecond Pulses of High Voltage". Translation by Foreign Technology Div., Wright Patterson AFB, Ohio, Rep. FTD-HC-23-385-71 (1971)
[25] Graneau P. N., Rossi J. O. and Smith P. W. "The Operation and Modelling of Transmission Line Transformers using a Referral Method". *Rev. Sci. Instrum.* **70**(7) (1999) 3180–3185
[26] Pandian S. G. and Arockiasamy R. "Equivalent Representation of Broad-Band Transmission-Line Transformer with Vector Impedances". *Proc. IEEE* **69** (1981) 1368–1369
[27] Van Valkenburg M. E. "Introduction to Modern Network Synthesis". John Wiley & Sons Inc. (1960)
[28] Graneau P. N., Rossi J. O., Brown M. P. and Smith P. W. "A High-Voltage Transmission-Line Pulse Transformer with Very Low Droop". *Rev. Sci. Instrum.* **67**(7) (1996) 2630–2635
[29] Collin R. E. "Foundations for Microwave Engineering". McGraw Hill (1996) ISBN 0-07-085125-5
[30] Slater J. C. "Microwave Transmssion". McGraw Hill (1942)
[31] Schatz E. R. and Williams E. M. "Pulse Transients in Exponential Transmission Lines". *Proc. Inst. Radio Engrs* **39** (1951) 84–86
[32] Williams E. M. and Schatz E. R. "Design of Exponential Pulse Transformers". *J. I. E. E.* **93**(IIIA) (1946) 559–563

7

Pulse Generators using Capacitive and Inductive Energy Storage

7.1 INTRODUCTION

Perhaps the simplest form of pulse generator relies on the discharge of a capacitor or capacitor bank into a load. The method is not only simple but relatively cheap and the necessary switches are commonly available. In this chapter a range of pulse generators will be described which rely primarily on the discharge of energy stored either in capacitors or inductors rather than in transmission lines or pulse-forming networks, as described in Chapters 3 and 4. The only exception to this is the spiral generator, which is included in this chapter as its operation relies on "vector inversion" in a similar way to that of the LC generator, which is also described in this chapter. Perhaps the most important of the generators to be described is the Marx generator, which is and continues to be the most popular and versatile high-voltage pulse generator for many pulsed power applications, particularly those which require the highest voltages and pulsed power levels.

7.2 THE BASIC PRINCIPLES OF CAPACITIVE AND INDUCTIVE ENERGY DISCHARGE

The discharge of capacitor, charged to an initial potential $V(0)$, into a load through a stray inductance L has already been analysed in Chapter 1 in Example 1.10. Although the circuit behaviour was analysed with two capacitors in the circuit the analysis is easy to modify for the basic circuit shown in Figure 7.1(a). The current flowing in the circuit, when the switch is closed, is given by the three Equations (7.1) which depend on whether the circuit is underdamped, critically damped and overdamped. From these equations it is easy to deduce that for a fast discharge the stray inductance in the circuit L must be small such that the circuit becomes underdamped, with the penalty that there will be ringing on the pulse

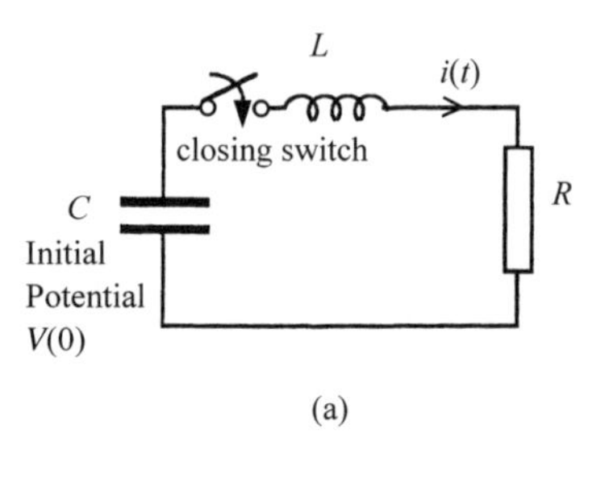

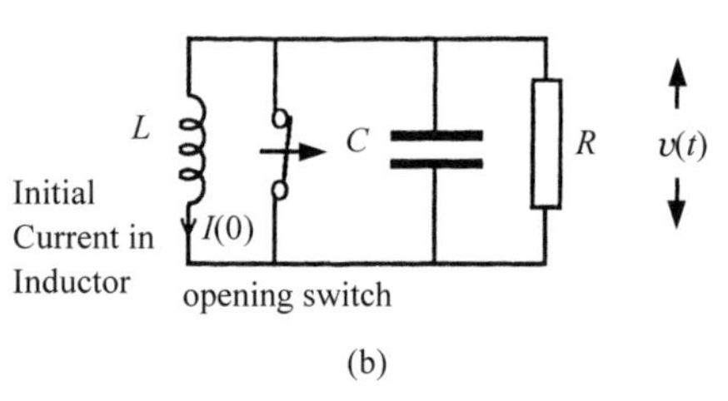

Figure 7.1 Basic capacitive and inductive discharge circuits

generated at the load (see for example Figure 1.11).

$$\left.\begin{aligned} i(t) &= \frac{V(0)}{\omega_d L}\exp(-\alpha t)\sin\omega_d t \qquad \omega_d^2 > 0 \quad \text{underdamped} \\ i(t) &= \frac{V(0)t}{L}\exp(-\alpha t) \qquad \omega_d^2 = 0 \quad \text{critically damped} \\ i(t) &= \frac{V(0)}{\beta L}\exp(-\alpha t)\sinh\beta t \qquad \omega_d^2 < 0 \quad \text{overdamped} \\ \alpha &= \frac{R}{2L}, \quad \omega_0 = \sqrt{\frac{1}{LC}} \quad \text{and} \quad \omega_d^2 = \omega_0^2 - \alpha^2 = -\beta^2 \end{aligned}\right\} \tag{7.1}$$

The rise-time of the current to its peak amplitude t_r is also easy to determine from these equations, and is found to be given by:

$$\left.\begin{aligned} t_r &= \frac{1}{\omega_d}\tan^{-1}\frac{\omega}{\alpha} \qquad \text{underdamped} \\ t_r &= \frac{1}{\alpha} \qquad \text{critically damped} \\ t_r &= \frac{1}{\omega_d}\tanh^{-1}\frac{\omega}{\alpha} \qquad \text{overdamped} \end{aligned}\right\} \tag{7.2}$$

A similar analysis results from the discharge of an inductor, with an initial current $I(0)$ flowing through it, into a load with stray capacitance C, as shown in Figure 7.1(b). In this circuit it is assumed that $I(0)$ has been provided by some current source (not shown) and at a time $t = 0$ a switch, which short-circuits the load, is opened so that the energy stored in the inductor is discharged into the load. Clearly the circuit, as drawn, is not realisable and additional switches are required to isolate the inductor from the charging current source, when the opening switch is opened, and to isolate the load from the inductor during the current charging phase.

The analysis of this circuit is also quite straightforward using the Laplace transformed circuit with initial current $I(0)$ flowing through the inductor. The voltage on the load is again found to be given by three Equations (7.3) depending on whether the circuit is underdamped, critically damped or overdamped. Again, for a fast discharge the stray capacitance C must be small such that the circuit is underdamped, with the penalty that the pulse on the load will ring.

$$\left.\begin{aligned} v(t) &= \frac{-I(0)}{\omega_d C}\exp(-\alpha t)\sin(\omega_d t) && \omega_d^2 > 0 \quad \text{underdamped} \\ v(t) &= \frac{-I(0)t}{C}\exp(-\alpha t) && \omega_d^2 = 0 \quad \text{critically damped} \\ v(t) &= \frac{-I(0)}{\beta C}\exp(-\alpha t)\sinh(\beta t) && \omega_d^2 < 0 \quad \text{overdamped} \\ \alpha &= \frac{1}{2CR}, \quad \omega_0 = \sqrt{\frac{1}{LC}} \quad \text{and} && \omega_d^2 = \omega_0^2 - \alpha^2 = -\beta^2 \end{aligned}\right\} \tag{7.3}$$

A similar analysis was also given in Example 1.12.

7.2.1 Pulse Generators based on Inductive Energy Storage

The main advantage of inductive energy storage compared to capacitive storage is size. The energy density storage capability of inductors is much higher than that of capacitors and indeed can be at least an order of magnitude greater. Its main disadvantage is that inductive discharge circuits must contain an opening switch which has to, in some circuits, turn off large currents very quickly for the circuit to work efficiently [1]. Unfortunately such opening switches are difficult to construct, particularly in circuits which are required to operate repetitively. Thus pulse generators using capacitive energy storage are far more common than their inductive counterparts.

The constraints on the opening switch performance can be best understood by the analysis of the two inductive energy discharge circuits shown in Figure 7.2. The load in the first circuit is resistive and in the second it is inductive, and in both circuits the opening switch is represented by a time-varying resistor $R(t)$. The initial current in the inductor $I(0)$ is accounted for by a voltage source $LI(0)$ (see Figure 1.8).

The simplest analysis for circuit (a) in Figure 7.2 is when the behaviour of the opening switch is ideal, i.e. $R(t)$ changes instantly from 0 to infinity. In this case the voltage on the load $\mathbf{V(s)}$ is given by

$$\mathbf{V(s)} = I(0)R_l\frac{1}{\mathbf{s}+\alpha}, \qquad \alpha = \frac{R_l}{L} \tag{7.4}$$

Hence the voltage on the load as a function of time is given by

$$v(t) = I(0)R_l\exp(-\alpha t) \tag{7.5}$$

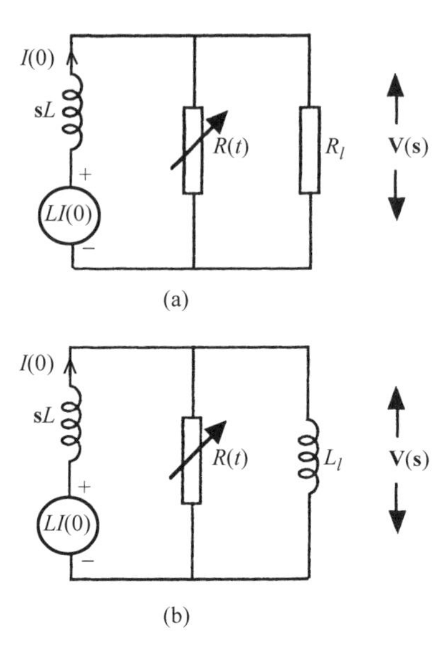

Figure 7.2 Two basic inductive discharge circuits with time-varying opening switches $R(t)$

If instead the resistance of the opening switch changes instantly from 0 to some finite value R_s, then the voltage on the load is found to be

$$v(t) = I(0)R_t \exp(-\alpha t), \qquad \alpha = \frac{R_t}{L}, \quad R_t = R_s||R_l = \frac{R_s R_l}{R_s + R_l} \tag{7.6}$$

A more interesting and realistic case is given when the opening switch resistance changes in some way with time so that the resistance of the switch is time-dependent, i.e. $R(t)$. The current in the load can be determined using the equivalent circuit given in Figure 7.3.

It is assumed that initially the storage inductor L_1 has a current $I(0)$ flowing through it and the load impedance Z is resistive, i.e. $Z = R_l$. Applying Kirchhoff's laws to this circuit

$$\left.\begin{aligned} &i_1(t) = i_2(t) + i_3(t) \\ &L_1 \frac{di_1(t)}{dt} + i_2(t)R(t) = 0 \\ &i_2(t)R(t) = i_3(t)Z = i_3(t)R_l \end{aligned}\right\} \tag{7.7}$$

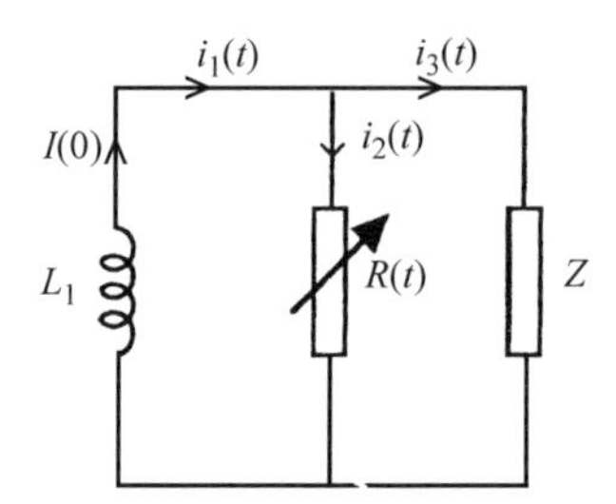

Figure 7.3 Equivalent circuit for the discharge of an inductor into a load impedance Z

From these equations a differential equation can be derived for $i_1(t)$, i.e.

$$\frac{di_1(t)}{dt} + \frac{R(t)R_l i_1(t)}{L_1(R(t)+R_l)} = 0 \tag{7.8}$$

The integrating factor $\phi(t)$ for this equation [2] is given by

$$\phi(t) = \exp \int \frac{R(t)R_l}{L_1(R(t)+R_l)} dt \tag{7.9}$$

Hence $i_1(t)$ is given by

$$i_1(t) = A \exp - \int \frac{R(t)R_l}{L_1(R(t)+R_l)} dt, \quad A \text{ is a constant} \tag{7.10}$$

From the boundary condition

$$\text{at} \qquad t = 0, \quad i_1(0) = I(0) \tag{7.11}$$

The constant A is clearly equal to $I(0)$. Hence $i_1(t)$ is given by

$$i_1(t) = I(0) \exp - \int \frac{R(t)R_l}{L_1(R(t)+R_l)} dt \tag{7.12}$$

and the current in the load $i_3(t)$ is given by

$$i_3(t) = I(0)\frac{R(t)}{R(t)+R_l} \exp - \int \frac{R(t)R_l}{L_1(R(t)+R_l)} dt \tag{7.13}$$

If $R(t)$ simply varies linearly with time then $R(t) = mt$, where m is a constant. Putting this expression into Equation (7.13) and carrying out the integration gives the current in the load $i_3(t)$ as

$$i_3(t) = I(0)\frac{mt}{mt+R_l} \exp\left\{\frac{R_l^2}{mL_1} \ln\left[\frac{(mt+R_l)}{R_l}\right] - \frac{R_l t}{L_1}\right\} \tag{7.14}$$

Alternatively, if $R(t)$ increases exponentially with time from some initially small value $R(0)$ then $R(t)$ can be written as

$$R(t) = R(0)\exp(at), \quad a \text{ is a constant} \tag{7.15}$$

and the current in the load $i_3(t)$ can be found to be

$$i_3(t) = I(0)\frac{R(0)\exp(at)}{R(0)\exp(at)+R_l} \exp\left\{-\frac{R(0)R_l}{L_1}\left[\frac{t}{R(0)} + \frac{1}{aR(0)} \ln\left(\frac{R(0)+R_l \exp(-at)}{R(0)+R_l}\right)\right]\right\} \tag{7.16}$$

In the case where the resistance of the opening switch increases linearly with time, using Equation (7.13) it is possible to determine the maximum power delivered to the load from the maximum load current. This can be done if the load is capacitive or inductive as well as resistive [1]. In the case of a resistive load the maximum power delivered to the load is at a time τ given by

$$\tau = \sqrt{\frac{L_1}{m}} \tag{7.17}$$

and the maximum power is

$$P_{\max} = \left(i_3(t)_{\max}\right)^2 R_l = \frac{I(0)^2 R_l \alpha^2}{(1+\alpha)^2} \exp\left[\left(\frac{-2}{\alpha}\right) + \left(\frac{2}{\alpha^2}\right) \ln(1+\alpha)\right]$$

$$\text{where} \qquad \alpha = \sqrt{\frac{mL_1}{R_l^2}} \tag{7.18}$$

From Equation (7.18) and the equivalent expressions for capacitive and inductive loads it can be shown that the greater the value of m, i.e. the faster the switch opens, the greater will be the efficiency of energy transfer from the storage inductor to the load. This highlights the main problem with inductive storage, i.e. the practical difficulty of designing an opening switch that can turn off large inductive storage currents very rapidly. The effect of switch opening time on peak current can be seen in Figure 7.4, where the load current given by Equation (7.14) is plotted for different m values for the circuit given in Figure 7.3. For illustrative purposes, in the calculation L was chosen to be 1 μH, the load was chosen to be 50 Ω and $I(0)$ was 5000 A.

7.2.2 The Efficiency of Energy Transfer from Inductive Energy Stores

The efficiency issue outlined above is of considerable interest in the case of inductive loads [3] such as those presented by imploding plasmas. If the load Z in Figure 7.3 is an inductor L_2, and it is assumed that the switch opens in a finite time Δt, then by conservation of magnetic flux and energy the initial energy in the storage inductor E_0 is divided between the storage inductor L_1 and the load inductor L_2. If these energies are given by E_1

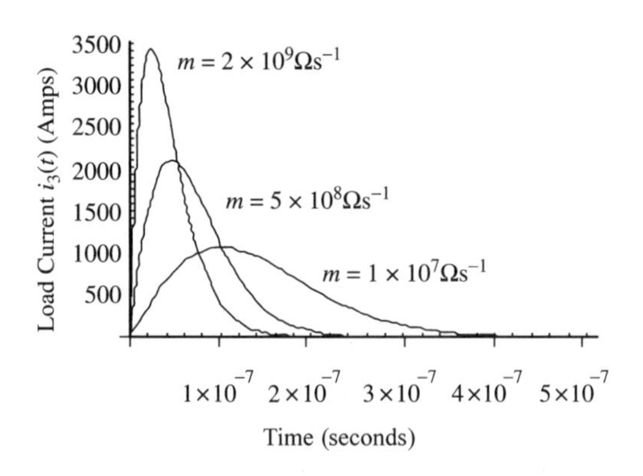

Figure 7.4 Example plot of Equation (7.14) showing the effect of opening switch time

and E_2 then

$$E_1 = \left[\frac{L_1^2}{(L_1 + L_2)^2}\right] E_0 \quad \text{and} \quad E_2 = \left[\frac{L_1 L_2}{(L_1 + L_2)^2}\right] E_0$$

$$\text{where} \qquad E_0 = \frac{1}{2} L_1 I(0)^2 \tag{7.19}$$

The energy dissipated in the switch E_s is given by

$$E_s = \left[\frac{L_2}{L_1 + L_2}\right] E_0 \tag{7.20}$$

Thus the efficiency η by which energy is transferred from L_1 to L_2 can be found to be

$$\eta = \frac{E_2}{E_0} = \frac{L_2 L_1}{(L_1 + L_2)^2} \tag{7.21}$$

This efficiency has a maximum value of only 25% when L_1 equals L_2. The voltage V induced across the switch during the opening phase is

$$V \cong \frac{L_2 I_2}{\Delta t} = \left[\frac{L_1 L_2}{L_1 + L_2}\right] \frac{I(0)}{\Delta t} \tag{7.22}$$

where $I(0)$ is the initial current in L_1 and I_2 is the load current after switching. From Equation (7.22) it is clear that if the opening time of the switch is fast it will have a large voltage induced across it if the initial storage current is large. Note also that the energy dissipated in the switch, using Equation (7.20) for the case when L_1 and L_2 are equal, is 50% of the energy originally stored in the storage inductor L_1. Both facts exacerbate the opening switch problem.

A similar inefficiency problem is encountered when a mutually coupled inductor or transformer is used as an inductive energy store in a circuit such as that shown in Figure 7.5(a). In this circuit an initial current $I(0)$ is supplied to the storage inductor L_1 which is coupled to a second inductor L_2 with mutual inductance M. When the storage inductor is charged, switch SW1 is opened and switch SW2 is closed so that the energy stored in L_1 can be transferred to the load inductor L_3. A current I_1 flows in the secondary circuit after switching is completed. The energy in the storage inductor and the energy in the load inductor just before and after operation of the switches is given by

$$E_0 = \frac{1}{2} L_1 I(0)^2 \quad \text{and} \quad E_3 = \frac{1}{2} L_3 I_1^2 \tag{7.23}$$

Conservation of magnetic flux requires

$$M I(0) = I_1 (L_1 + L_2) \tag{7.24}$$

Hence the efficiency η by which energy is transferred from the storage inductor L_1 to the load inductor L_3 is given by

$$\eta = \frac{E_3}{E_0} = \frac{L_3 M^2}{L_1 (L_2 + L_3)^2} \tag{7.25}$$

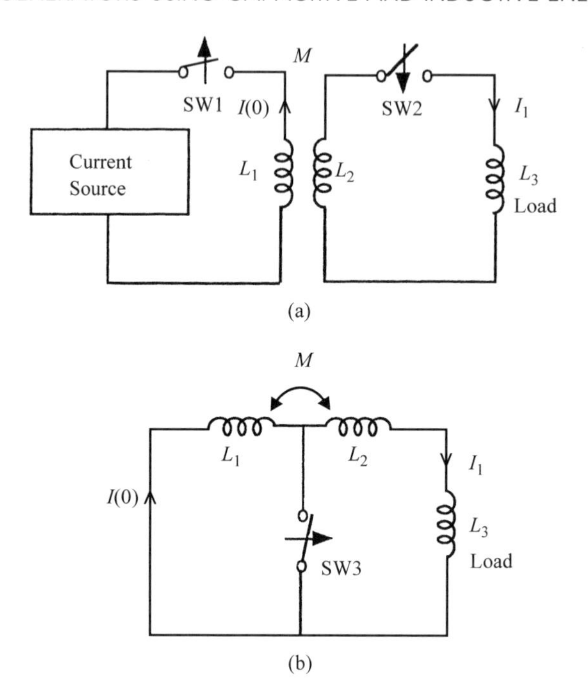

Figure 7.5 Two more inductive energy storage circuits

If perfect coupling is assumed between L_1 and L_2, then $M^2 = L_1L_2$ and the efficiency is given by

$$\eta = \frac{L_2L_3}{(L_2 + L_3)^2} \tag{7.26}$$

which has a maximum value of just 25% when L_2 equals L_3.

This efficiency problem can be improved using the so called "meat-grinder" circuit shown in Figure 7.5(b) [4,5]. Here an additional energy transfer inductor L_2 is included between the storage inductor L_1 and the load inductor L_3. L_1 and L_2 are mutually coupled, and after the storage inductor L_1 is charged to an initial current $I(0)$ the switch SW3 is opened. It can be shown that the efficiency of energy transfer from the storage inductor to the load inductor can be at least doubled, depending on the values of the ratios L_1/L_3 and L_2/L_3. Even better transfer efficiencies, approaching 100%, can be achieved in multi-stage "meat-grinder" circuits where a series of energy transfer inductors are sequentially switched between the storage inductor and load inductor [6,7].

An alternative circuit in which the energy transfer from a storage inductor to a load inductor can be, theoretically, 100% is given in Figure 7.6. In this circuit a capacitor C is placed in parallel with the load inductor L_2 in the basic inductor circuit described earlier (Figure 7.3) and energy is discharged into it from the storage inductor L_1 when the switch is opened. The circuit has been transformed into its Laplace equivalent and for ease of analysis the opening switch behaviour is assumed to be ideal.

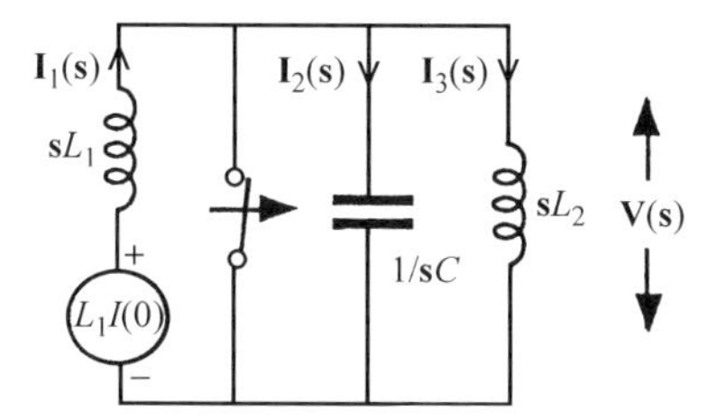

Figure 7.6 Laplace equivalent of an efficient inductive energy storage circuit

The current in the load inductor can be found by determining $\mathbf{V(s)}$ and then dividing by the load impedance $\mathbf{s}L_2$, i.e.

$$\mathbf{V(s)} = L_1 I(0) \frac{\mathbf{s}L_2 || \dfrac{1}{\mathbf{s}C}}{\mathbf{s}L_2 || \dfrac{1}{\mathbf{s}C} + \mathbf{s}L_1}$$

$$= \frac{I(0)}{C} \frac{1}{\mathbf{s}^2 + \omega^2}, \qquad \omega = \sqrt{\frac{(L_1 + L_2)}{CL_1 L_2}} \tag{7.27}$$

The current in the load inductor L_2 is therefore

$$\mathbf{I}_3(\mathbf{s}) = \frac{I(0)}{\mathbf{s}} \frac{L_1}{L_1 + L_2} \frac{\omega^2}{\mathbf{s}^2 + \omega^2} \tag{7.28}$$

Taking the inverse transform gives $i_3(t)$ as

$$i_3(t) = I(0) \frac{L_1}{L_1 + L_2} (1 - \cos(\omega t)) \tag{7.29}$$

Similarly the current $i_2(t)$ through the capacitor C can be found to be

$$i_2(t) = I(0) \cos(\omega t) \tag{7.30}$$

The operation of the circuit can be best understood by plotting the three currents $i_1(t)$, $i_2(t)$ and $i_3(t)$ as functions of time, recognising that the current in the storage inductor is just the sum of the currents in the capacitor and the load inductor. From this plot, as shown in Figure 7.7, it can be seen that the current initially stored in the storage inductor is completely transferred to the load inductor with a return path back through the capacitor. For the case where the storage inductor and load inductor are equal, it is a simple matter to show that the efficiency of energy transfer is now 100%. Clearly, in a practical circuit stray resistance in the inductors, the capacitor and the circuit itself will reduce the transfer efficiency below the 100% figure. The other potential problem with the circuit, at high powers, is finding a suitable capacitor which could handle the large circulating current through the load inductor.

It is interesting to note that the behaviour of this circuit is similar to the capacitive discharge circuit described in Example 1.10. Without the resistor, the circuit given in Figure 1.10 would be the dual of the inductive discharge circuit described, and similarly

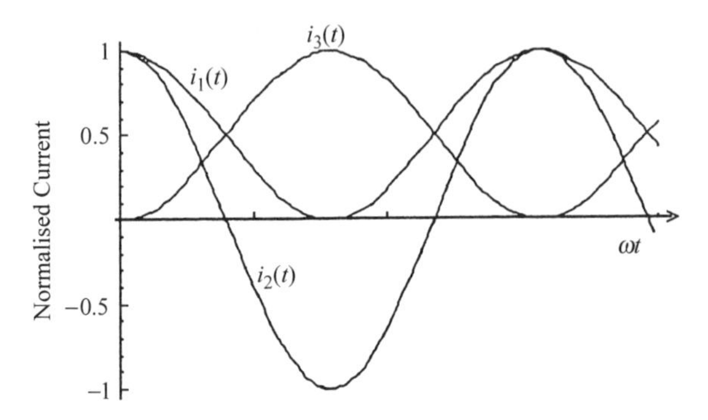

Figure 7.7 Plots of the currents in the circuit shown in Figure 7.6 as functions of time

energy stored in the storage capacitor C_1 can be transferred with 100% efficiency to the capacitor C_2, provided they are equal in value.

7.2.3 *Flux Compression Circuits*

The storage and discharge of energy in an inductor is also used in magnetic flux compression generators [8,9,10,11] but, in this case, the bulk of the energy supplied to the load comes from the explosive used to compress the magnetic flux produced in a current-carrying inductor.

A wide variety of different types of device have been described, but the basic method of operation can best be understood by reference to the simplified diagram of a helical flux compression generator shown in Figure 7.8.

Magnetic flux is generated between a coil and conducting tube which runs along the centre of the coil by connecting one end of the tube and coil to a current source as shown. A load is connected at the other end of the coil and tube. When the explosive contained in the tube is detonated, at the left hand end of the tube as shown, the explosion causes the tube to expand from left to right thereby compressing the magnetic flux between the coil and the tube. By comparison of this device to conventional rotating electrical machinery, the tube is

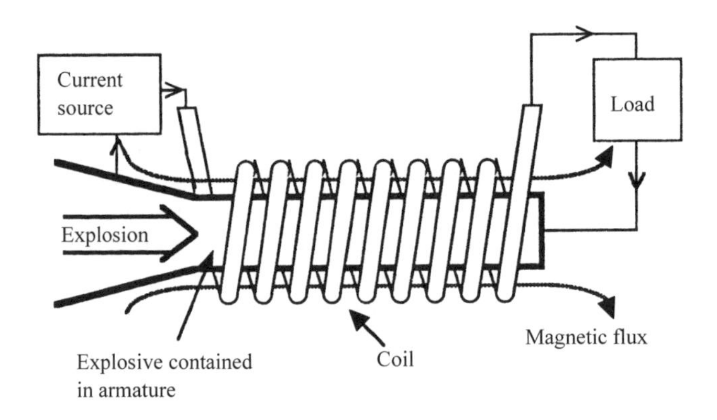

Figure 7.8 Basic diagram of a helical magnetic flux compressor

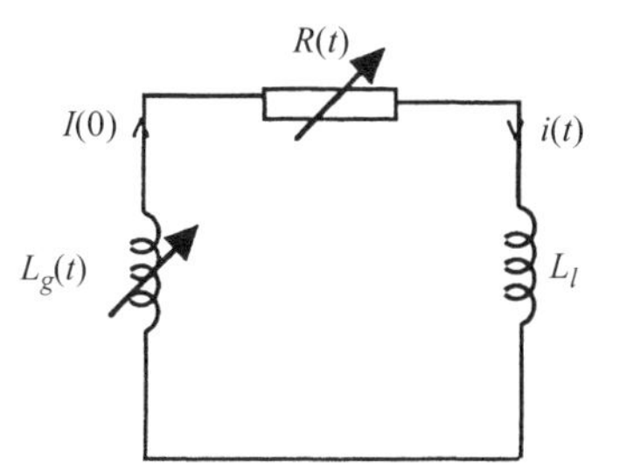

Figure 7.9 Equivalent circuit of a flux compressor

usually referred to as the armature and the coil as the stator. In practice it is difficult to maintain electrical contact during the explosion so a double ended tube is often used, which is detonated at both ends, with the load connected to a central tap on the coil [10]. Although these devices can generate extremely high pulsed powers, the efficiency by which the chemical energy of the explosive is converted to electrical energy is usually quite low. An equivalent electrical circuit for the compressor is given in Figure 7.9, where the armature/stator arrangement is represented by an inductor whose size decreases with time. It is assumed in the analysis of the behaviour of the circuit that the load is purely inductive and that loss of flux which is "switched out" of the circuit, due to the explosive process [12], can be ignored. The total resistance of the generator $R(t)$ is also assumed to be time-dependent.

If an initial current $I(0)$ is flowing in the storage inductor at a time $t = 0$ then the current in the circuit $i(t)$ will be the solution of the differential equation

$$(L_g(t) + L_l)\frac{di}{dt} + \frac{dL_g(t)}{dt}i + iR(t) = 0$$
$$\text{or} \qquad \frac{di}{dt} + \frac{\left(\frac{dL_g(t)}{dt} + R(t)\right)i}{(L_g(t) + L_l)} = 0 \tag{7.31}$$

The integrating factor [2] for this equation is given by

$$\phi(t) = \exp\left[\int \frac{\frac{dL_g(t)}{dt}}{(L_g(t) + L_l)}dt + \int \frac{R(t)}{(L_g(t) + L_l)}dt\right] \tag{7.32}$$

Thus, by applying the boundary condition given above, the solution to Equation (7.31) is given by

$$i(t) = I(0)\left[\frac{L_l + L_g(0)}{L_l + L_g(t)}\right] \exp - \int_0^t \frac{R(t)}{L_l + L_g(t)} dt \tag{7.33}$$

where $L_g(0)$ is the value of the generator inductance at time $t = 0$. For this equation to be solved it is clearly necessary to be able to determine the functions $R(t)$ and $L_g(t)$, and this can only be done from knowledge of the mechanics of the explosive process and the physical construction of the generator.

The fraction of the magnetic flux that remains in the circuit after a time t is known as the flux efficiency $\eta(t)$ and is given by

$$\eta(t) = \exp - \int_0^t \frac{R(t)}{L_l + L_g(t)} dt \tag{7.34}$$

Clearly if

$$\left[\frac{L_g(0) + L_l}{L_l}\right] \eta(\tau)^2 > 1, \qquad \text{where } L_g(\tau) = 0 \tag{7.35}$$

then the generator will contain more electromagnetic energy after it has been operated than before, the increase being the result of the conversion of explosive energy into electrical energy.

7.3 MARX GENERATORS

The basic capacitive discharge circuit, described in the first section of this chapter, is limited in operation to roughly 100 kV. This is because of the difficulty of making reliable switches that will operate above this potential, the physical size of the components needed to build the circuit, the high DC voltages needed from the power supply used to charge the discharge capacitor and the difficulties in suppressing corona discharges from the parts of the circuit raised to high DC potentials. These problems can be circumvented by the Marx generator [13] in which a number of capacitors are charged in parallel, usually via resistors with large values of resistance, and then discharged in series through a number of high-voltage switches, usually spark gaps. The generator therefore has a voltage gain approximately equal to the number of capacitors used. The Marx generator is one of the most important circuits used in the field of pulsed power, particularly when very high voltages and powers are required. As such the circuit is widely used for applications such as lightning and EMP simulation, streamer chambers, flash X-ray and electron beam generators and high-power gas lasers. Many different types of Marx generator have been described in the literature and their use is not solely limited to single-shot high-power pulse generation. For example, the circuit can be made to operate at low voltages and high repetition rates using solid state switches to produce high repetition rate generators [14], trigger generators [15], and a novel type of DC converter [16].

The basic circuit of a three-stage generator is given in Figure 7.10. In this circuit the switches are simple two-electrode spark gaps which break down when the potential across them exceeds their break down or sparking potential.

The capacitors are each charged via the charging resistors R to a potential V by a DC charging supply such that the potentials at points A, C, E are V volts and the potentials at points B, D, F are zero. It is assumed that the values of the charging resistors are sufficiently high that they can be neglected once the circuit is operated. If spark gap SG1 breaks down first, then the potential at point B will rise to V and the potential at point C will rise to 2 V. Assuming that the break down potential of all three spark gaps is around V volts, then SG2 will be over-volted and will also break down. This then causes the potential at point D to rise to 2 V and the potential at point E to rise to 3 V. SG3 then breaks down and the voltage on the

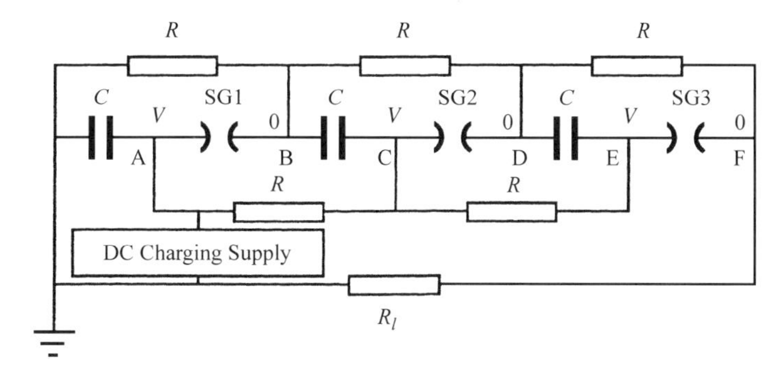

Figure 7.10 Basic circuit of a three-stage Marx generator

load resistor R_l rises to 3 *V*. Thus the voltage gain of the circuit is equal to the number of stages, i.e. 3.

In practice the operation of Marx generators is far more complex than this simple description, due mainly to the existence of stray capacitance from various points in the circuit to ground C_g, in the switches C_s and between stages C_i [17], as illustrated in Figure 7.11. Two basic physical constructions of five-stage Marx generators are given in this figure, namely a basic stacked structure and a folded structure which is both more compact and has a lower inductance than the stacked generator.

The way in which the switches themselves operate in the circuit, particularly if they are triggered, also add to the problem of fully understanding the way in which a particular generator "erects". Figure 7.12 gives the circuit of a six-stage generator in which the

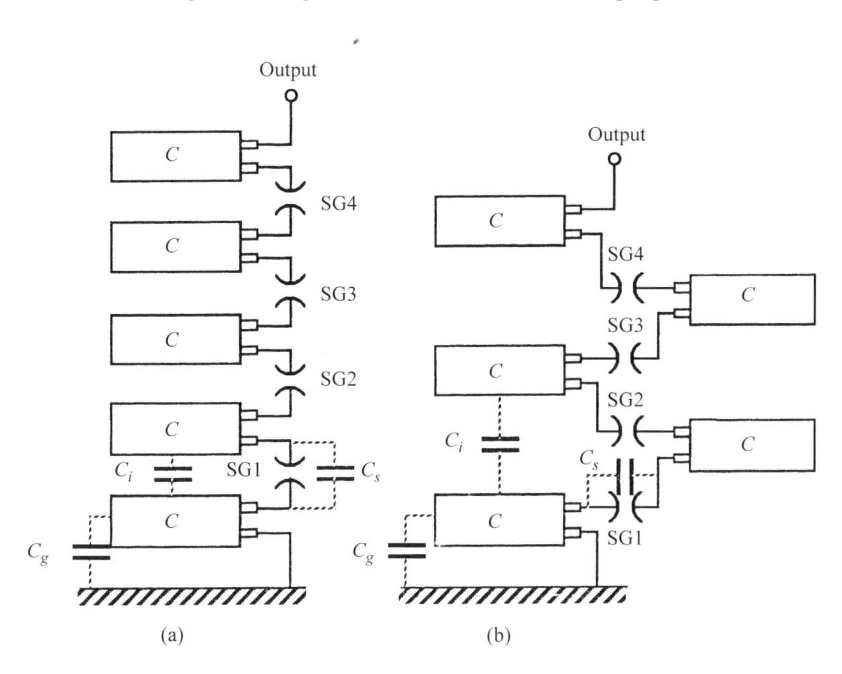

Figure 7.11 Two basic physical constructions of a Marx generator: (a) a linear stack, (b) a folded structure

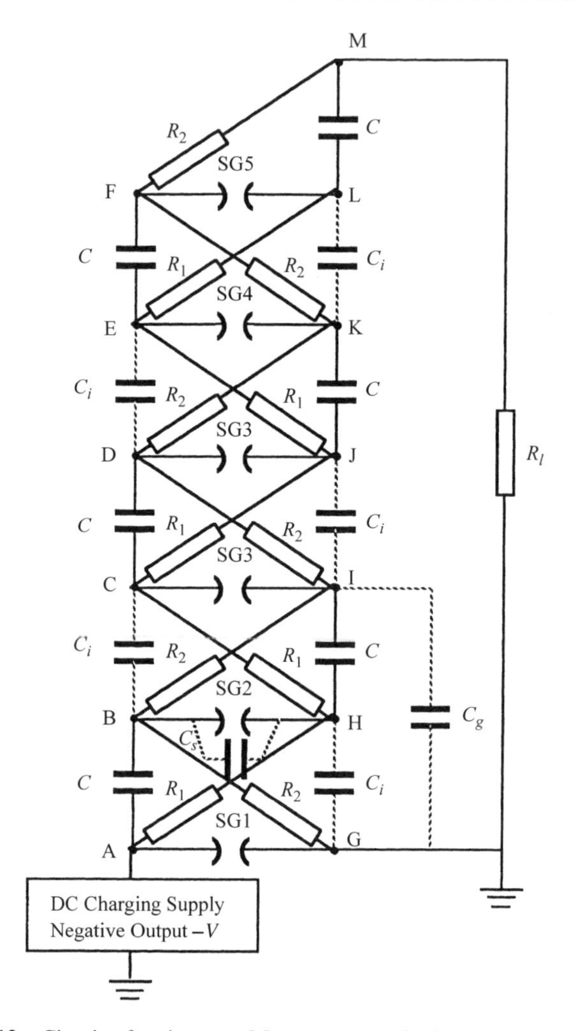

Figure 7.12 Circuit of a six-stage Marx generator including stray capacitances

various stray capacitances have been included. Note that to simplify the diagram not all the switch and stray ground capacitances have been included. In this circuit two sets of charging resistors R_1 and R_2 have been used, as is common practice. R_1 is chosen to be much greater than R_2 so that the points in the circuit, at the end of the charging period, which should be at ground potential do not move significantly away from ground potential during the charging period. This circuit also uses a negative charging supply, although when the Marx generator is operated (or "erected") the output is positive.

Although, from the simple description of the operation of a Marx generator given above, one might expect that the breakdown potential of the spark gaps could be increased successively along or up the circuit due to the successive increase in overvoltage on the gaps, in practice it is found that that the breakdown potential of SG2 can only be set a little above that of SG1 for reliable operation. If this is not the case there tends to be a finite delay before SG2 breaks down. The cause of this is the stray capacitances in the circuit. If it is assumed, as before, that during the operation of the circuit the charging resistors can be ignored, then the potential at point H will be set by the sizes of the stray capacitances to which it is connected. Assuming that the sizes of the stray capacitances will be much less than the sizes of the storage capacitors C and that the interstage capacitance C_i between points B and C can be ignored, then the potential at point H is fixed by the potential divider formed by C_s and the parallel combination of C_g and C_i. Thus on the break down of SG1 point A will rise in potential from $-V$ to zero, and point B will rise from 0 to V, which causes the potential of point H to rise from a potential of $-V$ to

$$V_H = -V + V\left(\frac{C_s}{C_i + C_s + C_g}\right) = -V\left(\frac{C_i + C_g}{C_i + C_s + C_g}\right) \tag{7.36}$$

Therefore the potential difference across SG2 becomes

$$V_{BH} = V\left(1 + \frac{C_i + C_g}{C_i + C_s + C_g}\right) \tag{7.37}$$

If the stray capacitance in SG1 is zero then the potential across the gap SG2 will be $2V$. In practice this will never be the case, and depending on the relative sizes of the three stray capacitances C_i, C_g and C_s the potential across the gap could be reduced to a value which is only slightly greater than V. It should be remembered that this is only a transient effect as the stray capacitances have very small values usually not exceeding 10 pF. Taking typical values for the charging resistors, the stray capacitances will charge and discharge with time constants of the order of 10^{-8} to 10^{-7} s. This effect occurs at every stage in the Marx, leading to firing delays as the stray capacitances charge. Nonetheless the firing delays and jitter caused by this effect can be problematic. Transient effects of this type can be even more severe in the "erection" of compact, low inductance, Marx generators with folded structures (see Figure 7.11) where it has been suggested [18] that the transient voltage swing on the storage capacitors can reach up to four times the Marx generator's charging voltage.

However the coupling that exists between the stages of the Marx generator can be deliberately engineered, to some advantage, to actually improve the operation of the Marx [19,20]. With care it is possible to obtain overvolting of the switches of several times the charging potential of the generator. If the gap spacings, in the case of spark gap switches, are kept constant throughout the generator it is possible to operate some Marx generators reliably over a wide range of charging potentials, assuming that the bottom spark gap is triggered. The coupling need not necessarily be capacitive; resistive coupling can be used or a mix of resistive and capacitive coupling.

7.3.1 Circuit Analysis of the Marx Generator

An equivalent circuit of a Marx generator is given in Figure 7.13. The storage capacitors are lumped into a single capacitor C and the self-inductance and equivalent series resistance of the generator are represented by the components L and R, respectively. The load R_l is assumed to be partly capacitive so a component C_l is included in the circuit as shown.

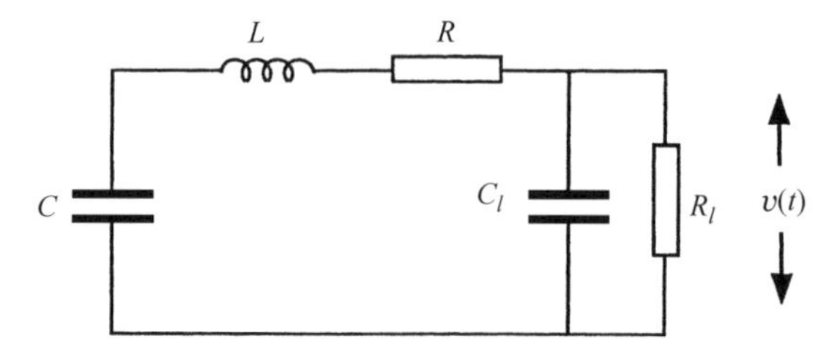

Figure 7.13 Equivalent circuit of a multi-stage Marx generator

If the voltage on the capacitor of the erected Marx generator is V the voltage on the load $\mathbf{V(s)}$ is given by

$$
\begin{aligned}
\mathbf{V(s)} &= \frac{V}{\mathbf{s}} \frac{\left(R_l \| \dfrac{1}{\mathbf{s}C_l}\right)}{\dfrac{1}{\mathbf{s}C} + \mathbf{s}L + R + R_l \| \dfrac{1}{\mathbf{s}C_l}} \\
&= \frac{V}{\mathbf{s}LC_l} \frac{1}{\mathbf{s}^3 + a\mathbf{s}^2 + b\mathbf{s} + c}
\end{aligned}
$$

$$\text{where} \quad a = \left(\frac{R}{L} + \frac{1}{C_l R_l}\right), \quad b = \left(\frac{1}{LC_l} + \frac{1}{LC} + \frac{R}{LR_l C_l}\right), \quad c = \frac{1}{LCC_l R_l} \tag{7.38}$$

Taking the inverse transform leads to an expression for $v(t)$ of the form

$$v(t) = \frac{V}{LC_l}\left[\frac{\exp(-\alpha_1 t)}{(\alpha_1 - \alpha_2)(\alpha_1 - \alpha_3)} + \frac{\exp(-\alpha_2 t)}{(\alpha_2 - \alpha_1)(\alpha_2 - \alpha_3)} + \frac{\exp(-\alpha_3 t)}{(\alpha_3 - \alpha_1)(\alpha_3 - \alpha_2)}\right] \tag{7.39}$$

where α_1, α_2 and α_3 are roots of the equation

$$\mathbf{s}^3 + a\mathbf{s}^2 + b\mathbf{s} + c = 0 \tag{7.40}$$

These roots can only be found if numerical values of the components of the circuit shown in Figure 7.13 are substituted. Should some of these roots be imaginary then the output waveform from the generator will be underdamped and will display oscillations. It should also be pointed out that L and R will be strongly time-dependent as a major contribution to these components comes from the series connected spark gaps used to construct the Marx generator. To achieve an accurate solution will then require knowledge of the time-dependent inductance and resistance of the spark gaps [21,22,23,24].

It is often necessary to shape the profile of a Marx generator to a given specification for application to the simulation of lightning strikes [25], switching transients in power lines and EMPs. This can be done by manipulation of the series resistor R and the load resistor R_l. To illustrate the method consider the analysis of the Marx circuit given with the series inductance set to zero. In this case the voltage on the load $\mathbf{V(s)}$ is given by

$$\mathbf{V(s)} = \frac{V}{RC_l} \frac{1}{\mathbf{s}^2 + \mathbf{s}\left[\dfrac{1}{R_l C_l} + \dfrac{1}{RC_l} + \dfrac{1}{RC}\right] + \dfrac{1}{RR_l CC_l}} \tag{7.41}$$

The denominator of this expression cannot be factorised as it stands. However, recognising that the load resistor will be much bigger than the series resistor ($R_l \gg R$) and that the load capacitance C_l will be much smaller than the capacitance of the "erected" Marx C, suggests that one of the roots β_1 will approximately be given by

$$\beta_1 \cong \frac{1}{RC_l} \tag{7.42}$$

Thus completing the square and taking the inverse transform suggests an approximate solution for $v(t)$ of the form

$$v(t) \cong \frac{V}{(\beta_1 - \beta_2)RC_l}[\exp(-\beta_2 t) - \exp(-\beta_1 t)],$$

$$\text{where} \quad \beta_1 = \frac{1}{RC_l} \quad \text{and} \quad \beta_2 = \frac{1}{R_l C} \tag{7.43}$$

Noting that $\beta_1 \gg \beta_2$, a plot of the expression given in Equation (7.43) is given in Figure 7.14.

From the plot it is observed that the front or leading edge of the waveform is controlled by the series resistance of the Marx generator R and the load capacitor C_l, and the tail is controlled by the capacitance of the erected Marx C and the load resistor R_l. Since the load capacitance and capacitance of the Marx generator are usually fixed, adjustment of the series resistance of the Marx and the load resistance is used to control the shape of the pulse. This can be done by deliberately adding series resistance to the circuit to increase the rise-time of the leading edge of the pulse, or by increasing/decreasing the load resistance to increase or shorten the length of the tail of the pulse.

Using resistance to alter the shape of the pulse tends to be rather inefficient and, for example, the addition of inductance to the load resistance to increase tail length is more efficient [26]. To increase the length of the leading edge of the pulse, the series resistance which must be added is usually distributed within the generator. This avoids having to add a series resistor at the output of the Marx generator which has to withstand the full rated output voltage of the generator. Such a resistor tends to be both bulky and expensive.

7.3.2 Fast Marx Generators

For some applications very fast rise-times are required in the output waveform from Marx generators. As any Marx generator will have an inherent minimum rise-time due to its

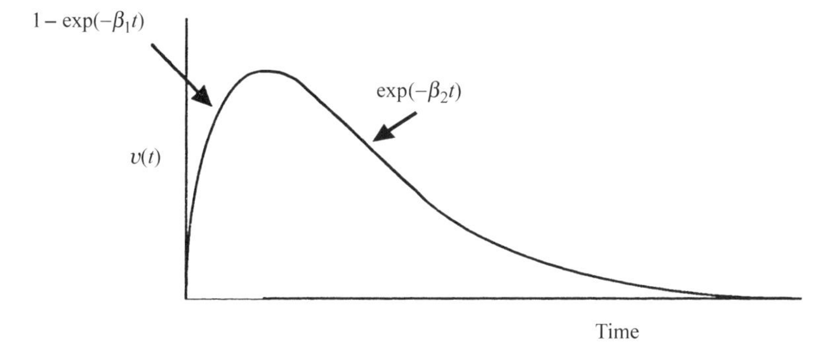

Figure 7.14 Approximate output waveform from a Marx generator

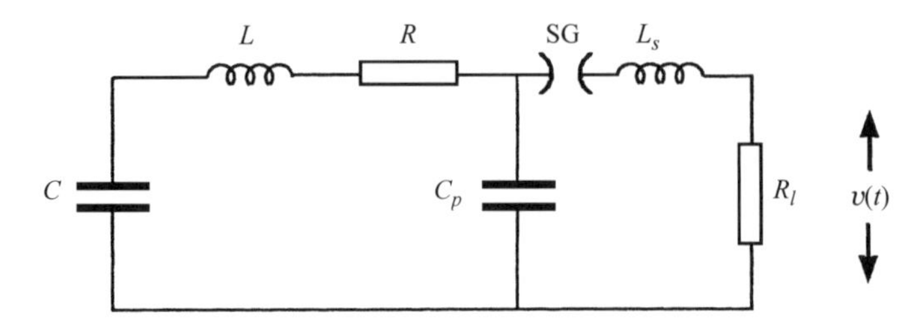

Figure 7.15 Marx generator with peaking capacitor and switch

equivalent series resistance and inductance, a circuit modification is needed if this rise-time is to be reduced. This can be done by using a peaking capacitor and switch arrangement, as shown in Figure 7.15.

The peaking capacitor C_p is chosen such that its value is much less than the capacitance of the "erected" Marx C. The breakdown potential of the spark gap is set so that it breaks down at the maximum voltage to which the peaking capacitor is charged by the Marx generator. There are now two discharge circuits operating. The first is a *CLRC* circuit identical to that described in Example 1.10, and the second is a simple but fast discharge circuit where the rise-time is primarily limited by the stray inductance L_s in the circuit (see the first section of this chapter). Not only can very fast rise-times be achieved in this way but also the peaking capacitor can be charged to almost twice the output voltage of the Marx generator, if its value is small. If a shaped pulse is required from the generator the peaking capacitor can be replaced by a pulse-forming line or network or a Blumlein pulse-forming line or network. The lines are effectively pulse-charged by the Marx generator and then switched to initiate pulse-forming action at the maximum voltage to which the lines are charged. This technique of generating high-voltage shaped pulses is very commonly used in the pulsed power field.

The peaking capacitor technique is very effective and has been extended by Kekez [27] to construct a Marx generator with a rise-time of just 50 ps. In his generator each stage of the Marx is effectively switched into a peaking capacitor which is formed between one of the electrodes of each spark gap and the metallic enclosure surrounding the generator. Thus the rise-time of the voltage is successively reduced at each stage of the Marx generator during the "erection" process.

7.3.3 Triggered Marx Generators

In standard Marx generators it is common practice for the first two or lowest spark gaps to be three-electrode devices or trigatrons. The Marx generator is then fired or "erected" by applying a strong trigger pulse to the trigger electrodes of these spark gaps. It is then assumed in a well-designed Marx generator that there will be significant overvolting of the spark gaps further up the Marx, leading to reliable operation. However in some applications, for example where a number of Marx generators have to be operated in parallel [28], it is critical that the firing jitters are very low, often down to a few ns. To achieve such performance it is necessary to trigger all the gaps in the Marx generator. In large multi-stage Marx generators this may require a very powerful trigger generator to

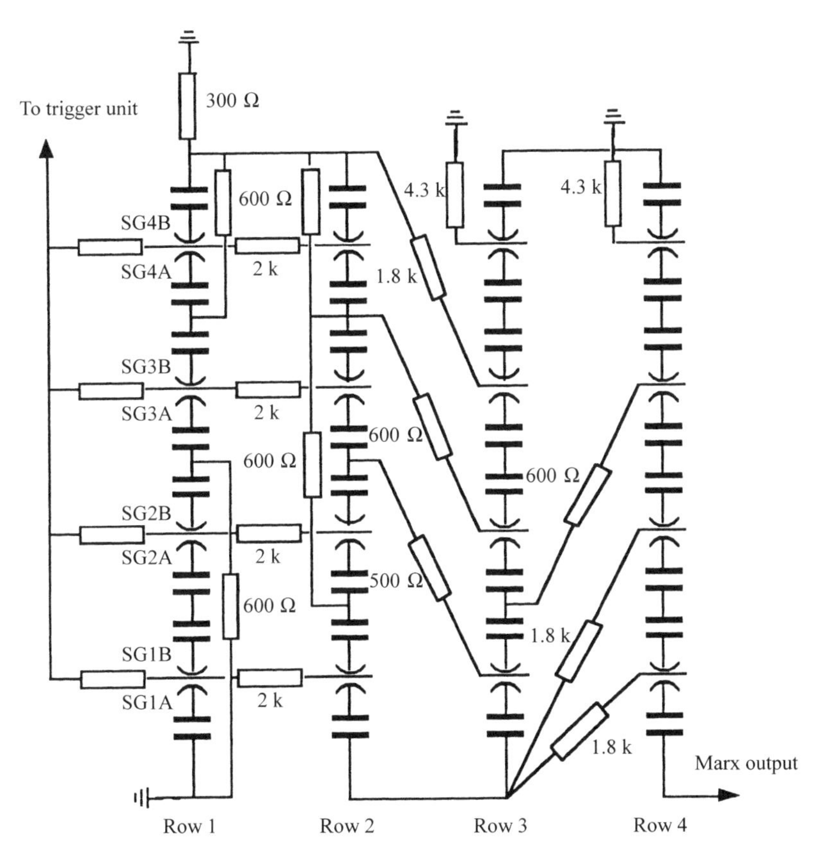

Figure 7.16 A 16-stage Marx generator with mixed external/internal triggering

provide sufficient current to all of the triggered gaps to ensure reliable and low jitter operation. As an alternative it is possible to use a mixed internal external trigger system, where some of the gaps are triggered by an external trigger generator and the rest by internal resistive coupling within the generator, as shown in Figure 7.16. This is a diagram of the PBFA I Marx generator developed at Sandia National Laboratories for plasma fusion studies [29].

To get the triggering system to work effectively it is necessary to cause a cascade erection or sequential breakdown of the spark gaps in the first row of the Marx generator. The spark gaps are midplane-triggered and the trigger pulse causes the first portion A of spark gaps 1 to 4 to break down first. This then allows the erected Marx voltage to break down the second half B of the gaps. Once the first row has erected it is then used to trigger subsequent gaps in the Marx with low impedance. Consequently very fast rising trigger pulses are coupled to the other gaps in the Marx generator and very low triggering jitter can be achieved. Stray capacitance within the generator can reduce the energy available to charge the stray capacitance of the gaps which can lead to faulty cascade "erection". However this can be avoided by connecting the trigger and ground resistors to decouple the detrimental stray capacitances.

7.4 VECTOR INVERSION GENERATORS

7.4.1 The LC Generator

The *LC* generator circuit [30], as with the Marx generator, relies on the "erection" of a set of capacitors connected in series. Although less popular than the Marx, it is possible to build quite low inductance generators [31] which can give very fast "erection" times. The basic circuit of a three-stage *LC* generator is given in Figure 7.17.

In this generator the capacitors are charged to a potential V with the polarities of adjacent capacitors opposing each other, as indicated by the arrows, so that the net potential across the series connected capacitors is zero. Once charged the switches are simultaneously closed, causing the potential on the capacitors, that are connected to the switches through inductors L, to ring and consequently reverse. Once reversed the potentials on all the capacitors lie in the same direction and the output voltage is $2nV$, where n is the number of stages in the generator. The erection time of the generator depends on the values of the capacitors, the inductors, and any stray series resistance in the switched *LC* circuits. This

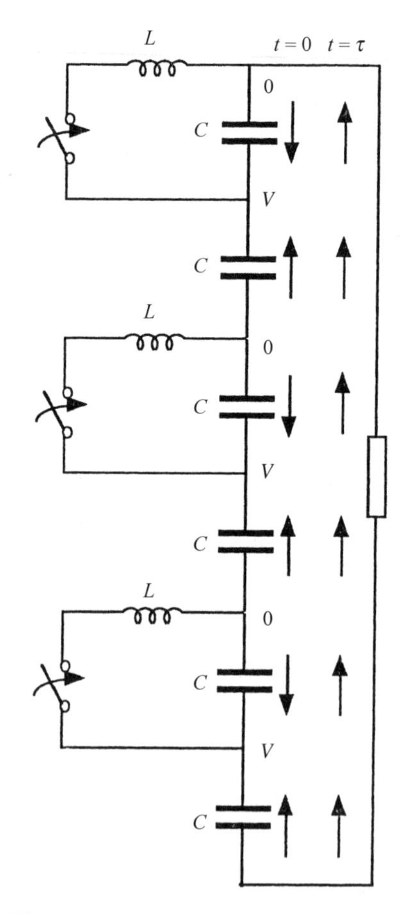

Figure 7.17 A three-stage *LC* generator

resistance will also lower the gain from the ideal value of $2n$, consequently it is important to try to minimise this resistance by minimising, for example, the ESR of the capacitors that are used. It is also important that the capacitors are capable of up to a 95% reversal in polarity without failure. Although triggering of the generator is less complicated than the Marx generator, as only half the number of switches are required, unlike the Marx the LC generator does not have the advantage of the overvolting of switches at stages further up the generator. Also, the large transient voltages that appear in the "erection" of Marx generators due to stray capacitance are much reduced in the LC generator, as any stray capacitance is charged relatively slowly.

As the series resistance R in the switched circuits is small, the circuits are clearly underdamped, and therefore using Equations (7.1) the output voltage will be given by

$$v_o(t) = nV[1 - \exp(-\alpha t)\cos(\omega_d t)],$$

$$\text{where} \quad \alpha = \frac{R}{2L}, \quad \omega_o = \sqrt{\frac{1}{LC}}, \quad \text{and} \quad \omega_d^2 = \omega_o^2 - \alpha^2 \tag{7.44}$$

The "erection time" τ will be approximately given by

$$\tau = \pi\sqrt{LC} \tag{7.45}$$

and the maximum output voltage will also, approximately, be

$$v_{o(\max)} = nV[1 + \exp(-\alpha\tau)] \tag{7.46}$$

An interesting variant of the LC generator has been described [32,33] in which just one switch is used to "erect" the generator. The circuit of a three-stage generator of this type is given in Figure 7.18. Once charged, as before, the circuit is triggered by closing the switch SW which is usually a spark gap. A second spark gap isolates the generator from the load during the "erection" process. The transformers are 1:1 pulse transformers which are usually wound on magnetic cores.

To understand the operation of the circuit consider the bottom stage only containing the capacitors C_a and C_b. Provided the sense of the winding of the pulse transformers is as shown, then on switch closure capacitor C_a is connected to an inductance which is simply the leakage inductance of the transformer, as any flux generated, as a result of current flowing in the primary winding, will be cancelled by the flux produced by the same current flowing in the secondary winding. Thus the "erection" time of the generator is determined by the ringing frequency of capacitor C_a with the leakage inductance of the pulse transformer. On the other hand capacitor C_b is connected across the primary winding of the pulse transformer and, provided the inductance of this winding is large enough, little current will flow through the winding during the "erection" of the generator. The normal "erection" process will be little affected by this connection. The pulse transformers are usually bifilar wound transformers to maximise flux linkage and minimise the leakage inductance for a fast "erection" time. It should also be noted that special care needs to be taken to adjust the amount of flux linkage in the transformers in multi-stage generators because, as the number of stages is increased, the leakage inductances of the transformers are connected partly in parallel and partly in series. For example, looking at capacitor C_a it can be seen that, as well as being connected across the bottom transformer in the generator,

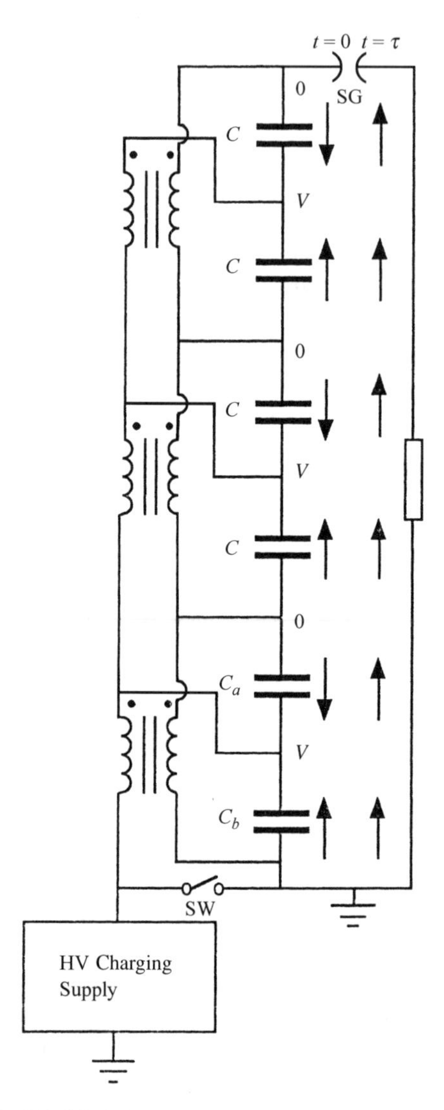

Figure 7.18 Modified three-stage LC generator with a single switch

it is also connected across the next transformer up and then via this transformer to the third transformer and so on.

7.4.2 The Spiral Generator

Although not in common use, the spiral generator is a very compact and ingenious device that can provide the means of generating very high voltage pulses (around 1MV) [30,34]. In basic terms the spiral generator is similar to a rolled up Blumlein pulse-forming line made from strip lines. The main disadvantage of the generator is that, unlike the Blumlein pulse-forming line, it generates a triangular output pulse rather than a rectangular pulse. Each layer of the spiral consists of an active line, which is shaded in Figure 7.19, and a passive unshaded line. The generator is charged to a potential V as shown and a switch SW, usually a low inductance spark gap, is used to initiate the output pulse generation process. In the generator shown the switch is on the outside of the spiral although, in practice there is no reason why the switch should not be placed at the mid-point of the spiral which leads to a faster "erection" time. The load is connected from the centre of the generator to the outside (see Figure 7.19).

Once the structure is charged, prior to triggering, the direction of the resulting electric fields on the strip lines is shown in the figure by the small arrows. Looking at the top left hand diagram, which represents the state of the generator just after switch closure, it can be seen that the fields in the lines oppose each other and therefore the output voltage V_{out} from

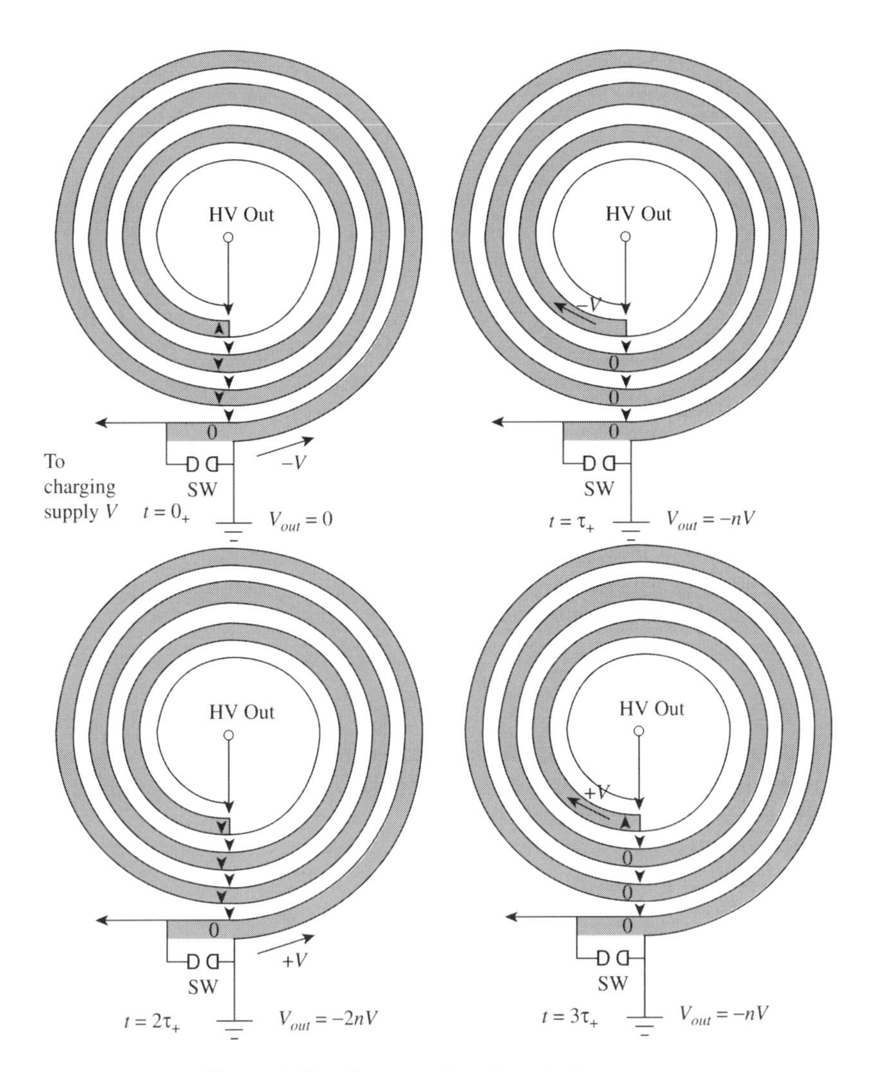

Figure 7.19 The operation of a spiral generator

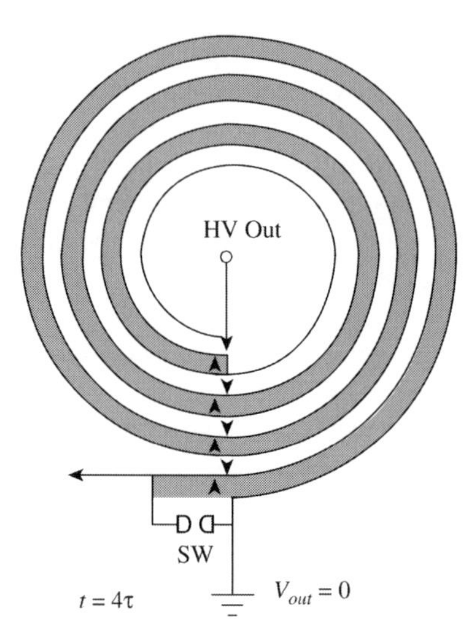

Figure 7.19 (*Continued*)

the generator is zero. When the switch fires a step of amplitude $-V$ is propagated from the switch into the shaded strip line as shown. As the step propagates, the potential on the shaded line is reduced to zero, and after a time $t = \tau$ the step reaches the open-circuited end of the line in the centre of the spiral. Looking at the direction of the fields, on the unshaded strip lines in the diagram at the top right hand side of the figure, it can be seen that the potential on the output terminal has reached $-nV$, where n is the number of turns in the spiral. The velocity of propagation of the step is given by

$$v_p = \frac{c}{\sqrt{\varepsilon_r}} \tag{7.47}$$

where c is the velocity of light and ε_r is the dielectric constant of the insulator separating the conductors in the lines. Thus the time τ, which is the time taken for the step to propagate from the switch to the central output terminal of the generator, is given by

$$\tau = \frac{n\pi d}{v_p} \tag{7.48}$$

where d is the mean diameter of the spiral and n is the number of turns. Thus as the step propagates the output voltage ramps to the potential $-nV$, and during this period the output voltage of the generator is given by

$$v_{out}(t) = \frac{-nVt}{\tau} \qquad 0 < t < \tau \tag{7.49}$$

When the propagating step reaches the centre of the spiral it "sees" an open circuit and reflects with a reflection coefficient $\rho = +1$. The step therefore reflects, propagating back towards the outside of the spiral, and charges the shaded line to a potential $-V$ as it propagates, as shown in the diagram on the top right of Figure 7.19. The output voltage from the generator, therefore, continues to ramp up in voltage until at a time $t = 2\tau$ the step reaches the switch, which is still in a closed condition. During this part of the "erection" process of the generator the output voltage is given by

$$v_{out}(t) = -nV - \frac{nV(t-\tau)}{\tau} = -nV\frac{t}{\tau} \qquad 0 < t < 2\tau \tag{7.50}$$

When the step reaches the switch it now "sees" a short circuit, with reflection coefficient $\rho = -1$, as the switch is closed, and reverses polarity. It now starts to return to the centre of the spiral, as shown in the middle diagram on the left of the figure. As it propagates it now reduces the potential of the shaded line to zero causing the output voltage of the generator to start to reduce. After a further reflection at the centre of the spiral, the step continues to cause the output voltage to ramp back to zero, and when it finally reaches the switch again the generator returns to its original state. Thus the output voltage during the final phases of pulse generation is given by

$$v_{out}(t) = -nV\left(4 - \frac{t}{\tau}\right) \qquad 2\tau < t < 4\tau \tag{7.51}$$

The shape of the output pulse therefore is as shown in Figure 7.20.

Although multiplication factors of the charging potential of 90% of the theoretical maximum have been achieved, in practice the performance of the generator falls short of the ideal for several reasons.

The above description of the operation of the generator assumed that it was connected to an open circuit load. When it is connected to a resistive load the output waveform tends to skew about the zero axis, giving rise to a positive triangular shaped tail.

The inductance of the switch must be kept low otherwise the step launched into the line will have an exponential rise-time with time constant τ_s given by

$$\tau_s = \frac{L_s}{Z_0} \tag{7.52}$$

where L_s is the switch inductance and Z_o is the impedance of the striplines in the generator. Clearly this rise-time should be short compared to the propagation time τ around the spiral, so that the generator "erects" correctly and there is not a consequent serious loss in gain.

Resistive losses in the lines also degrade the shape of the pulse as the distance travelled by the propagating step can be quite long in a high-gain generator. This can be particularly severe if skin effect losses (see Chapter 2) in the lines are significant, and it leads to an error function shaped loss of pulse rise-time in the propagating step.

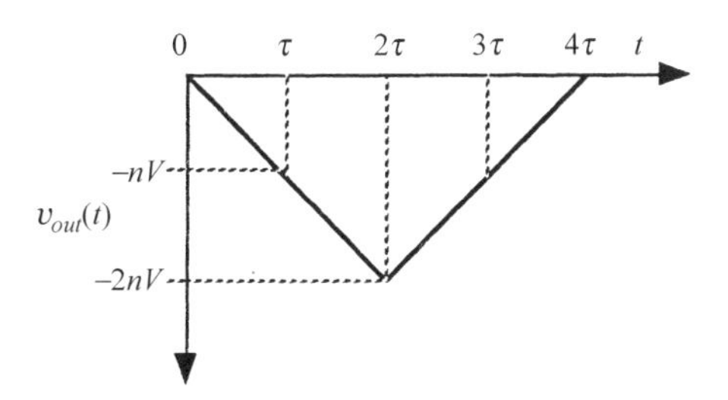

Figure 7.20 Shape of the output pulse from a spiral generator into an open circuit load

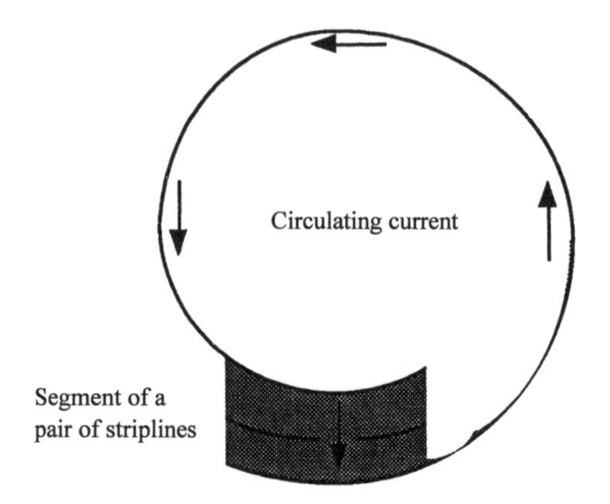

Figure 7.21 Circulating currents in a spiral generator

It has been pointed out that the simple description of the operation of the generator based on the propagation of a voltage step around the spiral is really only applicable to generators which have a very large diameter [35]. The main problem is the fact that the generator is threaded with a pair of continuous conductors, which interconnect regions of different potential, causing currents to flow across the generator that tend to reduce these potential differences. This is illustrated in Figure 7.21 which is a diagram of a segment of the spiral consisting of two layers of stripline linked by one turn of conductor. Current flowing along this closed loop passes across the striplines, changing the potentials on the lines. This circulating current must therefore also be taken into account, as well as the current flowing along the striplines, if an accurate description of the operation of the spiral generator is to be achieved. This particular problem has been analysed in detail in [35].

REFERENCES

[1] Schoenbach K. H., Kristiansen M. and Schaefer G. "A Review of Opening Switch Technology for Inductive Energy Storage". *Proc. IEEE* **72** (1984) 1019–1040

[2] Stevenson G. "Mathematical Methods for Science Students" (Longman Scientific and Technical (1973) 393, ISBN 0-582-44416-0

[3] Maisonnier Ch., Linhart J. G. and Gourlan C. "Rapid Transfer of Magnetic Energy by Means of Exploding Foils" *Rev. Sci. Instrum.* **37** (1966) 1380–1384

[4] Giorgi D., Lindner K., Long J., Navapanich T. and Zucker O. "A Circuit to Enhance the Transfer of Inductive Energy to Imploding Plasma Loads". Proceedings of the 5th IEEE Pulsed Power Conference (1985) 619–622

[5] Pai S. T. and Zhang Qi. "Introduction to High Power Pulse Technology". World Scientific, Advanced Series in Electrical and Computer Engineering, Vol 10 (1995) ISBN 981-02-1714-5

[6] Giorgi D., Lindner K., Long J., Navapanich T. and Zucker O. "The Design and Analysis of a Multi-Stage meatgrinder Circuit". Proceedings of the 5th IEEE Pulsed Power Conference (1985) 615–618

[7] Zucker O., Long J., Lindner K., Giorgi D. and Navapanich T. "Inductive Energy Transfer Circuit Proof of Principle Experiment". *Rev. Sci. Instrum.* **57** (1986) 859–862

[8] Conger R. L. "Large Electric Power Pulses by Explosive Magnetic-Field Compression". *J. Appl. Phys.* **38** (1967) 2275–2277
[9] Conger R. L., Johnson J. H., Long L. T. and Parks J. A. "Production of Large Electric Pulses by Explosive Magnetic Field Compression". *Rev. Sci. Instrum.* **38** (1967) 1608–1610
[10] Shearer J. W., Abraham F. F., Aplin C. M., Benham B. P., Faulkner J. E., Ford F. C., Hill M. M., McDonald C. A., Stephens W. H., Steinberg D. J., and Wilson J. R., "Explosive-Driven Magnetic-Field Compression Generators". *J. Appl. Phys.* **39** (1968) 2102–2116
[11] Crawford J. C. and Damerow R. A. "Explosively Driven High-Energy Generators". *J. Appl. Phys.* **39** (1968) 5224–5213
[12] Cowan M. and Kaye R. J. "Finite-Element Circuit Model of Helical Explosive Generators". Proceedings of the 4th IEEE Pulsed Power Conference (1983) 105–108
[13] Marx E. "Versuche uber die Prufing von Isolatoren mit Spannungsstossen". *Elektro-technishnishe Zeitschrift* **45** (1925) 625
[14] Mallik K. "The Theory of Operation of Transistorized Marx Bank Circuits". *Rev. Sci. Instrum.* **70** (1999) 2155–2160
[15] Pronko S. G. E., Ngo M. T. and Germer R. F. K. "A Solid-State Marx-Type Generator". Proceedings of the 18th Power Modulator Symposium (1988) 211–214
[16] Donaldson P. E. K. "The Mosmarx Voltage Multiplier". *Electronics and Wireless World* **94** (1988) 748–750
[17] Kuffel E. and Zaengl W. S. "High Voltage Engineering". Pergamon Press (1984) ISBN 0-08-024213-8
[18] Fitch R. A. "Marx and Marx-Like High - Voltage Generators". *IEEE Trans. on Nucl. Sci.* **NS18** (1971) 190–198
[19] Prestwich K. R. and Johnson D. L. "Development of an 18MV Marx Generator". *IEEE Trans. on Nucl. Sci.* **NS16** (1969) 64–69
[20] Morrison R. W. and Smith A. M. "Overvoltage and Breakdown Patterns of Fast Marx Generators". *IEEE Trans. on Nucl. Sci.* **NS19** (1972) 20–31
[21] Johnson D. L., Van Devender J. P. and Martin T. H. "High Power Density Water Dielectric Switching". *IEEE Trans.* **PS-8** (1980) 204
[22] Kushner M. J., Kimura W. D. and Byron S. R. "Arc Resistance of Laser-Triggered Spark Gaps". *J. Appl. Phys.* **58** (1985) 1744
[23] Martin T. H., Guenther A. H. and Kristianson M. "J. C. Martin on Pulsed Power". *Advances in Pulsed Power Technology, Plenum* **3** (1996) ISBN 0-306-45302-9
[24] Schaefer G., Kristianson M. and Guenther A. H. "Gas Closing Switches". *Advances in Pulsed Power Technology, Plenum* **2** (1990) ISBN 0-306-43619-1
[25] Carrus A. and Funes L. E. "Very Short Tailed lightning Double Exponentil Wave Generation Techniques Based on Marx Circuit Standard Configurations". *IEEE Trans. on Power App. and Sys.* **PAS-103** (1984) 782–787
[26] Carrus A. "An Inductance on the Marx Generator Tail Branch. New Technique for High Efficiency Laboratory Reproduction of Short Time to Half Value Lightning Impulses". *IEEE Trans. on Power Del.* **4** (1989) 90–94
[27] Kekez M. M. "Simple sub-50-ps Rise-Time High Voltage Generator". *Rev. Sci. Instrum.* **62** (1991) 2923–2930
[28] Lockwood G. J., Schneider L. X., Neyer B. T. and Ruggles L. E. "Diagnosing High-Reliability, Low Jitter Marx Generators". *Rev. Sci. Instrum.* **58** (1987) 1297–1302
[29] Schneider L. X. and Lockwood G. J. "Engineering high Reliability, Low Jitter Marx Generators" Proceedings of the 5th IEEE Pulsed Power Conference (1985) 780–783
[30] Fitch R. A. and Howell V. T. S. "Novel Principle of Transient High-Voltage Generation". *Proc. IEE.* **111** (1964) 849–855

[31] Harris N. W. and Milde H. I. "15 kJ *LC* Generator: Low Inductance Device for a 100 GW Pulsed Electron Accelerator". *J. Vac. Sci. Technol.* **12** (1975) 1188–1190
[32] Meyer B., Watson A., Engel T. G. and Kritianson M. "A Single Gap Transformer Coupled LC Generator with Resonant Frequency Compensation". Proceedings of the 7th IEEE Pulsed Power Conference (1989) 749–752
[33] Engel T. G. and Kritianson M. "A Compact High Voltage Vector Inversion Generator". Proceedings of the 10th IEEE Pulsed Power Conference (1995) 1389–1393
[34] Brau C. A., Rayburn J. L., Dodge J. B. and Gilman F. M., "Simple, Pulsed, Electron Beam Gun". *Rev. Sci. Instrum.* **48** (1977) 1154–1160
[35] Ruhl F. and Herziger G. "Analysis of the Spiral Generator". *Rev. Sci. Instrum.* **51** (1980) 1541–1547

8 Nonlinear Pulsed Circuits

8.1 INTRODUCTION

In the previous chapters of this book it has always been implicitly assumed that the components in the pulsed circuits described are linear in the sense that their values or sizes do not depend on the current passing through them or the voltage applied to them. The only exception to this is in the use of magnetic cores for the transformers described in Chapters 5 and 6. Since there is a nonlinear relation between the magnetic field H applied to a magnetic core and the resulting flux density B (see Figure 5.12) the inductance of any winding placed on the core depends on the current passing though the winding and is therefore not constant. This nonlinear relation, as explained, has to be taken into account in the design of any transformer or inductor wound on a magnetic core to ensure, for example, that the core does not saturate and cause the inductance of the windings on it to collapse.

Although this nonlinear property can be considered to be a handicap or nuisance in the design of magnetically cored inductors or transformers, it can also be exploited to produce a whole family of new circuits with characteristics that are not found in linear circuits. The nonlinear components of such circuits contain either inductors whose inductance decreases or increases with current ($L(I)$) or capacitors whose capacitance are voltage-dependent ($C(V)$) or a combination of both. The technology can also be extended to transmission lines that are insulated either by nonlinear magnetic materials or nonlinear dielectric materials.

Unfortunately the circuit equations for circuits containing such components are inevitably nonlinear and are either difficult or impossible to solve exactly. Thus circuit analysis is generally only feasible using numerical techniques or by the use of approximations [1].

8.2 MAGNETIC SWITCHING

Probably the first piece of work to exploit the use of the nonlinearity of inductors to achieve fast switching was that of Melville on the use of saturable reactors (inductors) in pulse generators [2]. The basic concept is to drive sufficient current through a winding on a magnetic core such that the applied field H produces a flux density B in the core in excess of the core's saturation flux. In doing so the inductance of the winding changes from a relatively high value to a very low value and the inductor behaves as a magnetic switch. Since Melville's work there has been a significant amount of work carried out on magnetic switches, particularly on the magnetic materials used to make the cores. Consequently magnetic switches are now in common use in many different types of pulsed circuit.

As described in Chapter 5, the change in flux density in a magnetic core resulting from the application of a rectangular voltage pulse of constant amplitude V_p applied to a winding on the core is given by

$$\Delta B = \frac{1}{NA}\int_0^{t_p} V_p\,dt = \frac{V_p t_p}{NA} \tag{8.1}$$

where N is the number of winding turns, A is the core area, and t_p is the duration of the applied pulse. If the amplitude of the applied pulse is large enough, the core will saturate after a time t_s given by

$$t_s = \frac{\Delta BAN}{V_p} = \frac{(B_{sat} \pm B_r)AN}{V_p} \tag{8.2}$$

where B_{sat} is the saturation flux density for the core and B_r is the residual or remanent flux density in the core at the start of the pulse (see Chapter 5). Typical values of B_s for a power ferrite are 300 to 500 mT, and for iron-based core materials 0.5 to 2.5 T. In order to maximise ΔB (which then minimises the area of core material required for a given switch) it is desirable to set the remanent flux density of the core at the start of the pulse. Referring to Figure 8.1(b), which shows the BH loop of a typical magnetic core, it can be seen that if the core is initially in position a then ΔB will be maximised. To set the core to this position a

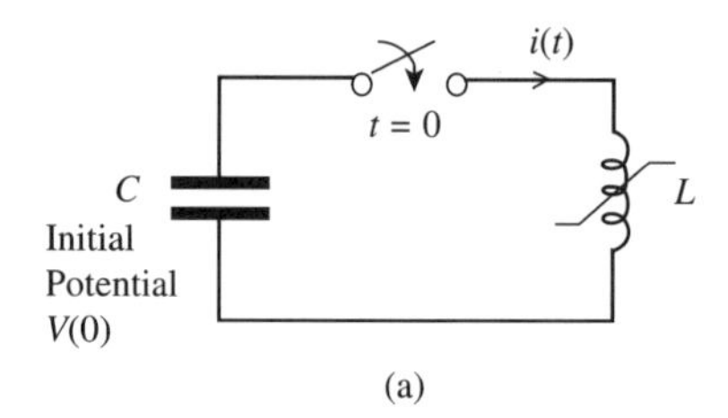

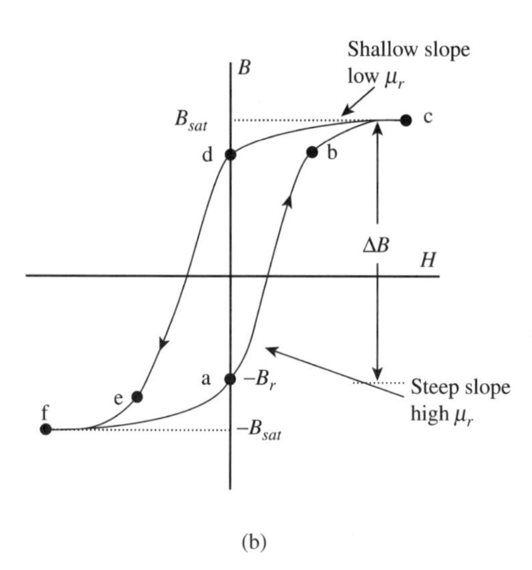

Figure 8.1 *LC* circuit and *BH* loop to describe magnetic switch behaviour

negative pulse can be applied to the winding before operation of the magnetic switch, or a pulse can be applied to a separate winding on the core that can also be used to reset the core before switching.

Assuming that the core is initially at position a, the slope of the loop at that point is steep and the relative permeability μ_r will be high. Conversely if an applied magnetic field H causes the core to move via point b to position c then the slope of the loop at that point is shallow and μ_r will be very low. For a typical core μ_r could change from say 1000 to 2 or 3 in moving from position a to position c. Therefore from Equation (5.46) it can be seen that the inductance L of a winding on a core such as this can vary substantially in moving around the BH loop in this way.

Referring to Figure 8.2, the variation of both current and voltage in the LC circuit of Figure 8.1(a) can be described as the core responds to a positive and then negative magnetic field H caused by the discharge of the capacitor C shown. L here is the inductance of a winding on the magnetic core. Starting again from position a, L will initially have a high value, as explained, and little current will flow in the circuit. As the current starts to rise, the magnetic flux in the core will move to point b, where the core starts to saturate, and then on to point c, where the core is in full saturation. The inductance of the winding will now have fallen to a very low value and consequently the impedance in the LC circuit has also dropped appreciably and the current will rise significantly. Since the circuit contains just a capacitor and an inductor, the current will have a sinusoidal dependence (see Equation (7.1)) and eventually start to fall and then reverse. As it does so the magnetic flux in the core moves to position d and then falls and also reverses. The slope of the loop is again steep so the inductance in the circuit rises sharply again and the current falls. Eventually the core saturates negatively as the flux in the core moves from point e to point f. Once again the current increases significantly but this time the current is negative. The current then peaks and starts to fall. As it does so the core once again comes out of saturation and eventually the flux returns to point a. The current and voltage in the circuit at the different positions marked on the BH loop in Figure 8.1(b) are shown in Figure 8.2. Notice that the final voltage $V(T)$ on the capacitor, after one compete cycle of duration T, is less that the original voltage on the capacitor $V(0)$. This is mainly due to loss in the circuit caused by driving the core around one circuit of the BH loop. The area of the BH loop can therefore be equated to the difference in the energy stored on the capacitor at the beginning and end of one cycle, i.e.

$$\oint B\,dH = \oint H\,dB \cong \frac{1}{2}C(V(0)^2 - V(T)^2) \tag{8.3}$$

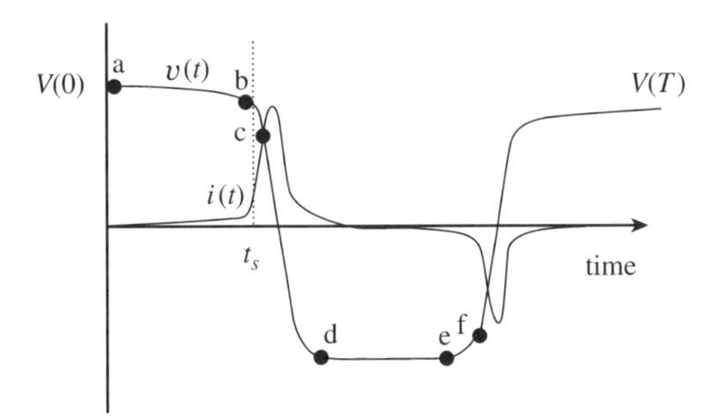

Figure 8.2 Variation of current and voltage in the LC circuit shown in Figure 8.1(a)

Magnetic switches have many advantages which can be exploited in pulsed circuits. They have a very long lifetime, as they do not suffer from erosion problems like gas switches, they can be operated at high repetition rates (1 kHz and above), and they are reliable and rugged. The main disadvantage is heating due to core loss which can lead to a change in the time taken for the switch to saturate. This is because the saturation flux density of most magnetic materials is temperature-dependent. It is therefore important to cool the cores and use well-regulated power supplies to supply any circuits containing magnetic switches. Typical power handling in magnetic switch circuits is 10 kW average and 800 kW in a burst mode, and with care it is possible to build power modulators whose output pulse rise-times can exceed that of more conventional gas-switched circuits.

8.2.1 Magnetic Pulse Compressors

If a series of magnetic switches is used, as shown in the circuit in Figure 8.3, it is possible to successively reduce the rise-time and compress a pulse propagating on the line at each stage. Such a circuit is often referred to as a magnetic pulse compressor or "Melville" line. The compression increases the peak power of the pulse and is achieved by using the magnetic switches to transfer energy between the capacitors. If the inductance of the switches is reduced successively along the line then each capacitor is charged and discharged faster than the previous capacitor. Furthermore, if the capacitors themselves are also successively reduced in size, the amplitude of the pulse grows as it propagates along the line and substantial voltage gain is possible.

Referring to Figure 8.3, on switch closure the capacitor C_0, initially charged to a potential $V(0)$, is discharged through the inductor L_0 into capacitor C_1. As the potential on C_1 rises a point is reached at which the first magnetic switch L_1 will saturate. C_1 then discharges rapidly into capacitor C_2. Again as the potential on capacitor C_2 rises a potential will be reached at which the second magnetic switch L_2 will also saturate, and providing this switch has a lower saturated inductance than L_1 then C_2 will discharge even more rapidly into the next capacitor along the line. This process continues until C_n discharges into the load.

This process is illustrated in Figure 8.4 where it has been assumed that the capacitors on the line all have the same value as C_0 and that there is no loss in the compressor circuit so that the potential reached on the capacitors is constant. Note also that if the capacitors are all of the same value and the saturated inductances of the magnetic switches is reduced progressively along the line, then the impedance of the line falls along the line and there can be considerable current gain.

As explained earlier a completely accurate analysis of the behaviour of the magnetic pulse compressor is very difficult because the circuit contains nonlinear components. However a good approximate analysis has been given by Melville [2] which has been

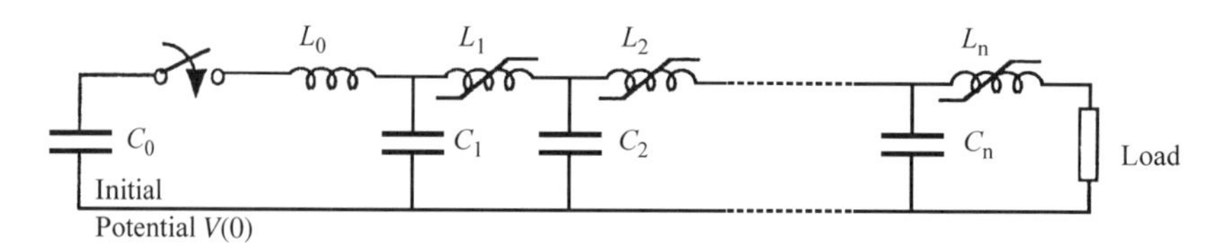

Figure 8.3 Basic magnetic pulse compressor circuit

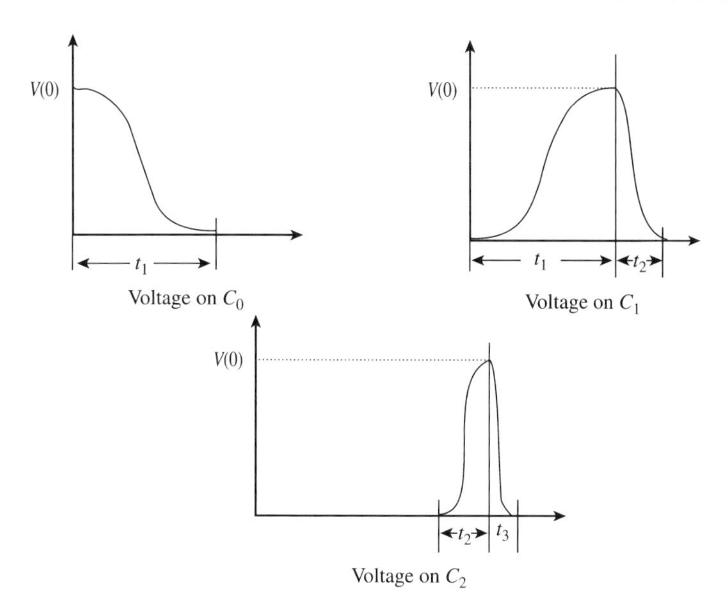

Figure 8.4 The operation of the basic magnetic pulse compressor

extended by Birx [3]. The starting point for the analysis is to look in more detail at the behaviour of two stages in the middle of a pulse compressor. A diagram of such a stage is given in Figure 8.5.

Consider the charging of capacitor C_n from the previous capacitor C_{n-1} via the saturated inductor $L_{(n-1)s}$ (note that the subscripts s and u refer to the saturated and unsaturated inductances, respectively). From Equation (1.129) the voltage $v_n(t)$ will be given by

$$v_n(t) = V(0)\left(\frac{C_n}{C_{n-1} + C_n}\right)(1 - \cos(\omega t)),$$
$$\omega = \sqrt{\frac{1}{\dfrac{L_{(n-1)s} C_n C_{n-1}}{C_n + C_{n-1}}}} \tag{8.4}$$

In the special case, when all the capacitors have the same value, this equation simplifies to

$$v_n(t) = \frac{V(0)}{2}(1 - \cos(\omega t)),$$
$$\omega = \sqrt{\frac{2}{L_{(n-1)s} C_0}} \tag{8.5}$$

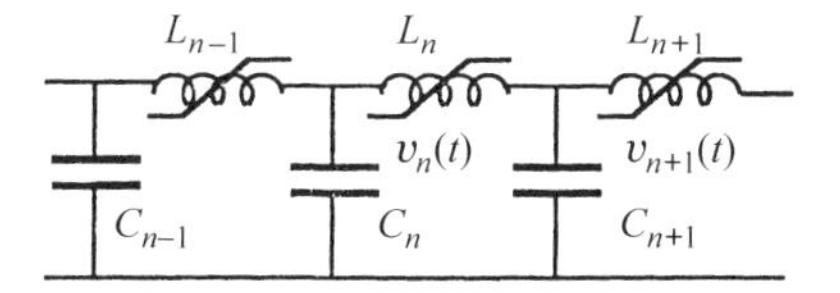

Figure 8.5 Intermediate stages in a magnetic pulse compressor

and the time T_n to charge capacitor C_n is given by

$$T_n = \pi\sqrt{\frac{L_{(n-1)s}C_0}{2}} \tag{8.6}$$

The compressor is designed such that the inductor, after the capacitor which is being charged from the previous stage, saturates when the capacitor is fully charged, i.e. inductor L_n is designed to saturate when capacitor C_n is fully charged. Thus from Equation (8.1) the change in flux $B_n(t)$ in the core of inductor L_n is given by

$$\begin{aligned} B_n(t) &= \frac{1}{N_nA_n}\int_0^{T_n} v_n(t)\,dt \\ &= \frac{V_0}{2N_nA_n}\int_0^{T_n} (1-\cos(\omega t))\,dt \end{aligned} \tag{8.7}$$

where T_n is the time for the inductor L_n to saturate. Carrying out the integration gives

$$B_n(T_n) = \Delta B_n = \frac{T_nV_0}{2N_nA_n} = B_{(sat)n} + B_{rn}$$

$$\text{or} \qquad T_n = \frac{2\Delta B_nA_nN_n}{V_0} \tag{8.8}$$

where B_{rn} is the assumed reset position of the core flux at the start of operation of the flux compressor. Equating this saturation time to the charge time of capacitor C_n (Equation (8.6)) gives

$$\frac{2\Delta B_nA_nN_n}{V_0} = \pi\sqrt{\frac{L_{(n-1)s}C_0}{2}}$$

$$\text{or} \qquad L_{(n-1)s} = \left[\frac{2\Delta B_nA_nN_n}{\pi V_0}\right]^2 \frac{2}{C_0} \tag{8.9}$$

From Equation (5.5) the saturated inductance of inductor L_n is given by

$$L_{ns} = \frac{\mu_0\mu_{rs}N_n^2A_n}{l} = \frac{\mu_0\mu_{rs}N_n^2A_n^2}{Vol_n} \tag{8.10}$$

where μ_{rs} is the saturated permeability and Vol_n is the core volume of inductor L_n. Therefore the ratio of the saturated inductances of successive inductors is

$$\frac{L_{ns}}{L_{(n-1)s}} = \mu_0\mu_{rs}\left(\frac{C_0V_0^2}{2}\right)\frac{1}{Vol_n}\frac{\pi^2}{4\Delta B_n^2} \tag{8.11}$$

Since the bracketed term in this equation is simply the pulse energy, and the ratio of the saturated inductances of successive inductors is fixed by the compressor design, it is clear that the core volumes in the inductors must be increased linearly in proportion to the pulse energy. As it has been assumed that the capacitors in the flux compressor all have the same value, then from Equation (8.6) the ratio of the charging times of successive stages, which is often known as the stage gain G of the compressor, can be found to be

$$G = \frac{T_n}{T_{n+1}} = \sqrt{\frac{L_{(n-1)s}}{L_{ns}}} \tag{8.12}$$

Combining this equation with Equation (8.11) and rearranging gives the useful design relationship

$$Vol_n = \frac{G^2(\text{pulse energy})\pi^2\mu_0\mu_{rs}}{4\Delta B_n^2} \tag{8.13}$$

For efficient operation of the compressor it is important that the saturated inductance of the inductor, at a given stage, is very much smaller than the unsaturated inductance of the inductor of the next stage. If this is not the case there will be significant current leakage through the inductor of the next stage on to the stage beyond that. Therefore, it is important that

$$L_{nu} \gg L_{(n-1)s} \tag{8.14}$$

According to Melville [2] if $\gg$ is interpreted as "twenty times", then an estimate for the maximum gain or compression per stage can be derived from the ratio of the unsaturated and saturated permeabilities of the magnetic cores, i.e.

$$G^2 = \frac{L_{(n-1)s}}{L_{ns}} = \frac{L_{nu}}{20L_{ns}} = \frac{\mu_{ru}}{20\mu_{rs}} \tag{8.15}$$

Typically, as stated earlier

$$\frac{\mu_{ru}}{\mu_{rs}} \cong 400 \tag{8.16}$$

Thus the maximum gain or compression per stage is around 4 to 5.

Finally, it is worth describing two important variants of the basic pulse compresor described. The first is to use a succession of transformers whose cores are allowed to saturate, as shown in Figure 8.6 [4].

Once the storage capacitor C is charged, the switch is closed and the capacitor discharges into next capacitor which has a value of $C/4$ via a pulse transformer. Since the referred value of this capacitor is equal to the value of the storage capacitor, total energy transfer takes place if losses are ignored. Once the $C/4$ capacitor has charged the core of the first transformer saturates, effectively isolating the first stage of the circuit, and the $C/4$ capacitor starts to charge the last capacitor which also has a matched value through the last transformer

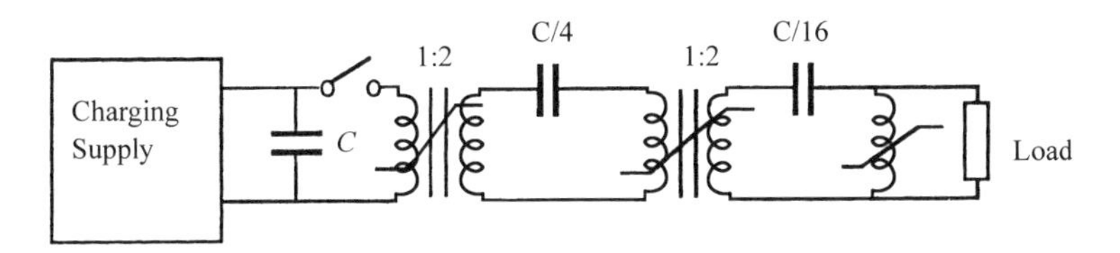

Figure 8.6 Magnetic pulse compressor using saturable transformers

of $C/16$. The magnetic switch in parallel with the load, is biased so that at this stage it is saturated and little current flows through the load. Once the capacitor $C/16$ is charged, the magnetic switch is designed to come out of saturation so that the capacitor then discharges into the load. As with the first stage, the second transformer is also designed to saturate at the same time in order to isolate its primary from the rest of the circuit. As the transformers have turns ratios of 1 : 2 and the values of the capacitors are reduced along the compressor, the circuit provides both pulse compression and voltage gain.

The other variant [5] is alternator-driven. A basic diagram is given in Figure 8.7 together with sketches of the voltage waveforms at different parts of the circuit. The advantages of this circuit are that there are no conventional switches, only magnetic switches, and the output is a repetitive train of pulses. Reset circuits, which are not shown, are necessary for correct operation of the circuit.

In this circuit the sinusoidal voltage waveform from the alternator is fed to a capacitor C_1. The core of the first magnetic switch SW1 is set such that it saturates at the peak of the first negative excursion of the waveform from the generator. Once SW1 has saturated the energy stored on C_1 is transferred via the transformer to capacitor C_2. Since $C_2 \ll C_1$, the rise-time of the voltage on C_2 is now much less than the period of the sinusoidal waveform from the

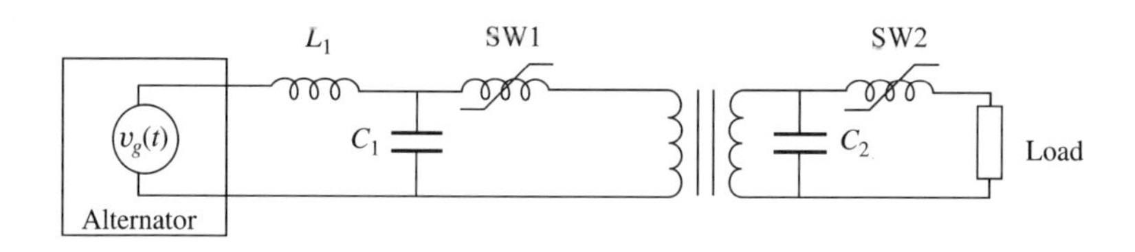

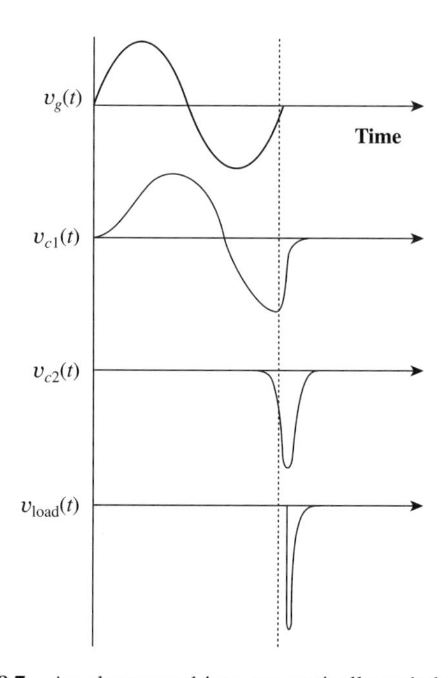

Figure 8.7 An alternator-driven magnetically switched modulator

generator. Once C_2 is fully charged, the second switch saturates and an even faster rising pulse is delivered to the load. Voltage gain is also possible in the circuit if the transformer has a turns ratio greater than unity. For efficient energy transfer between the two capacitors, it is again necessary to make the referred value of C_2 to the primary equal to the value of C_1.

8.3 PULSE SHARPENING USING NONLINEAR CAPACITORS

Nonlinear capacitors, such as those based on ferroelectric ceramics made from barium or strontium titanate, can also be used in circuits that are similar to the magnetic pulse compressor to speed up the rise or fall times of electrical pulses [6]. The nonlinear capacitors are used as the nonlinear elements in an *LC* ladder network instead of saturating nonlinear inductors as shown in Figure 8.8. In contrast to the standard magnetic pulse compressor, the ladder is usually uniform, i.e. each section is identical and comprises a nonlinear capacitor and an inductor.

Pulse sharpening is caused by the amplitude dependence of the phase velocity of signals propagating on such a line. For a linear *LC* ladder network, such as that described in Chapter 4, the propagation constant γ is given by Equation (4.15), i.e.

$$\cosh(\gamma) = 1 + \left(\frac{LC}{2}\right)\mathbf{s}^2 \tag{8.17}$$

For continuous signals on a loss-free ladder (see Chapter 2)

$$\left.\begin{aligned} &\gamma = j\beta,\ \cosh(j\beta) = \cos(\beta) \\ &\mathbf{s} = j\omega \end{aligned}\right\} \tag{8.18}$$

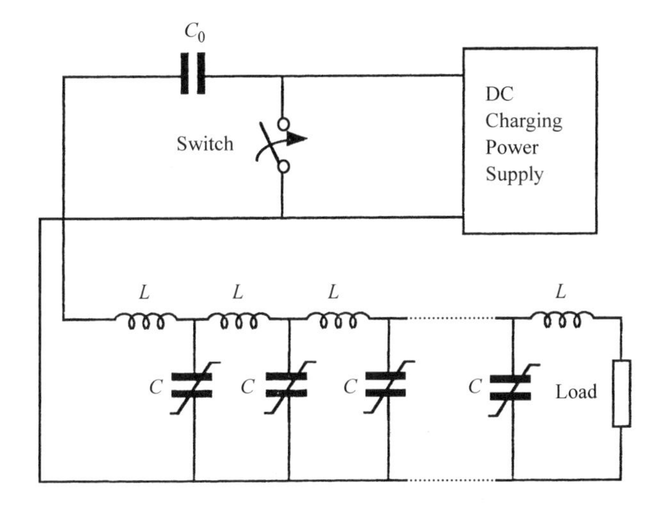

Figure 8.8 A pulse sharpener using nonlinear capacitors

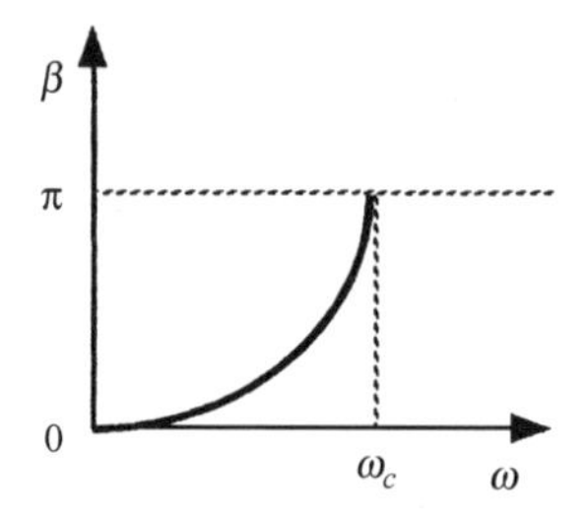

Figure 8.9 The variation of β with frequency ω from Equation (8.19)

Therefore

$$\cos(\beta) = 1 - \frac{\omega^2 LC}{2} = 1 - 2\left(\frac{\omega}{\omega_c}\right)^2 \tag{8.19}$$

where ω_c is the cut-off frequency of the ladder [7]. A plot of the relationship between β and ω is given in Figure 8.9.

For small values of β corresponding to low frequencies

$$\cos(\beta) \cong 1 - \frac{\beta^2}{2} = 1 - \frac{\omega^2 LC}{2} \tag{8.20}$$

Hence the phase velocities of low-frequency signals propagating on the ladder (see Equation (2.29)) are given by

$$v_p = \sqrt{\frac{1}{LC}} \tag{8.21}$$

Since the capacitors in the ladder are nonlinear, such that the capacitance falls with applied voltage $C(V)$, it is clear that the phase velocity will increase with amplitude. If a pulse is injected on to the ladder, the peak of the pulse will travel at a faster velocity than the lower amplitude portions of the pulse, and the propagation delay per section of the ladder will be at a minimum at the peak of the pulse. This causes the peak of the pulse to catch up with the low-amplitude portion of the leading edge, which leads ultimately to the formation of a shock-wave. A typical circuit for demonstrating this effect is shown in Figure 8.8. A charged storage capacitor is discharged into the nonlinear ladder by closing the switch. The resulting pulse then propagates along the ladder and the rise-time of its leading edge reduces as it propagates towards the load.

This effect can also be demonstrated on lines where the nonlinear capacitors are reverse-biased varactor diodes [8,9]. Although the pulse sharpening effect is similar to the operation of magnetic pulse compressors, it is more difficult to carry out a circuit analysis, as the nonlinearity of capacitors made from ferroelectric ceramic dielectrics such as barium titanate is relatively weak in comparison to the large change in inductance seen in saturating inductors, which leads to the magnetic "switching" effect described in the last section. A typical characteristic for a commercial 1.2 nF 20 kV barium titanate capacitor is shown in Figure 8.10.

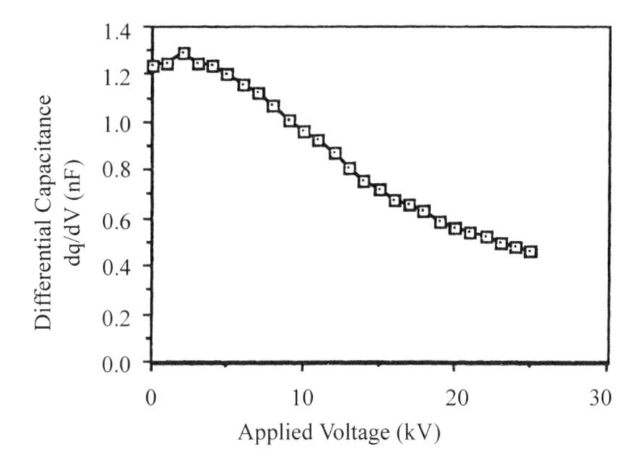

Figure 8.10 The differential capacitance of a typical nonlinear capacitor as a function of applied voltage. © 1991 IEEE

This characteristic is measured by biasing a capacitor from a DC voltage source and then measuring the capacitance of the series connection of the capacitor under test and a much larger linear capacitor [10]. Hence, strictly the differential capacitance is measured. The reason for this nonlinearity is that some ferroelectric materials such as barium titanate have a nonlinear polarization (P) versus electric field (E) characteristic which is analogous to the BH loops observed in magnetic materials.

An accurate estimation of the amount of sharpening that will be found in a given ladder is difficult, not only because the mathematical analysis is nonlinear but also because the phase velocity on the ladder is frequency-dependant. A rough estimate can be made by calculating the difference in time delay of each section of the ladder between the low-amplitude portion of the pulse and its peak. The total rise-time reduction ΔT from the input to the ladder to its output will then be given by the relationship

$$\Delta T = n(\sqrt{LC_{us}} - \sqrt{LC_s}) \tag{8.22}$$

where n is the number of sections in the ladder, C_{us} is the value of the capacitors at zero applied voltage, and C_s is the value of the capacitors at an applied voltage equal to the peak amplitude of the propagating pulse. Clearly, the shortest rise-time that can be achieved by the ladder network, as the number of sections is increased, will ultimately be limited by the cut-off frequency of the ladder ω_c, assuming that the ladder is loss-free. Loss in the circuit will further increase the minimum rise-time that is achievable.

Another rather surprising effect that is observed in ladders of this type is the appearance of high-frequency oscillations which are superimposed on the pulse and are particularly strong at the middle sections of the ladder. This oscillation is caused by the pulse breaking up into a soliton array, a phenomenon which has been observed previously in nonlinear ladders which use varactor diodes as the nonlinear capacitors [11]. The oscillation frequency is found to be close to the cut-off frequency of the ladder. An example of this effect is shown in Figure 8.11. Here a pulse has been sharpened on a ladder with 15 sections, and the rise-time has been reduced from 280 ns at the input to 50 ns at the load.

Since the value of the capacitors in the ladder depends on the amplitude of the pulse which is propagating, it is clear that the characteristic impedance of the ladder will also

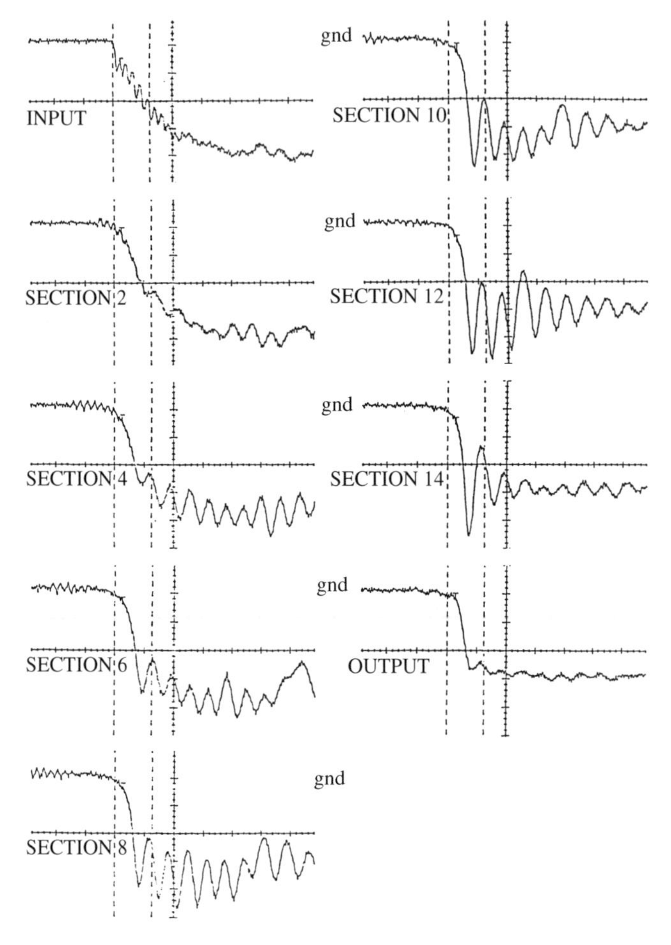

Figure 8.11 Example of pulse sharpening on a nonlinar capacitor ladder network. The horizontal time scale is 100 ns per division. © 1991 IEEE

be dependant on the amplitude V of the pulse and will decrease with increasing amplitude, i.e.

$$Z_0(V) = \sqrt{\frac{L}{C(V)}} \tag{8.23}$$

Finally it is worth noting that if a nonlinear line such as that shown in Figure 8.8 is biased, so that the signal propagation velocity *decreases* with amplitude, it is possible to generate a sharpening effect on the tail of a pulse rather than on the leading edge. A suitable experimental circuit is shown in Figure 8.12. If switches SW1 and SW2 are closed, with switch SW3 open, then the line will sharpen the leading edge of the pulse as described earlier. If, however, switch SW3 is closed and switches SW2 an SW2 are left open, then it is possible to apply a DC bias to the line so that the nonlinear capacitors C are stressed and

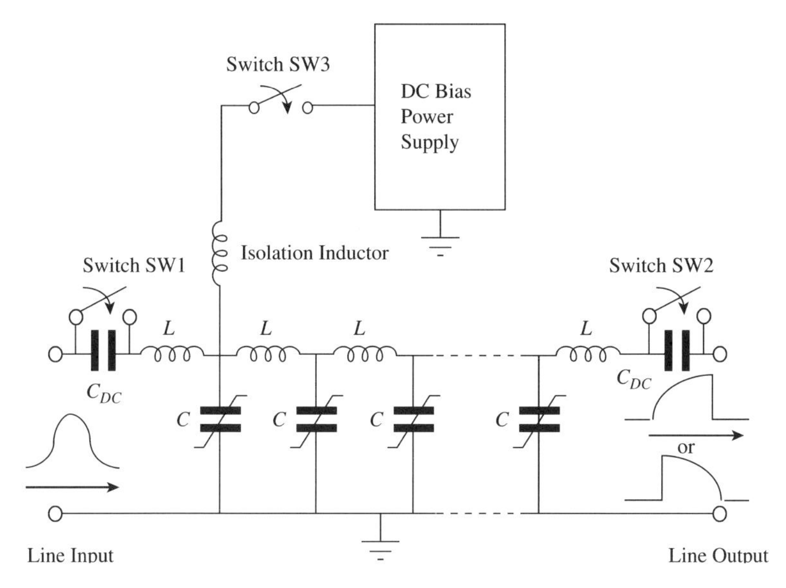

Figure 8.12 A biased line for pulse leading edge sharpening or tail sharpening

have a lower value than when they are unstressed. The input pulse then causes their capacitance to increase, as the amplitude of the pulse increases, and the propagation velocity then slows. This causes a sharpening effect on the tail of the pulse. The two capacitors C_{DC} simply isolate the DC bias from the input and output of the line, and the isolation inductor ensures that the DC power supply is isolated during the pulse sharpening process.

8.3.1 The Analysis of Pulse Sharpening on LC Ladders with Nonlinear Capacitors

The rise-time of a pulse, sharpened by the amplitude dependence of its phase velocity (amplitude dispersion), remains finite, in practice, because of frequency dispersion or frequency-dependent loss. In a distributed transmission line frequency dispersion is caused, as explained in Chapter 2, either by loss in the dielectric used to insulate the line or by the skin effect, which results in resistive loss in the conductors at high frequencies [12]. In a ladder network, frequency dispersion results from the periodic structure, which causes frequencies close to the cut-off frequency of the ladder to propagate more slowly than lower frequencies (see the last section). Providing that the nonlinear line or network is long enough, the rise-time of a sharpened pulse will depend on the balance between amplitude and frequency dispersion.

To fully understand and hence design nonlinear pulse shaping circuits, it is important to be able to mathematically model and analyse their behaviour. This is a difficult problem because the analysis of nonlinear networks and transmission lines results in nonlinear differential-difference equations or nonlinear partial differential equations, the solutions to which may not be known. It may be impossible to derive exact time-dependent transfer and response functions for such networks and lines, and approximations may be needed. In this section an approximate analytical treatment of electrical signal propagation on nonlinear ladder networks is given. Despite being approximate, this analysis provides an important insight into the mechanism of the pulse sharpening process, and some useful data suitable for circuit design can be obtained.

As will be evident from the analysis of ladder networks given in Chapter 4, in a linear loss-free LC ladder network the equations governing the currents and voltages at the kth mesh are

$$C\frac{dv_k(t)}{dt} = i_{k-1} - i_k \tag{8.24}$$

$$L\frac{di_k(t)}{dt} = v_k(t) - v_{k+1}(t) \tag{8.25}$$

By substituting into these equations the forward voltage and current travelling waves

$$\left.\begin{aligned} v_k(k,t) &= V_0 \exp[j(\omega t - \beta k)] \\ i_k(k,t) &= I_0 \exp[j(\omega t - \beta k)] \end{aligned}\right\} \tag{8.26}$$

the dispersion relation given in the last section is obtained

$$LC\omega^2 = 2[1 - \cos(\beta)] \tag{8.27}$$

The shortest wavelength λ for which a real solution of this equation exists is $\lambda=2$ sections, equivalent to $\beta=\pi$. This wavelength corresponds to a cut-off frequency ω_c given by

$$\omega_c^2 = \frac{4}{LC} \tag{8.28}$$

Hence there are no travelling wave solutions of Equations (8.24) and (8.25) for frequencies above the cut-off frequency. From Equations (8.24) to (8.26) the impedance of the line can be deduced to be

$$Z = \frac{V_0}{I_0} = Z_0 \exp(j\phi) \tag{8.29}$$

where

$$\sin(\phi) = \frac{\omega}{\omega_c} \quad \text{and} \quad Z_o = \sqrt{\frac{L}{C}} \tag{8.30}$$

If the network is constructed from 'T' sections [7], the matched terminating impedance Z_T is

$$Z_T = Z_0\left(1 - \frac{\omega^2}{\omega_c^2}\right)^{1/2} \tag{8.31}$$

For the case of a Π section, the matched terminating impedance is

$$Z_\Pi = Z_0\left(1 - \frac{\omega^2}{\omega_c^2}\right)^{-1/2} \tag{8.32}$$

In the long-wavelength limit $\beta \ll 1$ or $\omega \ll \omega_c$, the current and voltage are in phase and the frequency dispersion is negligible, so the phase velocity v_p is given by

$$v_p = \frac{\beta}{\omega} = \sqrt{\frac{1}{LC}} \tag{8.33}$$

as before. Therefore in the low-frequency limit the ladder cannot be distinguished from an ideal distributed loss-free transmission line of characteristic impedance Z_0.

An approximate continuous wave equation for the lumped line can be derived from Equations (8.24) and (8.25), which may be combined to give

$$LC\frac{d^2 v_k(t)}{dt^2} = v_{k+1} - 2v_k + v_{k-1} \tag{8.34}$$

$v_k(t)$ can be approximated by a continuous function $v(x,t)$, such that $v_k(t) = v(k\delta x,t)$, where δx is a constant. By expressing v_{k+1} and v_{k-1} in terms of Taylor expansions of $v(x,t)$ about $v(k\delta x,t)$ it can be shown that

$$L'C'\frac{\partial^2 v(x,t)}{\partial t^2} - \frac{\partial^2 v(x,t)}{\partial x^2} \cong \frac{\delta x^2}{12}\frac{\partial^4 v(x,t)}{\partial x^4} \tag{8.35}$$

where $L'\delta x = L$ and $C'\delta x = C$. In the case where the capacitance is nonlinear, this equation can be written

$$L'\frac{\partial^2 C'(v)v(x,t)}{\partial t^2} - \frac{\partial^2 v(x,t)}{\partial x^2} \cong \frac{\delta x^2}{12}\frac{\partial^4 v(x,t)}{\partial x^4} \tag{8.36}$$

On the assumption that

$$C'(v)v(x,t) = C_0'(v(x,t) + \varepsilon f(v)) \tag{8.37}$$

where C_0' is a constant, $f(v)$ is a function representing the nonlinearity and ε is a constant which determines the strength of the nonlinearity, Equation (8.36) becomes

$$\frac{\partial^2 v(x,t)}{\partial t^2} + \varepsilon\frac{\partial^2 f(v)}{\partial t^2} \cong \frac{1}{L'C_0'}\left(\frac{\partial^2 v(x,t)}{\partial x^2} + \frac{\delta x^2}{12}\frac{\partial^4 v(x,t)}{\partial x^4}\right) \tag{8.38}$$

Following a perturbation method introduced by Washimi and Taniuti [13], $v(x,t)$ is now expressed as a power series expansion in the small parameter ε,

$$v(x,t) = \varepsilon v^{(1)}(x,t) + \varepsilon^2 v^{(2)}(x,t) + \cdots \tag{8.39}$$

where $v^{(1)}$ is the first-order approximation, $v^{(2)}$ is the second-order approximation, etc. It is mathematically convenient to change coordinates to a system moving at velocity c by substituting

$$\xi = \varepsilon^{1/2}(x - ct) \quad \text{and} \quad \tau = \varepsilon^{3/2}ct \tag{8.40}$$

where

$$c = \sqrt{\frac{1}{L'C_0'}} \tag{8.41}$$

Making these substitutions, and collecting terms in the same powers of ε, an equation for $v^{(1)}$ is obtained

$$2\frac{\partial^2 v^{(1)}(x,t)}{\partial\xi\partial\tau} - \frac{\partial^2 f(v)}{\partial\varepsilon^2} + \frac{\delta x^2}{12}\frac{\partial^4 v^{(1)}(x,t)}{\partial\xi^4} = 0 \tag{8.42}$$

Integrating this equation once with respect to ξ, and assuming a quadratic nonlinearity of the form $f(v) = -\alpha v^2$, results in

$$\frac{\partial v^{(1)}(x,t)}{\partial\tau} - \alpha v^{(1)}(x,t)\frac{\partial v^{(1)}(x,t)}{\partial\xi} + \frac{\delta x^2}{24}\frac{\partial^3 v^{(1)}(x,t)}{\partial\xi^3} = 0 \tag{8.43}$$

which is a form of the now well-known Korteweg–de Vries (KdV) equation [14]. A great deal is known about the character of the solutions of this equation [15]. A particularly important result is that an arbitrary initial voltage waveform $v(\xi,0)$ will separate into an array of solitons travelling in the forward direction (i.e. in the direction $\xi \to +\infty$) and an oscillatory dispersive wave travelling in the reverse direction ($\xi \to -\infty$). The relevance of this result is that a shock-wave front propagating along a nonlinear ladder network subject to Equation (8.43) will split up into an array of solitons. This can cause a modulation of the power delivered to the load. Furthermore, as the wavefront evolves into a soliton array an increasing fraction of the propagating power is coupled into frequencies close to the cut-off frequency of the line. These frequencies cannot all be coupled out simultaneously because of the rapid variation of the matched value of the terminating impedance of the line near the cut-off frequency (Equations (8.31) or (8.32)). Any resistive termination of the line will therefore reflect a fraction of the incident power. The implication is that there is an upper limit to the useful length of a lumped element shock line, above which no further reduction in pulse rise-time occurs and the overall pulse shape may be degraded by the instability of the waveform.

8.3.2 Soliton Generation

A soliton is effectively a solitary wave solution of a travelling wave equation. For a signal propagating in a one-dimensional medium to be defined as travelling wave it must be propagating at a constant velocity c and its shape must not change. That is, the signal will be constant in a frame of reference propagating with a velocity c. If the translation $\zeta = x - ct$, where c is constant and ζ is the new frame coordinate, is used then the wave function $\phi(x,t) = \phi(\zeta)$ for a travelling wave. One type of travelling wave is a localised travelling wave, or solitary travelling (ST) wave $\phi_{\mathrm{ST}}(\zeta)$. This is more precisely defined as a travelling wave whose transition from one constant asymptotic state to another, as ζ changes from $+\infty$ to $-\infty$, is essentially localised in ζ. To describe a soliton it is only necessary to consider the case where the solitary wave function ϕ_{ST} and all its derivatives tend to 0 as $\to \pm\infty$.

The solitary wave was initially considered an unimportant curiosity in nonlinear wave theory and simply a special solution of a number of nonlinear partial differential equations. Originally it was thought that where two such waves were launched on a collision course, the nonlinear interaction of the collision would destroy the identity of the waves. The advent of high-speed computers made testing of such theories possible and in 1962 Perring and

Skyrme [16] investigated such interactions numerically. Their computer simulations indicated that solitary waves were not destroyed on collision, but emerged with the same shapes and velocities as they had before they collided. Shortly afterward Zabusky and Kruskal [17] published results from an independent computer model to study the propagation of waves in plasmas. They also concluded that solitary waves emerged unaltered from a collision, and coined the term soliton as a result of this remarkable property. The formal definition of a soliton $\phi_{ST}(\zeta)$, first suggested by Scott *et al.* [18], is a solitary wave solution of a wave equation which asymptotically preserves its shape and velocity on collision with other solitary waves. Although a more detailed discussion of solitons is beyond the scope of this book, interested readers are referred to the books by Drazin and Johnson [19] and Lonngren and Scott [11].

The study of nonlinear dispersive transmission lines or ladder networks is a fairly new subject and several papers have been published on soliton generation and interaction on nonlinear, dispersive transmission lines, consisting of nonlinear capacitors and inductors [11]. As described earlier, nearly all these lines have used varactor diodes as nonlinear capacitors, and consequently the soliton amplitude and frequency, in the experiments reported to date, are typically several volts and a few MHz, respectively. In 1988 Ikezi [20] suggested that the same technique could be scaled to produce soliton bursts with microwave frequencies, at GW power levels, by using modulated strip lines which are periodically loaded with nonlinear blocks of dielectric made from ferroelectric ceramics such as barium or strontium titanate. This technique has now been shown to work experimentally both by Ikezi [21] and others [22].

The KdV equation, derived in the last section from an analysis of the signal propagation characteristics on nonlinear LC ladder networks, proves to be one of many equations which are known to have soliton solutions. Its particular solution takes the form

$$v(x, t) = -\frac{3c}{\beta}\,\mathrm{sech}^2\left(\sqrt{\frac{c}{\alpha}}(x - ct)\right) \tag{8.44}$$

where α and β are constants which depend on the geometry of the transmission line and the strength of the nonlinearity, respectively. Two important properties of solitons can be deduced from this expression, namely that the amplitude of the soliton is proportional to its velocity, and that the width of the soliton is inversely proportional to the square root of its velocity.

A circuit of a high-power soliton generator is given in Figure 8.13. The heart of the generator is a "modulated" strip line in which blocks of a nonlinear ferroelectric ceramic made from strontium titanate are periodically bonded between two parallel plate conductors. This construction essentially forms an LC ladder network where the blocks are the capacitors and the spaces between the blocks are the inductors. The inductance L and the unstressed capacitance C of each section of the modulated line are given by

$$\left.\begin{aligned} L &= \mu_0\left(\frac{lh}{w}\right)\left(1 - \frac{d}{l}\right)\mathrm{H} \\ C &= \varepsilon_r\varepsilon_0\frac{wd}{h}\mathrm{F} \end{aligned}\right\} \tag{8.45}$$

It is assumed that the width w of the conductors is much larger than the separation h, so that end effects can be neglected. d is the length of the blocks, l is the length of the space

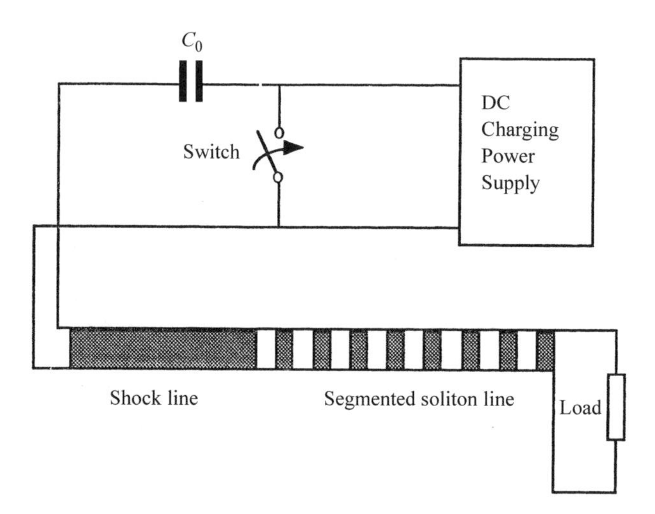

Figure 8.13 A circuit of a high-power soliton generator

between the blocks and ε_r is the unstressed permittivity of the blocks. The frequency ω_s of the solitons produced on such a line is given by

$$\omega_s = \sqrt{\frac{1}{LC(V_{max})}} \tag{8.46}$$

where V_{max} is the peak amplitude of the solitons that are produced. To propagate a soliton on such a line, an input pulse of duration t_p must be such that $t_p \leq 1/\omega_s$. The easiest way to produce a high-frequency soliton burst from the line is to inject a shock wave on to the line, where the rise-time of the shock wave is $t_r \leq 1/\omega_s$. In the circuit shown, the shock wave is generated by closing the switch which discharges a charged capacitor C_0 into a shock line (discussed in the next section). The shock line is then connected directly into the segmented soliton line which is terminated in a resistive load.

It is possible to carry out a numerical analysis of the behaviour of such a circuit [22] and investigate how the modulation depth and frequency of the solitons are determined by the amplitude and rise-time of the input shock wave and the number of sections in the

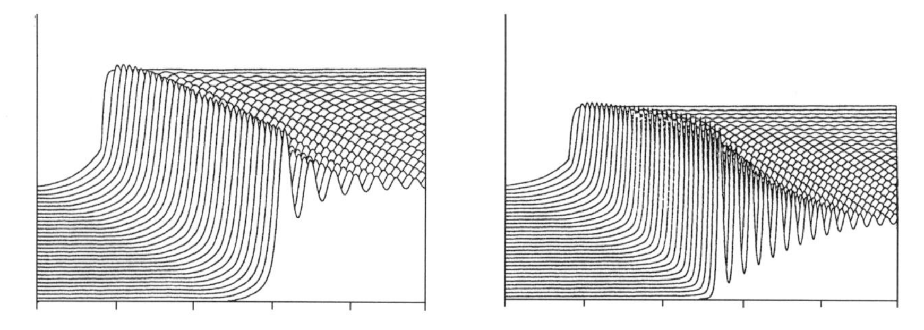

Figure 8.14 Numerical simulation of the effect of increasing the amplitude of the input pulse to a soliton line from 10 kV (left) to 30 kV (right). © 1977 IEEE

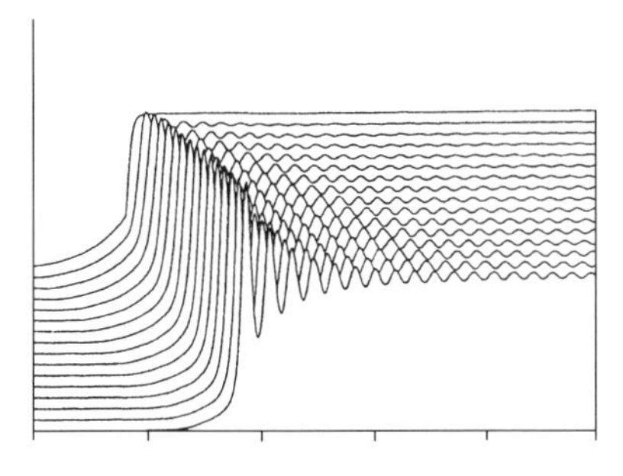
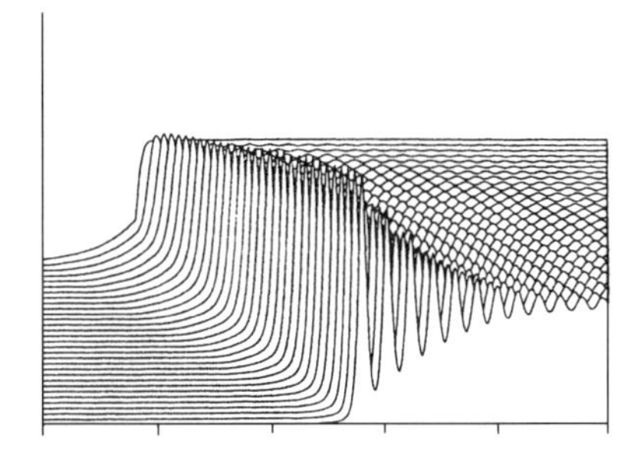

Figure 8.15 Numerical simulation of the effect of increasing the number of sections in a soliton line from 15 (left) to 30 (right). © 1977 IEEE

segmented soliton line. In Figure 8.14 two numerical simulations are shown, where the amplitude of the input shock wave is increased by a factor of three on an experimental circuit with 30 sections, as described in [22]. The simulation shows the soliton activity along the line from the input (at the back of the diagram) to the output (at the front). It can be seen that as the amplitude is increased from 10 kV to 30 kV, the modulation depth is also increased. In Figure 8.15 two further simulations are given where the number of sections in the line has been doubled from 15 to 30. In this case, as the number of sections is increased, the modulation depth is also increased.

Finally it is also possible to study, by numerical simulation, the problem of matching the soliton line with a resistive load. As described earlier the line will have a varying impedance which depends on the amplitude of the soliton burst. Rather surprisingly, and in agreement with Ikezi [23], it is found that a high matching efficiency can be achieved which is close to 100%.

8.4 ELECTROMAGNETIC SHOCK WAVE GENERATION IN NONLINEAR TRANSMISSION LINES

The generation of high-voltage, high-power, electrical pulses with very fast rise-times (<1 ns) is difficult to achieve directly because high-voltage switches with very short closing times do not satisfy simultaneously the requirements of being reliable, being able to switch at high power levels, and the capability of being operated repeatedly at high frequencies. One way around this problem is to feed a relatively slowly rising voltage pulse into a nonlinear circuit or component, such as a magnetic switch which, as has been described earlier, will sharpen the rising edge of the pulse, thereby reducing its rise-time. This reduces the closing time requirements on the primary switch, generating the pulse, so that reliable long lifetime devices such as thyratrons or thyristors can be used. However there is a limit to the rise-time reduction that is possible using magnetic switches or ladder networks with nonlinear capacitors due to the stray inductance in the connections to the load or, in the case of a magnetic switch, its finite saturated inductance.

This problem can be overcome by the use of a nonlinear transmission line rather than a circuit which contains discrete nonlinear elements. The nonlinearity may be due to changes in resistance, capacitance or inductance distributed along the transmission line [24] but transmission lines containing ferrites with nonlinear magnetic properties have received the greatest attention [25–30]. It is also possible to use a line insulated with a nonlinear ferroelectric material, although little work has been published on such lines [31,32].

The principle behind the basic sharpening mechanism is fairly simple although, again, detailed mathematical analysis of the effect is difficult as the sharpening process is based on nonlinearity. The speed of propagation or phase velocity of a signal propagating on a transmission line depends, as explained in Chapter 2, on the relative permittivity and permeability of the dielectric used to insulate the line. If the permittivity or permeability depends on the voltage V or current I, respectively, flowing on the line then the velocity v_p will be given by

$$v_p = \frac{c}{\sqrt{\varepsilon_r(V)\mu_r(I)}} \tag{8.47}$$

where c is the velocity of light. It can be seen from this equation that the velocity will depend on the amplitude of any signal propagating on the line. If a pulse is injected on to the line then, if either the relative permittivity or permeability fall with increasing pulse amplitude, the velocity of the peak of the pulse is greater than the lower amplitude portions of the leading edge of the pulse. Thus the peak of the pulse catches up with its rising edge, causing a reduction in rise-time, in a similar way to that observed on nonlinear LC ladder networks.

This process of rise-time reduction cannot continue indefinitely since, if the rise-time were to reach zero, the voltage or current flowing on the front edge of the pulse would become multivalued. Instead a shock wave develops. At the shock wave front the transmission line undergoes a sudden transition, either between being uncharged or charged or unmagnetised and magnetised, the duration of which is determined by the relaxation characteristics of the particular material used to insulate the line. The rise-time of the pulse cannot be reduced beyond the rise-time of this shock transition as it will be limited by the dispersive characteristics of this material. Once the shock wave is fully developed, so that the rise-time is limited by the material properties, it will continue to propagate with constant rise-time. In the case of a step-like wave, it is envisaged that a steady state will be achieved in which the wave no longer changes shape as it propagates along the transmission line. For a real pulse waveform, however, a true steady state cannot be achieved since some of the pulse energy is always dissipated at the shock front. The time taken for the shock wave to develop depends on the rise-time of the pulse injected at the input to the line, and can be roughly estimated using the velocities of the low- and high-voltage or current portions of the pulse. The shock wave is expected to be fully developed when the peak of the pulse has caught up to the front by an amount which is just less than the initial rise-time. This sharpening process can also be thought of as a harmonic generation method as high-frequency components are added to the pulse in the shock wave front.

8.4.1 Shock Wave Formation on Ferrite Loaded Transmission Lines

The construction of a standard a ferrite line pulse sharpener is illustrated in Figure 8.16. Ferrite rings are threaded on to a cylindrical conductor and this assembly is then inserted into a tightly fitting tubular conductor which is usually lined with a plastic dielectric material. The line, therefore, has a coaxial geometry. A DC current is usually passed along the line to bias it, so that the ferrite ring cores are magnetised (often into saturation) in a

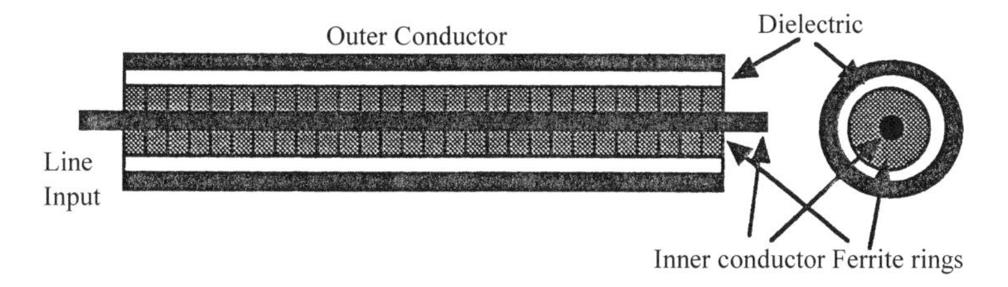

Figure 8.16 The construction of a typical ferrite line sharpener

sense which is opposite to that which results when the pulse, to be sharpened, is injected on to the line. On arrival of the pulse, the magnetisation of the cores is subjected to a large opposing magnetic field, thus the cores are forced to switch from the initial reverse biased state to an opposed saturated state. The transition from one state to the other is called the "spin reversal region", which takes a finite time corresponding to the time it takes for the magnetisation in the core to switch from its initial position on the *BH* loop to its final position. It has been observed that if the lines are not fully biased into saturation, i.e. a full *BH* loop reversal is not used, then there are fewer *BH* loop losses and faster switching times are achievable [27]. Indeed pulse sharpening will occur in unbiased lines provided that the magnetisation of the lines is returned to zero, using a reset pulse, before the arrival of a subsequent pulse to be sharpened. Furthermore it has been found that the lines can be biased magnetically using permanent magnets which obviates the need for current biasing [28]. Pulse sharpening in nonlinear ferrite lines can be quite impressive and the rise-time of 10 kV pulses rising in 5 ns has been successfully reduced to 150 ps [28].

In the case of current-biased lines it is the finite relaxation or switching time of the ferrite which limits the sharpening process. The standard analysis of the this switching time is based on the magnetisation reversal mechanism described by the Gilbert form of the Landau Lifschitz equation [25,30,33] which gives the time dependence of the magnetisation of a magnetic material. However, more recently, it has been suggested that the extremely fast rise-times observed on magnetically biased ferrite lines could be caused by a different mechanism relating to the stimulation of gyromagnetic precession behaviour in the ferrite cores, which is similar to that exploited in microwave ferrite devices such as circulators and isolators. This suggests that a model based on the damped precession of the magnetic dipoles at an atomic level, using the Landau–Lifshitz form of the magnetisation equation would be more appropriate. Such a model has been shown to be quite effective in accounting for the experimental results observed on a magnetically biased line [28].

8.4.2 Shock Wave Generation on Nonlinear Ferroelectric Lines

As an alternative to ferrites, ferroelectric ceramic materials can be used as the dielectric insulator in transmission lines. Since the dielectric constant of such materials falls with increasing applied electric field, then again the phase velocity of the peak of a pulse travelling on such a line will be greater than the low-amplitude part of the rising edge. Hence the rise-time of the pulse will be decreased and a shock wave will form. If $C_{x\%}$ represents the capacitance of the transmission line when it is stressed to x% of the full pulse amplitude, t_{ir} is the 10–90% rise-time of the input pulse and t_{sr} is the 10–90% rise-time of the shock wave, then the length l of transmission line required for the shock wave to form is estimated as:

$$l = \left(\frac{v_1 v_2}{v_2 - v_1}\right)(t_{ir} - t_{sr}) \tag{8.48}$$

where

$$v_1 = \sqrt{\frac{1}{LC_{10\%}}} \quad \text{and} \quad v_2 = \sqrt{\frac{1}{LC_{90\%}}} \tag{8.49}$$

A transmission line with a nonlinear dielectric can be represented by a series of infinitesimal circuit elements, as shown in Figure 8.17. Here L is the inductance per unit length and $U(Q)$ is the voltage which must be applied to the nonlinear dielectric in a length dx for it to store a charge Q. R represents the loss in the dielectric due to the relaxation of the nonlinear polarisation, and its value is chosen to reflect the relaxation frequency of the dielectric for reasons which will be explained later.

C_{hf} is an additional capacitance which represents the combined capacitance between the transmission line conductors without the dielectric and other polarisation mechanisms in the dielectric besides the nonlinear response. For example, in the case of a ferroelectric ceramic dielectric based on barium or strontium titanate, as well as the strongly nonlinear polarization in such a material, other linear electronic polarisation also occurs which does not play a part in the development of a shock wave. In this type of dielectric the contribution of C_{hf} to the charge stored on the transmission line is negligible compared to the charge stored due to the nonlinear polarisation, and in the following analysis its effect is ignored.

The change of current i and voltage v on the transmission line with distance x can thus be approximated by the equations

$$\frac{\partial i}{\partial x} = -\frac{\partial Q}{\partial t} \qquad \frac{\partial v}{\partial x} = -L\frac{\partial i}{\partial t} \tag{8.50}$$

and

$$v = U(Q) + R\frac{\partial Q}{\partial t} \tag{8.51}$$

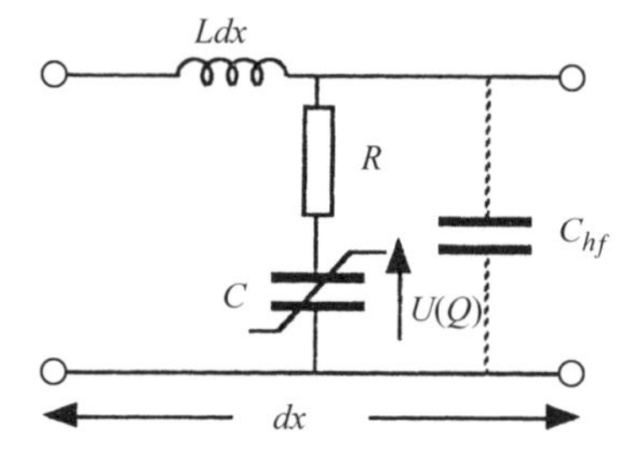

Figure 8.17 Circuit element of a transmission line with a nonlinear dielectric

Following the analysis of Peng and Landauer [34] consider a nonlinear dielectric response of the form

$$U(Q) = \frac{Q}{C_0} + \eta \frac{Q^3}{C_0^3} \tag{8.52}$$

where C_0 is the value of capacitance on the line when no charge is stored and η is a constant. This equation implies that the capacitance on the line tends to zero as the stored charge becomes infinitely large, whereas strictly it should tend to C_{hf}.

It is now necessary to find a steady-state solution to the set of Equations (8.50) to (8.52) in order to find an estimate of the rise-time that might be achieved on the nonlinear transmission line. A steady-state solution is one for which all parts of a wave or pulse propagate at the same constant velocity, so that as it moves along the transmission line its shape remains unchanged. For such a solution, the profile can be described by

$$Q = Q(\vartheta) \qquad \vartheta = x - ut \tag{8.53}$$

where u is the velocity of the steady-state shock profile and ϑ is a phase variable which is constant for a profile moving with this velocity. Because the wave has a constant profile, u is independent of the charge at any particular point on the transmission line. Under these conditions Equations (8.50) and (8.51) can be rewritten as

$$\frac{\partial i}{\partial \vartheta} = u \frac{\partial Q}{\partial \vartheta} \qquad \frac{\partial v}{\partial \vartheta} = uL \frac{\partial i}{\partial \vartheta} \tag{8.54}$$

and

$$v = U(Q) - uR \frac{\partial Q}{\partial \vartheta} \tag{8.55}$$

and integrated directly to give

$$v = u^2 LQ \tag{8.56}$$

The constant of integration has been set to zero since, for an initially uncharged transmission line, v and Q must both be zero as $\vartheta \to \infty$. Equations (8.52), (8.55) and (8.56) can now be combined to give

$$u\tau_0 \frac{dQ}{d\vartheta} = \eta \frac{Q^3}{C_0^2} - Q\left(\frac{u^2}{v_0^2} - 1\right)$$
$$\text{where} \qquad v_0 = \sqrt{\frac{1}{LC_0}} \quad \text{and} \quad \tau_0 = RC_0 \tag{8.57}$$

where τ_0 is the relaxation time of the nonlinear dielectric. Equation (8.57) is a separable, first-order ordinary differential equations which can be rewritten as

$$u\tau_0 \frac{dQ}{Q^3 - Q_s^2 Q} = \frac{d\vartheta}{Q_s^2 w_0} \tag{8.58}$$

where

$$Q_s^2 = \frac{C_0^2}{\eta}\left(\frac{u^2}{v_0^2} - 1\right) \quad \text{and} \quad w_0 = \frac{\eta Q_s^2}{u\tau_0 C_o^2} \tag{8.59}$$

Partialising the left hand side of Equation (8.58) gives

$$-\frac{1}{2Q_s}\left(\frac{1}{Q(Q_s-Q)}+\frac{1}{Q(Q_s+Q)}\right)dQ=\frac{d\vartheta}{Q_s^2 w_0} \tag{8.60}$$

using the indefinite integral

$$\int\frac{dx}{x(ax+b)}=\frac{1}{b}\ln\frac{x}{ax+b} \tag{8.61}$$

Equation (8.60) can be integrated to give, after some rearrangement

$$-\frac{1}{2}\ln\left(\frac{Q_s^2-Q^2}{Q^2}\right)=\frac{\vartheta-x_0}{w_0} \tag{8.62}$$

where x_0 is the constant of integration. It can be shown that

$$\ln\left(\frac{Q_s^2-Q^2}{Q^2}\right)=\tanh^{-1}\left(\frac{Q_s^2-2Q^2}{Q_s^2}\right) \tag{8.63}$$

Using this result, and writing Equation (8.62) in terms of the original coordinates (x,t), the shock wave profile is given by

$$Q^2(x-ut)=\frac{1}{2}Q_s^2\left[1-\tanh\left(\frac{x-x_0-ut}{w_0}\right)\right] \tag{8.64}$$

The integration constant x_0 is determined by the position of the shock wave at $t=0$.

Equation (8.64) is plotted in Figure 8.18. The tanh function is centred on $\vartheta=0$ and the square of the shock wave amplitude reaches half its maximum value at this point. The charge on the transmission line never reaches either 0 or its maximum value Q_s as ϑ tends to $\pm\infty$, indicating that this ideal shock wave can never be realised. The points x_1 and x_2

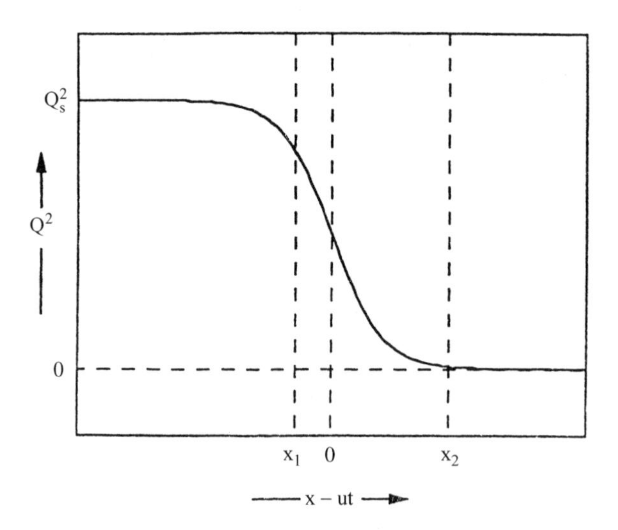

Figure 8.18 A plot of the shock profile given by Equation (8.64)

marked in the figure are where $Q^2 = 0.01\ Q_s^2$ and $0.81\ Q_s^2$, i.e. where $Q = 0.1\ Q_s$ and $0.9\ Q_s$, respectively. The voltage at any point on the transmission line must according to Equation (8.56) be proportional to the charge, so the 10–90% voltage rise-time can be determined if these two positions and the shock wave velocity are known. The wave reaches 90% of its full voltage at a point x_1 when

$$\tanh \frac{x_1 - x_0}{w_0} = -0.62 \tag{8.65}$$

or, setting x_0 to be zero when $t = 0$, when $x_1 = -0.73\ w_0$.

Similarly x_2 is found by setting

$$\tanh \frac{x_2 - x_0}{w_0} = 0.98 \tag{8.66}$$

giving $x_2 = 2.30\, w_0$. The shock wave velocity can be determined by rearranging Equation (8.59) to give

$$u = \sqrt{\frac{1}{LC_0}} \left(1 + \frac{\eta Q_s^2}{C_0^2}\right)^{1/2} \tag{8.67}$$

τ_0 is experimentally determined as the reciprocal of the relaxation frequency of the dielectric, and then from these results the 10–90% rise-time of the voltage profile can be shown to be

$$t_r = \frac{x_2 - x_1}{u} = 3.03 \frac{C_0^2}{\eta Q_s^2} \tau_0 \tag{8.68}$$

Notice that t_r is proportional to the relaxation time of the dielectric, and inversely proportional to the nonlinearity factor η.

To produce a shock wave with the minimum possible rise-time it is important that

$$\frac{\eta Q_s^2}{C_0^2} > 1 \tag{8.69}$$

but then t_r in this region will be largely determined by the relaxation time of the dielectric, as shown in Figure 8.19.

8.4.3 *Ferroelectric Shock Lines: Some Practical Considerations*

There are a number of dielectrics which can be made in the form of ceramics which will stand high levels of electric stress and are nonlinear. The most strongly nonlinear, that is those with the greatest voltage coefficient of permittivity, are based upon ferroelectrics such as lead zirconate, barium titanate, barium-strontium titanate or lead-magnesium niobate [10,35]. To maximise the nonlinearity, they should be used at a temperature near to the Curie temperature, where a ferroelectric undergoes a phase transition and becomes paraelectric. In this region the permittivities of the compounds listed above can be several thousands, and may change by an order of magnitude when the dielectrics are stressed to their full voltage rating.

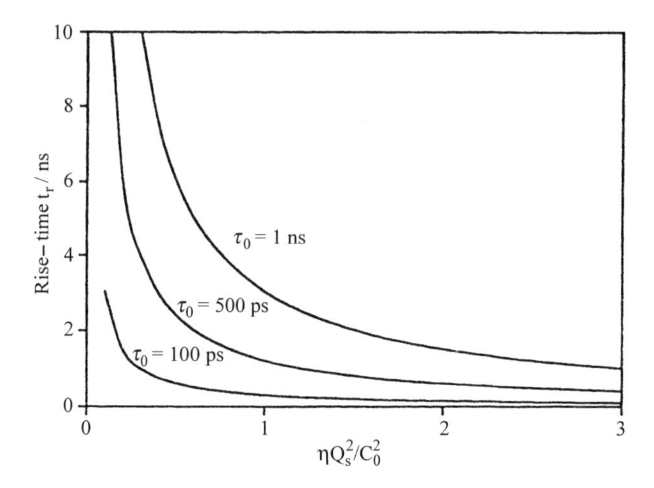

Figure 8.19 Variation of shock wave rise-time with nonlinearity for different dielectric relaxation times

In addition to a strong nonlinearity, a dielectric must also have a high relaxation frequency if it is to be used for the generation of shock waves with fast rise-times. The relaxation frequency is the highest frequency with which the nonlinear polarisation mechanism can respond to a changing electric field. The permittivity and relaxation frequency of a ferroelectric dielectric is dependent on additives and the processing of the ceramic during manufacture. Of the materials listed above, barium-strontium titanate has the highest relaxation frequency at over 1 GHz. The nonlinearity of this material can be enhanced by the addition of various chemical dopants but, unfortunately, this reduces the relaxation frequency and any possible advantage of the added dopant is lost.

As stated earlier, the nonlinearity of ferroelectric dielectrics is not as strong as that of ferrites. When a ferrite reaches its saturation magnetisation, its relative permeability will fall from a few hundreds or even above a thousand to around 2 or 3. Ferroelectrics, on the other hand, still have relative permittivities of several hundreds when stressed to their maximum working voltage. This has two practical consequences for a practical shock wave generator. Firstly, because the permittivity of the dielectric remains high, at the full pulse voltage, the impedance of the shock line can be quite low. In the case of a shock line with a parallel plate geometry the impedance (see Chapter 2) is given by

$$Z_0 \approx \frac{377}{\sqrt{\varepsilon_r}} \frac{h}{w} \tag{8.70}$$

where is the ε_r relative permittivity of the dielectric, and h and w are the separation and width of the conductors of the transmission line, respectively. If $h = 0.1\,w$ and the permittivity of the dielectric remains at around 1000, at the full pulse voltage, then Z_0 is of the order of 1.2 Ω. This means that at relatively modest voltages of a few kV shock waves with current levels of several kiloamps and power levels exceeding 10 MW can be produced.

The second effect of the low saturated impedance of the dielectrics is that the physical wavelength of signals propagating on the shock line is relatively short. For instance a 1 GHz signal in a dielectric with a relative permittivity of 1000 has a wavelength of only about 1 cm. This has important implications for the design of wide bandwidth probes, which must be incorporated into the transmission line if the rise-time of the shock wave is to be accurately measured.

The other important consideration for ferroelectric shock lines is loss. In a typical dielectric there are electrical losses due to several causes, including conduction losses, losses due to hysteresis in the polarisation (P), electric field (E) characteristic and relaxation losses [10]. The combined effect of all these mechanisms is dependent on electric field and frequency and is extremely difficult to measure under the conditions in which shock waves form. However conduction losses should be very small, and hysteresis loss is not expected to affect the profile of the shock front as the area of the hysteresis loop is small at the Curie temperature. Hence relaxation losses play the dominant role in determining the rise-time of the electromagnetic shock wave, and the value of the relaxation frequency is the best measurement of this loss that can be achieved.

The most convenient way to build a ferroelectric shock line is to use tiles or blocks of an appropriate ferroelectric ceramic which are metallised on opposite faces and then bonded to two strip conductors to form a parallel plate type transmission line. A photograph of this type of line is given in Figure 8.20 [32]. The line is seen as a long black strip which runs from one end of the oil tank to the other. A parallel plate geometry also has the advantage that the electric field between the two conductors is uniform. In a coaxial transmission line, the electric field is stronger near to the inner conductor than it is at the outer. This leads to self-focusing effects in the transmission line which degrade the rise-time of the shock wave.

The circuit used to operate the line is identical to that shown in Figure 8.13, with the exception that the soliton line is removed. The output of the shock line is connected directly to a very low inductance load constructed from two copper plates immersed in an aqueous copper sulphate solution. Figure 8.21 shows typical voltage profiles at the input and output

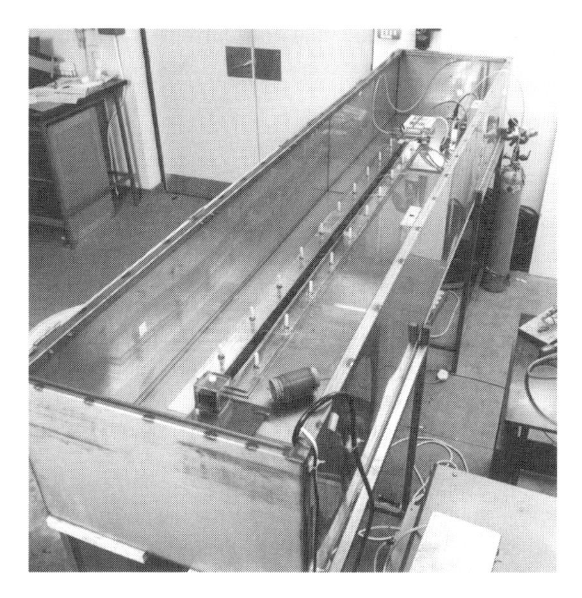

Figure 8.20 Photograph of a nonlinear ferroelectric shock line

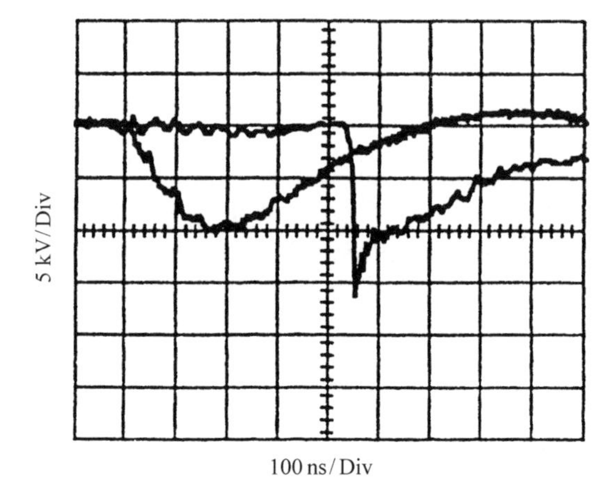

Figure 8.21 Typical voltage profiles at the input and output of the nonlinear transmission line shown in Figure 8.20

of a line. The rise-time of the input pulse appears to have been decreased from 130 ns to 10 ns at the load. In fact the rise-time at the load is very much less than 10 ns and the measurement shown is bandwidth-limited.

To enable higher bandwidth measurements to be made, an *I* dot probe can be built into the stripline. This is done by cutting a small slot in one of the conductors of the transmission line and then connecting a length of coaxial transmission across the slot. Provided the length of the slot is less than the wavelength of signals propagating on the transmission line, the slot acts like a series inductor, and a voltage proportional to the rate of change of current (*I* dot) is developed across the slot. This voltage signal is then simply integrated to determine the rise-time of the current in the shock wave. By this means, in the experiment described, it was possible to determine that the current was rising in just 410 ps [32] corresponding to a current rise-time in excess of 10^{13} A s^{-1}.

REFERENCES

[1] Kouril F. and Vrba K. "Non-linear and Parametric Circuits". Ellis Horwood series in Electrical and Electronic Engineering, Halstead Press (1988) ISBN 0-85312-606-2

[2] Melville W. S. "The Use of Saturable Inductors as Discharge Devices for Pulse Generators". *Proc. IEE* **98** (1951) 185–207

[3] Birx D. L., Lauer E. J., Reginato L. L., Schmidt J. and Smith M. "Basic Principles governing the Design of Magnetic Switches". Lawrence Livermore Laboratory Report No. UCID-18831 (1980)

[4] Nunnally W. C. "Stripline Magnetic Modulators for Lasers and Accelerators". Proceedings of the 3rd IEEE Pulsed Power Conference (1981) 210–213

[5] Kihara R. and Kirbie H. C. "An Alternator-Driven Magnetically Switched Modulator". Proceedings of the 17th Power Modulator Symposium (1986) 246–249

[6] Wilson C. R., Turner M. M. and Smith P. W. "Pulse Sharpening in a Uniform LC Ladder Network Containing Nonlinear Ferroelectric Capacitors". *IEEE Trans. on Electron Devices* **38** (1991) 767–771

[7] Bleaney B. I. and Bleaney B. "Electricity and Magnetism". Oxford University Press (1976) ISBN 0-19-851172-8
[8] Owens A. R. and White G. Generation of High-Speed Waveforms using Nonlinear Delay Lines *Proc. IEE* **113** (1966) 1763–1768
[9] Fallside F. and Bickly D. T. Nonlinear Delay Line with a Constant Characteristic Impedance *Proc. IEE* **113** (1966) 263–270
[10] Burfoot J. C. and Taylor G. W. "Polar Dielectrics and their Applications". Macmillan (1979)
[11] Lonngren K. E. and Scott A. (editors). "Solitons in Action". Academic Press (1978)
[12] Dreher T. "Cabling Fast Pulses? Don't Trip on the Steps". Electronic Engineer (1979) 71–75
[13] Washimi H. and Taniuti T. "Propagation of Ion-Acoustic Solitary Waves of Small Amplitude". *Phys. Rev. Lett.* **17** (1966) 996
[14] Korteweg D. T. and de Vries G. "On the Change of Form of Long Waves Advancing in a Rectangular Channel and on a New Type of Long Stationary Waves". *Phil. Mag.* **39** (1896) 422
[15] Newall A. C. "Solitons in Mathematics and Physics". CBMS-NSF Regional Conference Series in Applied Mathematics Vol 48 (1985)
[16] Perring J. K. and Skyrme T. H. R. "A Model Unified Field Equation". *Nuclear Physics.* **31** (1962) 550–555
[17] Zabusky N. J. and Kruskal M. D. "Interaction of Solitons in a Collision-Less Plasma and the Recurrence of Initial States". *Physical Review Letters* **15** (1965) 240–243
[18] Scott A. C., Chu F. Y. F. and McLaughlin D. W. "The Soliton—A New Concept in Applied Science". *Proceedings of the IEEE* **61** (1973) 1443–1483
[19] Drazin P. G. and Johnson R. S. "Solitons: An Introduction". Cambridge Texts in Applied Mathematics (1989) ISBN 0-521-33389-X
[20] Ikezi H., Wojtowicz S. S., Waltz R. E., deGrassie J. S. and Baker D. R. "High Power Soliton Generation at Microwave Frequencies". *J. Appl. Physics* **64** (1988) 3277–3281
[21] Ikezi H., deGrassie J. S. and Drake J. "Soliton Generation at 10 MW Level in the Very High Frequency Band". *Appl. Phys. Lett.* **58** (1991) 986–987
[22] Brown M. P. and Smith P. W. "High Power Pulsed Soliton Generation at Radio and Microwave Frequencies". Proceedings of the 11th IEEE International Pulsed Power Conference (1997) 346–354
[23] Lin-Liu Y. R., Chan V. S., deGrassie J. S. and Ikezi H. "Characteristic Impedance of a Soliton-Bearing Nonlinear Transmission Line". *Microwave and Optical Letters* **4** (1991) 468–471
[24] Katayev I. G. "Electromagnetic Shock Waves". Iliffe Books Ltd. (1966)
[25] Weiner M. "Pulse Sharpening in Ferrite Transmission Lines". Proceedings of the 2nd IEEE International Pulsed Power Conference (1979) 91–95
[26] Weiner M. and Silber L. "Pulse Sharpening Effects in Ferrites". *IEEE Trans. on Magnetics* **MAG-17** (1981) 1472–1477
[27] Pouladian-Kari R., Benson T. M., Shapland A. J. and Parkes D. M. "The Electrical Simulation of Pulse Sharpening by Dynamic Lines". Proceedings of the 7th IEEE International Pulsed Power Conference (1989) 178–181
[28] Dolan J. E., Bolton H. R. and Shapland A. J. "Analysis of Ferrite Pulse Sharpeners". Proceedings of the 9th IEEE International Pulsed Power Conference (1993) 308–311
[29] Seddon N. and Thornton E. "A High-Voltage Short Risetime Pulse Generator Based on a Ferrite Pulse Sharpener". *Rev. Sci. Instrum.* **59** (1988) 2497–2498
[30] Mesyats G. A. and Baksht R. B. "Deformation of Intense Waves Traversing a Ferrite Discontinuity in a Transmssion Line". *Soviet Physics-Technical Physics* **10** (1965) 685–689
[31] Branch G. and Smith P. W. "Electromagnetic Shock-Waves in Distributed delay Lines with Nonlinear Dielectrics". Proceedings of the 20th IEEE Power Modulator Symposium (1992) 355–359

[32] Branch G. and Smith P. W. "Fast-Rise-Time Electromagnetic Shock Waves in Nonlinear, Ceramic Dielectrics". *J. Phys. D: Phys.* **29** (1996) 2170–2178
[33] Kikuchi K. "On the Minimum of Magnetisation Reversal Time". *J. Appl. Phys.* **27** (1956) 1352–1357
[34] Peng S. T. and Landauer R. "Effects of Dispersion on Steady State Electromagnetic Schock Profiles". *IBM J. Res. Develop.* (1973) 299–306
[35] Hench L. L. and West J. K. "Principles of Electronic Ceramics". John Wiley and Sons (1990) ISBN 0-471-61821-7

APPENDIX: TABLE OF LAPLACE TRANSFORMS

$\mathscr{L}[f(t)]$ is defined by $\int_0^\infty f(t)\exp(-\mathbf{s}t)\,dt$ and is written as $\mathbf{F}(\mathbf{s})$

$f(t)$ from $t = 0_+$	$\mathbf{F}(\mathbf{s}) = \mathscr{L}[f(t)]$
Basic Properties	
Linearity	
1. $\alpha f_1(t) + \beta f_2(t)$	$\alpha\mathbf{F}_1(\mathbf{s}) + \beta\mathbf{F}_2(\mathbf{s})$
Differentiation	
2. $(d/dt)f(t)$	$\mathbf{sF}(\mathbf{s}) - f(0)$
3. $(d^n/dt^n)f(t)$	$\mathbf{s}^n\mathbf{F}(\mathbf{s}) - \mathbf{s}^{n-1}f(0) - \mathbf{s}^{n-2}f'(0)\cdots - f^{n-1}(0)$
Integration	
4. $\int_0^t f(u)\,du$	$\dfrac{\mathbf{F}(\mathbf{s})}{\mathbf{s}}$
5. $\int_0^t \cdots \int_0^t f(u)\,du^n = \int_0^t \dfrac{(t-u)^{n-1}}{(n-1)!}f(u)\,du$	$\dfrac{\mathbf{F}(\mathbf{s})}{\mathbf{s}^n}$
Shifting	
6. $\exp(-\alpha t)f(t)$	$\mathbf{F}(\mathbf{s}+\alpha)$
Translation	
7. $f(t-\alpha)u(t-\alpha) = f(t-\alpha) \quad t \geq \alpha$; $= 0 \quad t < \alpha$	$\exp(-\alpha\mathbf{s})\mathbf{F}(\mathbf{s})$
Transformation	
8. $\alpha f(\alpha t)$	$\mathbf{F}(\mathbf{s}/\alpha)$
Differentiation in the **s** domain	
9. $tf(t)$	$-\mathbf{F}'(\mathbf{s})$
10. $(-1)^n t^n f(t)$	$\mathbf{F}^n(\mathbf{s})$
Integration in the **s** domain	
11. $f(t)/t$	$\int_{\mathbf{s}}^\infty f(u)\,du$
Convolution	
12. $\int_0^t f_1(u)f_2(t-u)\,du = f_1 \otimes f_2$	$\mathbf{F}_1(\mathbf{s})\mathbf{F}_2(\mathbf{s})$
Tables of Transforms	
1. Unit Impulse Function δ	1
2. Unit Step Function $u(t)$	$\dfrac{1}{\mathbf{s}}$
3. Delayed Unit Step Function $u(t-\alpha)$	$\dfrac{\exp(-\alpha\mathbf{s})}{\mathbf{s}}$
4. Rectangular Pulse (width α)	$\dfrac{1-\exp(-\alpha\mathbf{s})}{\mathbf{s}}$
5. Ramp Function t	$\dfrac{1}{\mathbf{s}^2}$
6. $\dfrac{t^{n-1}}{(n-1)!}, 0! = 1$	$\dfrac{1}{\mathbf{s}^n} \quad n = 1, 2, 3$
7. $\dfrac{1}{\sqrt{\pi t}}$	$\dfrac{1}{\sqrt{\mathbf{s}}}$

8. $\exp(-\alpha t)$	$\dfrac{1}{s+\alpha}$
9. $1-\exp(-\alpha t)$	$\dfrac{\alpha}{s(s+\alpha)}$
10. $t\exp(-\alpha t)$	$\dfrac{1}{(s+\alpha)^2}$
11. $\dfrac{t^{n-1}}{(n-1)!}\exp(-\alpha t)$, $0!=1$	$\dfrac{1}{(s+\alpha)^n}$ $n=1,2,3\ldots$
12. $\dfrac{\exp(-\alpha t)-\exp(-\beta t)}{\beta-\alpha}$	$\dfrac{1}{(s+\alpha)(s+\beta)}$
13. $\dfrac{\alpha\exp(-\alpha t)-\beta\exp(-\beta t)}{\alpha-\beta}$	$\dfrac{s}{(s+\alpha)(s+\beta)}$
14. $\sin(\omega t)$	$\dfrac{\omega}{s^2+\omega^2}$
15. $\cos(\omega t)$	$\dfrac{s}{s^2+\omega^2}$
16. $1-\cos(\omega t)$	$\dfrac{\omega^2}{s(s^2+\omega^2)}$
17. $\omega t-\sin(\omega t)$	$\dfrac{\omega^3}{s^2(s^2+\omega^2)}$
18. $\dfrac{t}{2\omega}\sin(\omega t)$	$\dfrac{s}{(s^2+\omega^2)^2}$
19. $t\cos(\omega t)$	$\dfrac{s^2-\omega^2}{(s^2+\omega^2)^2}$
20. $\sin(\omega t)-\omega t\cos(\omega t)$	$\dfrac{2\omega^3}{(s^2+\omega^2)^2}$
21. $\sin(\omega t)+\omega t\cos(\omega t)$	$\dfrac{2\omega s^2}{(s^2+\omega^2)^2}$
22. $\exp(-\alpha t)\sin(\omega t)$	$\dfrac{\omega}{(s+\alpha)^2+\omega^2}$
23. $\exp(-\alpha t)\cos(\omega t)$	$\dfrac{s+\alpha}{(s+\alpha)^2+\omega^2}$
24. $\exp(-\alpha t)\left(\cos(\omega t)-\dfrac{\alpha}{\omega}\sin(\omega t)\right)$	$\dfrac{s}{(s+\alpha)^2+\omega^2}$
25. $\exp(-\alpha t)+\left(\frac{\alpha}{\omega}\right)\sin(\omega t)-\cos(\omega t)$	$\dfrac{\alpha^2+\omega^2}{(s+\alpha)(s^2+\omega^2)}$
26. $\sin(\omega t+\phi)$	$\dfrac{s\sin(\phi)+\omega\cos(\phi)}{s^2+\omega^2}$
27. $\sinh(\beta t)$	$\dfrac{\beta}{s^2-\beta^2}$
28. $\cosh(\beta t)$	$\dfrac{s}{s^2-\beta^2}$
29. $\operatorname{erf}\left(\dfrac{\alpha}{2\sqrt{t}}\right)$	$\dfrac{1-\exp(-\alpha\sqrt{s})}{s}$
30. $J_0(\alpha t)$	$\dfrac{1}{\sqrt{s^2+\alpha^2}}$
31. $\dfrac{1}{t}J_n(\alpha t)$	$\dfrac{\alpha^n}{n[s+\sqrt{s^2+\alpha^2}]^n}$

Index